T0211710

Lecture Notes in Computer Science   9360

Commenced Publication in 1973
Founding and Former Series Editors:
Gerhard Goos, Juris Hartmanis, and Jan van Leeuwen

More information about this series at http://www.springer.com/series/7407

Roland Meyer · André Platzer
Heike Wehrheim (Eds.)

Correct System Design

Symposium in Honor of Ernst-Rüdiger Olderog
on the Occasion of His 60th Birthday
Oldenburg, Germany, September 8–9, 2015
Proceedings

 Springer

Editors
Roland Meyer
University of Kaiserslautern
Kaiserslautern
Germany

Heike Wehrheim
University of Paderborn
Paderborn
Germany

André Platzer
Carnegie Mellon University
Pittsburgh, PA
USA

ISSN 0302-9743   ISSN 1611-3349   (electronic)
Lecture Notes in Computer Science
ISBN 978-3-319-23505-9   ISBN 978-3-319-23506-6   (eBook)
DOI 10.1007/978-3-319-23506-6

Library of Congress Control Number: 2015947414

LNCS Sublibrary: SL1 – Theoretical Computer Science and General Issues

Printed on acid-free paper

Springer International Publishing AG Switzerland is part of Springer Science+Business Media
(www.springer.com)

Ernst-Rüdiger Olderog
Photograph taken by Wilfried Golletz.

Preface

Ernst-Rüdiger Olderog turned 60 in 2015. Congratulations, Ernst-Rüdiger!

On the occasion of his 60th birthday, this Festschrift features contributed papers dedicated to Ernst-Rüdiger Olderog. Besides a brief laudatio, the Festschrift includes contributions from colleagues who have accompanied Ernst-Rüdiger Olderog through his scientific life in one way or another, be it in joint projects, research articles or even the writing of complete books. The contributions in this Festschrift formed the major part of the scientific program of the associated Festschrift Symposium, which took place September 8–9, 2015.

We would like to thank all the contributors to this Festschrift for their hard work and for their scientifically interesting as well as individual articles, which fit very well to Ernst-Rüdiger Olderog's own research agenda in the area of formal methods. These contributed articles make this Festschrift a very suitable surprise to celebrate Ernst-Rüdider Olderog's contributions to science on the occasion of his birthday.

Our thanks furthermore go to all the reviewers whose support has made excellent articles even better. Many thanks also to Alfred Hofmann at Springer for his helpful interactions and advice on how to approach a Festschrift and for his help in making it manifest, and to Anna Kramer from Springer for her fast responses in all matters big or small. In addition, we are much indepted to the Local Organizing Committee at Oldenburg, especially Manuel Gieseking, Stephanie Kemper, Heinrich Ody, and Maike Schwammberger, for numerous hours of work organizing the Festkolloquium. And, last but not least, we appreciate the financial support of the AVACS project and its director, Werner Damm, as well as the support of the symposium coordinator, Jürgen Niehaus.

July 2015

Roland Meyer
André Platzer
Heike Wehrheim

Organization

Organized by the group Correct System Design of the University of Oldenburg and AVACS:

Local Organization Chairs

Manuel Gieseking	University of Oldenburg, Germany
Stephanie Kemper	University of Oldenburg, Germany
Heinrich Ody	University of Oldenburg, Germany
Maike Schwammberger	University of Oldenburg, Germany

Additional Reviewers

Furbach, Florian
Henriques, David
Hüchting, Reiner
Jakobs, Marie-Christine
Jeannin, Jean-Baptiste
Ji, Ran
Krämer, Julia Désirée

Martins, João G.
Mitsch, Stefan
Müller, Andreas
Travkin, Oleg
Walther, Sven
Wolff, Sebastian

Contents

Logic

Analysis

Synthesis

Laudationes

From Program Verification to Time and Space: The Scientific Life of Ernst-Rüdiger Olderog

Roland Meyer[1] and Heike Wehrheim[2](✉)

[1] Department of Computer Science, University of Kaiserslautern,
Kaiserslautern, Germany
meyer@cs.uni-kl.de
[2] Department of Computer Science, University of Paderborn, Paderborn, Germany
wehrheim@upb.de

The Festschrift and associated symposium celebrate Ernst-Rüdiger Olderog's 60th birthday with invited contributions of colleagues, all touching the theme of formal modeling and correctness in system design. Here, we would like to say some words about Ernst-Rüdiger Olderog himself and his contributions to formal methods research over the years.

Geographically, Ernst-Rüdiger Olderog started his scientific life in Kiel, Germany. This start location Kiel (or more generally, Schleswig Holstein) has in a sense remained the geographical fixpoint in his life throughout the years, new years eve meetings with old friends in Kiel being an annually reoccuring event. In Kiel, he studied Mathematics, Computer Science and Logic finishing with a "Diplom" degree in 1979. This was followed by a record-breaking short time for his PhD, supervised by Hans Langmaak, in 1981. In what today would be called a post-doc phase, he then started to visit the world outside Kiel, most notably Oxford, Amsterdam, Edinburgh, Yorktown Heights, and Saarbrücken. For his habilitation he returned back to Kiel. Shortly after the habilitation, he received his first "Ruf" for a professorship in Oldenburg, Germany, to which he then moved and where he has since then been working.

Scientifically, Ernst-Rüdiger Olderog has gone through different phases in his research. This started with his PhD thesis *Charakterisierung Hoare'scher Systeme für ALGOL-ähnliche Programmiersprachen* about Hoare-style proof systems for imperative programs. The main result, characterizing the existence of sound and relatively complete proof systems in the presence of recursion, settled an important problem in program verification in the beginning of the 1980s and was received with considerable interest by the scientific community, leading to publications in prestigious venues like ICALP, Acta Informatica, and STOC. The work on Hoare logics also introduced Ernst-Rüdiger Olderog and the logician Krzysztof Apt to each other, resulting in a lifelong friendship, several joint articles, and the standard textbook on program verification. After Hoare logics, which Ernst-Rüdiger Olderog remains interested in until today, the (consequently) next topic to study was CSP, naturally together with Tony Hoare. This led his research into the area of concurrency and gave his work a shift towards system design. In his book *Nets, Terms and Formulas: Three Views of Concurrent Processes and Their Relationship* Ernst-Rüdiger Olderog explains the worries he had developed about *a posteriori* program verification and how he would favor a method that would design

© Springer International Publishing Switzerland 2015
R. Meyer et al. (Eds.): Olderog-Festschrift, LNCS 9360, pp. 3–4, 2015.
DOI: 10.1007/978-3-319-23506-6_1

correct systems right from the start. The outcome of this contemplation is the first ever formal process of synthesizing machine-level operational models of concurrent systems from high-level declarative specifications. Starting from a logical specification of the desired system characteristics, decisions for the system design are made explicit in terms of a formal derivation of a process-algebraic model. The algebraic model is then translated into a Petri net which shows how to implement the required control. The work, which was summarized in his habilitation thesis, was later awarded the *Leibniz Preis*, the highest award the German research association (DFG) has to give across its subject areas. The combination of different formalisms and the idea of designing a system via transformation became a constant in his research. With the project ProCoS, two more aspects of system design moved into Ernst-Rüdiger Olderog's focus: real time and data. Real time came about in the form of Duration Calculus, data aspects were formalized in the specification language Z. During ProCos, Ernst-Rüdiger Olderog met a second lifelong friend, Anders Ravn, with whom he shares several common interests, including the study of real-time systems. In the 1990s, the languages CSP, Z, and DC were integrated into a modeling method with the interesting name CSP-OZ-DC — an abbreviation all of the project members of the flagship initiative AVACS soon had to learn. AVACS finally brought the insight that it is not only CSP, Object-Z, and Duration Calculus what is required for formal modeling, but that, in addition to a model of time, an explicit model of space (e.g. on roads) is necessary as well.

While the list of topics reflects the broad spectrum of research, every new topic is well-connected to previous work. All topics center around the theme of correctness and have evolved along two dimensions. The work on modeling and semantics evolved towards expressiveness and applicability, starting from imperative programs over recursion to concurrency, real-time behavior, support of data, and spatial information. The modeling aspect is always accompanied by verification techniques. They evolved from manual correctness proofs to manual synthesis, from there to an automated verification approach, and today we find Ernst-Rüdiger Olderog working on automatic synthesis. "Correct system design" has become the name of his research group, the title of this Festschrift, and could also well be the headline of his scientific research.

In all his work, Ernst-Rüdiger Olderog's leading principle can be characterized as *elegance through simplicity*. This principle is not only reflected in his research articles, but also in his presentations at conferences and lectures for students. Numerous Oldenburg undergraduate students complement their concern of not being particularly good in theory with the observation "but when Professor Olderog explains it, I can understand it". This shows Ernst-Rüdiger Olderog's enormous interest in teaching and his opinion of teaching being equally important to research: Like research, teaching needs time, detailed preparation and careful thinking about examples. Indeed, a precious advice to students is to study concepts on examples — examples that are instructive but still small enough to fit on a single sheet of paper. Ernst-Rüdiger Olderog's deep interest in teaching can also been seen in his passion for writing books, books which soon become classics and are being used in courses at universities all over the world. We hope for much more in the future!

Ernst-Rüdiger Olderog: A Life for Meaning

André Platzer[(⊠)]

Computer Science Department, Carnegie Mellon University, Pittsburgh, USA
aplatzer@cs.cmu.edu

While it is indubitably impossible to summarize all the many important aspects of Ernst-Rüdiger Olderog's scientific contributions and to do justice to each and every one of the areas that he contributed to, it is remarkably easy to identify and characterize the common core behind his investigations. An important leitmotif in his research agenda is *semantics*, the study of meaning.

Even if a significant share of his research does not study semantics in and of itself, but rather as a means to an end, Ernst-Rüdiger Olderog stands out as having recognized the significance of semantics as an important foundation in the first place. He accurately observes how crucial the study of the meaning of a mathematical object of study is for the progress of a research area as a whole, as well as for achieving the individual results he is looking for. The most prominent results that Ernst-Rüdiger Olderog's research enables are various forms of correctness results, as are engraved in the name of his research group: "Correct System Design". The main analytic tool to obtain them, though, is Ernst-Rüdiger Olderog's dedication to semantics.[1] Be it to develop an understanding of concurrency by logical communication formulas to algebraic process terms and further to Petri nets [21], to understand sequential and concurrent programs for the purpose of verification [3,2], or specification and verification techniques for real-time systems [26]. A strong devotion to semantics is a pervasive aspect of his research throughout his career [27,4,28,5,21,3,22,12,26,11], whether on communicating processes [28,27], concurrency at large [5,4,21], real-time systems [26,22] and specification languages for richer realtime systems [15,12], traffic agents [9,8], or games for distributes systems synthesis [11]. Ernst-Rüdiger Olderog's attention to semantics is quite prominently featured already in the first column of a paper on his first research results [20]:

> To explore the applicability of this idea, we have to formalize what "capturing the true partial correctness formulas" means. [20]

In that paper he investigates completeness of Hoare calculi and limits thereof for programs with procedures [20,19] based on his dissertation [18], refining earlier results of Clarke [7]. It is quite an achievement for a proper treatment of semantics to span such a wide range of topics, each with its own different conceptual and technical challenges.

Most of his Ph.D. students got accustomed to Ernst-Rüdiger Olderog's quest for semantics quite quickly. So quickly that there was ample opportunity to overhear the following question among the first raised during many random hallway

---

[1] As reflected in the former name of Ernst-Rüdiger Olderog's group: "Semantics".

© Springer International Publishing Switzerland 2015
R. Meyer et al. (Eds.): Olderog-Festschrift, LNCS 9360, pp. 5–9, 2015.
DOI: 10.1007/978-3-319-23506-6_2

conversations among students about their most recent (sometimes ingenious but certainly still immature) discovery:

"Was soll denn das heissen???"
which roughly translates to: "But what's that even supposed to mean???"

Of course, Ernst-Rüdiger Olderog himself used an infinitely more polite and modest way of rephrasing this question, albeit in a semantics-preserving way. But by carefully thought-out remarks and polite questions, Ernst-Rüdiger made sure his students ultimately understand that everything else crumbles apart unless the semantics holds it all neatly together. It takes the right semantics to start off a scientific development. If the semantics is inappropriate, the best theorems about it do not help. If the semantics expresses the right aspects, but exposes them in inapt ways, then proofs about the semantics become tedious.

Ernst-Rüdiger Olderog does not just ask whether a semantics is "right". He asks whether it is "exactly right". Does the semantics fit to the intuitive expectations for the domain of discourse? Does it lend itself to extrapolating appropriate verification and reasoning techniques? Does it pass muster under the respective principles of semantics, such as compositionality principles in denotational semantics or progress properties in operational semantics? Does it lead to simple and elegant proofs? A semantics that is exactly right will satisfy all these criteria, which leads to a spectrum of quite finely nuanced semantical choices that the uninitiated might struggle with.

Indeed, Ernst-Rüdiger's focus on getting the semantics right also makes all the sense in the world in another context. Higher-order proof assistants such as Isabelle [17], Coq [16], and Nuprl [1] make it possible to develop rich theories with machine-checked proofs. Since that makes mistakes in proofs impossible (if the prover kernel is implemented correctly), getting the semantics right as reflected in the basic definitions themselves is what probably matters the most.

Ernst-Rüdiger Olderog deserves particular admiration for his dauntless investigation of semantical nuances in ever more demanding challenges. Rather than settling for the (already very challenging) world of concurrent processes [21], he went on to a semantic study of programs [3,2], even concurrent programs, with all their extremely subtle semantic interactions. In shared-variable parallel programs and synchronization in parallel programs, a fine line separates a permissive semantics that allows all kinds of concurrent interactions but makes analysis and predictions impossible from a semantics that makes verification straightforward but is so overly restrictive that it hardly does justice to the intended purposes of concurrent programs. Rather than settling for the challenges of concurrency, however, Ernst-Rüdiger went on to study the semantical challenges of real-time systems [26], and did not shy away from subtle and challenging integrations of real-time reasoning with concurrency and rich data structures [15].

These directions of real-time systems as well as of semantically integrating real-time systems with complex systems aspects were one instrumental part of the AVACS project *Automatic Verification and Analysis of Complex Systems* [6]. Ernst-Rüdiger Olderog played a major role in AVACS, where he has been serving as the coordinator for the whole Real-Time Systems group R as well as for one of its 3 subprojects on real-time systems with complex aspects such as rich data [23] ever since its beginning in 2004 until today. AVACS is a Sonderforschungsbereich/Transregio funded by the German Science Foundation (DFG) spanning real-time systems (the R subprojects), hybrid systems (H), and systems of systems verification (S) in a major research initiative of researchers at the University of Oldenburg, the University of Freiburg, the University of Saarbrücken as well as the Max-Planck Institute in Saarbrücken. AVACS involved around 10 subprojects, each with around 5 principle investigators and even more researchers.

While real-time systems have received significant attention by the research community at large, Ernst-Rüdiger Olderog noticed early on that real-time systems are not just important in isolation but that real-time aspects arise in systems with a proper software structure [29,24,30,12,15,25,26,23]. These considerations include transformations from real-time specifications to program specifications [29], decompositions of Duration Calculus formulas into untimed systems communicating by timers [24], and a verification and design approach for real-time software in Programmable Logic Controllers (PLCs) [22], which was held together by a backbone of a semantic link via Duration Calculus between Constraint Diagrams and PLCs, and which had applications in railway signaling systems. This approach has been implemented in the verification tool Moby/PLC [10], which led to Moby/RT [25], a more general verification tool for real-time systems with a similar basis. Ernst-Rüdiger Olderog pursued his interest in combinations of specification techniques further in the development of CSP-OZ-DC [15], an integration of Communicating Sequential Processes with Object-Z and Duration Calculus, which formed an important foundation for the real-time efforts throughout AVACS to understand the interaction of concurrent process behavior, infinite data structures, and continuous real-time dynamics [23].

In addition to serving as scientific director for the real-time efforts in AVACS, Ernst-Rüdiger Olderog also contributed to efforts on the hybrid systems side, especially in verification efforts for traffic agents obeying certain cooperation principles, both in railway systems [8] as well as in car platooning [9]. The basic observation that made an analysis more feasible, was that such traffic applications can often be simplified by partitioning its operations into phases where traffic agents are still far away, then when they come closer and need to negotiate a safe action, as well as when they are correcting their actions to avoid or mitigate safety risks. Of course, the corresponding verification principles are backed up by a detailed semantical analysis how, generically, the components of such a system can interact to justify its correctness. Careful considerations exploiting the abstract structure of traffic also led to a dedicated logic for multilane scenario reasoning in cars by abstracting the motion qualitatively in spatial interval logic [14,13] again, of course, with a dedicated semantics suitable for the application domain.

I cannot say for sure whether Ernst-Rüdiger Olderog found the meaning of life yet, but with all his dedication to giving it all a semantics, I am confident that he is leading a life for meaning. Congratulations, Ernst-Rüdiger Olderog!

References

1. Allen, S.F., Constable, R.L., Eaton, R., Kreitz, C., Lorigo, L.: The nuprl open logical environment. In: McAllester, D. (ed.) CADE-17. LNCS, vol. 1831, pp. 170–176. Springer, Heidelberg (2000). http://dx.doi.org/10.1007/10721959_12
2. Apt, K.R., de Boer, F.S., Olderog, E.R.: Verification of Sequential and Concurrent Programs, 3rd edn. Springer (2010)
3. Apt, K.R., Olderog, E.R.: Verification of Sequential and Concurrent Programs. Texts and Monographs in Computer Science, 2nd edn. Springer (1997)
4. de Bakker, J.W., Meyer, J.C., Olderog, E., Zucker, J.I.: Transition systems, infinitary languages and the semantics of uniform concurrency. In: Proceedings of the 17th Annual ACM Symposium on Theory of Computing, Providence, Rhode Island, USA, May 6–8, pp. 252–262 (1985)
5. de Bakker, J.W., Meyer, J.C., Olderog, E., Zucker, J.I.: Transition systems, metric spaces and ready sets in the semantics of uniform concurrency. J. Comput. Syst. Sci. **36**(2), 158–224 (1988)
6. Becker, B., Podelski, A., Damm, W., Fränzle, M., Olderog, E., Wilhelm, R.: SFB/TR 14 AVACS - automatic verification and analysis of complex systems (der sonderforschungsbereich/transregio 14 AVACS - automatische verifikation und analyse komplexer systeme). IT - Information Technology **49**(2), 118–126 (2007)
7. Clarke, E.M.: Programming language constructs for which it is impossible to obtain good Hoare axiom systems. J. ACM **26**(1), 129–147 (1979)
8. Damm, W., Hungar, H., Olderog, E.-R.: On the verification of cooperating traffic agents. In: de Boer, F.S., Bonsangue, M.M., Graf, S., de Roever, W.-P. (eds.) FMCO 2003. LNCS, vol. 3188, pp. 77–110. Springer, Heidelberg (2004)
9. Damm, W., Hungar, H., Olderog, E.R.: Verification of cooperating traffic agents. International Journal of Control **79**(5), 395–421 (2006)
10. Dierks, H., Tapken, J.: Tool-supported hierarchical design of distributed real-time systems. In: 26th Euromicro Conference on Real-Time Systems. IEEE Computer Society, Los Alamitos (1998)
11. Finkbeiner, B., Olderog, E.: Petri games: synthesis of distributed systems with causal memory. In: Proceedings Fifth International Symposium on Games, Automata, Logics and Formal Verification, GandALF 2014, Verona, Italy, September 10–12, pp. 217–230 (2014)
12. Fischer, C., Olderog, E.-R., Wehrheim, H.: A CSP view on UML-RT structure diagrams. In: Hussmann, H. (ed.) FASE 2001. LNCS, vol. 2029, pp. 91–108. Springer, Heidelberg (2001)
13. Hilscher, M., Linker, S., Olderog, E.-R.: Proving safety of traffic manoeuvres on country roads. In: Liu, Z., Woodcock, J., Zhu, H. (eds.) Theories of Programming and Formal Methods. LNCS, vol. 8051, pp. 196–212. Springer, Heidelberg (2013)
14. Hilscher, M., Linker, S., Olderog, E.-R., Ravn, A.P.: An abstract model for proving safety of multi-lane traffic manoeuvres. In: Qin, S., Qiu, Z. (eds.) ICFEM 2011. LNCS, vol. 6991, pp. 404–419. Springer, Heidelberg (2011)
15. Hoenicke, J., Olderog, E.: CSP-OZ-DC: A combination of specification techniques for processes, data and time. Nord. J. Comput. **9**(4), 301–334 (2002)
16. The Coq development team: The Coq proof assistant reference manual. LogiCal Project (2004). http://coq.inria.fr (version 8.0)
17. Nipkow, T., Wenzel, M., Paulson, L.C.: Isabelle/HOL: A Proof Assistant for Higher-order Logic. Springer, Heidelberg (2002)

18. Olderog, E.R.: Charakterisierung Hoare-scher Systeme für Algol-ähnliche Programmiersprachen. Ph.D. thesis, Universität Kiel (1981)
19. Olderog, E.: Sound and complete Hoare-like calculi based on copy rules. Acta Inf. **16**, 161–197 (1981)
20. Olderog, E.: A characterization of Hoare's logic for programs with pascal-like procedures. In: Proceedings of the 15th Annual ACM Symposium on Theory of Computing, Boston, Massachusetts, USA, April 25–27, pp. 320–329 (1983)
21. Olderog, E.R.: Nets, Terms and Formulas: Three Views of Concurrent Processes and Their Relationship. Cambridge University Press (1991)
22. Olderog, E.-R.: Correct real-time software for programmable logic controllers. In: Olderog, E.-R., Steffen, B. (eds.) Correct System Design. LNCS, vol. 1710, pp. 342–362. Springer, Heidelberg (1999)
23. Olderog, E.-R.: Automatic verification of real-time systems with rich data: an overview. In: Agrawal, M., Cooper, S.B., Li, A. (eds.) TAMC 2012. LNCS, vol. 7287, pp. 84–93. Springer, Heidelberg (2012)
24. Olderog, E.-R., Dierks, H.: Decomposing real-time specifications. In: de Roever, W.-P., Langmaack, H., Pnueli, A. (eds.) COMPOS 1997. LNCS, vol. 1536, pp. 465–489. Springer, Heidelberg (1998)
25. Olderog, E., Dierks, H.: Moby/RT: A tool for specification and verification of real-time systems. J. UCS **9**(2), 88–105 (2003)
26. Olderog, E.R., Dierks, H.: Real-Time Systems: Formal Specification and Automatic Verification. Cambridge Univ. Press (2008)
27. Olderog, E., Hoare, C.A.R.: Specification-oriented semantics for communicating processes. In: Diaz, J. (ed.) Automata, Languages and Programming. LNCS, vol. 154, pp. 561–572. Springer, Heidelberg (1983)
28. Olderog, E., Hoare, C.A.R.: Specification-oriented semantics for communicating processes. Acta Inf. **23**(1), 9–66 (1986)
29. Olderog, E.R., Schenke, M.: Design of real-time systems: the interface between duration calculus and program specifications. In: Desel, J. (ed.) Structures in Concurrency Theory, pp. 32–54. WICS. Springer (1995)
30. Schenke, M., Olderog, E.: Transformational design of real-time systems part I: from requirements to program specifications. Acta Inf. **36**(1), 1–65 (1999)

Warmest Congratulations, Ernst-Rüdiger!

Willem Paul de Roever

Christian-Albrechts-Universität zu Kiel,
Ebereschenweg 70, 24161 Altenholz, Germany
corinnederoever@gmail.com

Dear Ernst-Rüdiger,

This is the place to express my admiration for the continued high quality and scientific integrity of your scientific work through all these years. Frank de Boer and I, when speaking about you some while ago, came to the conclusion that you are a light beacon for both of us, you are a true Professors' Professor, raising and maintaining the high standard of scientific endeavor within our field, that of Program Verification and Program Semantics within Computer Science. At the same time you have never developed any pretensions, but always remained your amiable, utterly reliable, friendly, modest and respectable self. That is why you have been our beacon throughout our lives not only in scientific respect but also as a respected human being, who maintains these simple values of human life — values which make our lives worthwhile also outside of and beyond our professional activities. Your students speak about you with warmth and respect, and consider themselves fortunate that you have been their mentor! All these qualities make you a truly worthy successor of your friend and Doctor-Vater Hans Langmaack in the very best tradition of German Science. What more can a University Professor aspire to be!

© Springer International Publishing Switzerland 2015
R. Meyer et al. (Eds.): Olderog-Festschrift, LNCS 9360, pp. 10–11, 2015.
DOI: 10.1007/978-3-319-23506-6_3

I hope from the bottom of my heart that the coming 5 years until your retirement will be as fruitful as your many previous years have been at the Carl von Ossietzky Universität Oldenburg, and that you will be offered a nice gradual transition from your professional life to what you would like to do once your retirement takes effect. I would have liked to dedicate a scientific contribution to your festschrift, but since I have withdrawn from research, that lays outside my possibilities.

The warmest wishes for a happy life from your friends,

Willem Paul de Roever

Semantics

Understanding Probabilistic Programs

Joost-Pieter Katoen[✉], Friedrich Gretz, Nils Jansen,
Benjamin Lucien Kaminski, and Federico Olmedo

RWTH Aachen University, Aachen, Germany
{katoen,friedrich.gretz,nils.jansen,benjamin.kaminski,
federico.olmedo}@cs.rwth-aachen.de

Abstract. We present two views of probabilistic programs and their relationship. An operational interpretation as well as a weakest precondition semantics are provided for an elementary probabilistic guarded command language. Our study treats important features such as sampling, conditioning, loop divergence, and non-determinism.

1 Introduction

Probabilistic programs are sequential programs with the ability to draw values at random from probability distributions. Probabilistic programs are not new at all. Seminal papers from the mid–eighties consider their formal semantics [16] as well as their formal verification [25]. Variations of probabilistic propositional dynamic logic [5] have been defined to enable reasoning about probabilistic programs. McIver and Morgan [17] generalized Dijkstra's weakest pre–conditions to weakest pre–expectations (wp) so as to formally analyze pGCL—the probabilistic guarded command language. Mechanized wp–reasoning has been realized [3,13].

In the last years the interest in probabilistic programs is rapidly growing [8]. This is mainly due to their wide applicability. Probabilistic programs are used in security to describe cryptographic constructions (such as randomized encryption) and security experiments [1], in machine learning to describe distribution functions that are analyzed using Bayesian inference, and naturally occur in randomized algorithms [18]. Other applications include [6] scientific modeling, information retrieval, bio–informatics, epidemiology, vision, seismic analysis, semantic web, business intelligence, human cognition, and more. The variety of probabilistic programming languages is immense. Almost each programming language, being it imperative, declarative, object–oriented or logical, has a probabilistic counterpart. Probabilistic C [21] extends C with sampling, Church is based on the λ–calculus, Figaro [22] is fully integrated in the Scala object–oriented language, and CHRiSM is a probabilistic version of Prolog. Probabilistic programs are not just of academic interest; they are highly relevant to industry; DARPA invests 48 million US dollar on probabilistic programming for advanced machine learning because:

This work was supported by the Excellence Initiative of the German federal and state government.

R. Meyer et al. (Eds.): Olderog-Festschrift, LNCS 9360, pp. 15–32, 2015.
DOI: 10.1007/978-3-319-23506-6_4

"probabilistic programming is a new programming paradigm for managing uncertain information. By incorporating it into machine learning, we seek to greatly increase the number of people who can successfully build machine learning applications, and make machine learning experts radically more effective".

Microsoft has recently started a large initiative to improve the usability of probabilistic programming. New languages and approaches such as Infer.NET (akin to C#), R2 [19] and Tabular [7] emerged.

What is special about probabilistic programs? They are typically just a few number of lines, but hard to understand and analyze, let alone algorithmically. For instance, the elementary question of almost–sure termination—for a given input, does a probabilistic program terminate with probability one?—is as hard as [14] the universal halting problem—does an ordinary program halt on *all* possible inputs? Loop invariants of probabilistic programs typically involve quantitative statements and synthesizing them requires more involved techniques than for ordinary programs [2,15]. Modern probabilistic programming languages do not just support sampling, but also have the ability to condition values of variables in a program through *observations*. Conditioning blocks all program runs violating its Boolean condition and prevents those runs from happening. Consequently, the likelihood of the remaining runs is normalized. The latter effect makes observations differ from program annotations like probabilistic assertions [24].

Conditioning of variables through observations is less well–understood and raises various semantic difficulties, in particular in the presence of possibly non–terminating loops and non–determinism[1]. Previous works on semantics for probabilistic programs with observations [12,19] do not consider these important features. In fact, many works on probabilistic programs ignore the notion of non–termination and assume that loops always terminate—a property that is unrealistic in practice and highly undecidable to establish. This paper sketches the semantic intricacies, and presents ideas of providing a formal semantics of pGCL treating conditioning in presence of possibly diverging loops and non–determinism.

Much in the vein of Olderog's view [20] that multiple semantic perspectives are useful for a full understanding of programs and systems, we provide *two* semantic views and study their relationship. We present an operational semantics in terms of infinite–state parametric Markov decision processes [23] as well as a weakest (liberal) precondition semantics à la McIver and Morgan [17] and Dijkstra [4]. The main result is a transfer theorem that establishes the relationship between the two semantics. A program transformation is described to remove conditioning and its correctness is established. The presentation is kept informal; full technical details can be found in [9–11].

---

[1] As stated in [8], "representing and inferring sets of distributions is more complicated than dealing with a single distribution, and hence there are several technical challenges in adding non–determinism to probabilistic programs".

2 Probabilistic Programs

This section introduces our programming language. Probabilistic programs are presented by means of examples that elucidate the key insights behind them.

Main Features. Roughly speaking, probabilistic programs are ordinary sequential programs with two additional features:

(i) The ability to *draw samples* from a probability distribution. For simplicity, we consider discrete probability distributions only, and model sampling by means of a probabilistic choice[2] of the form:

$$\{P_1\}\ [p]\ \{P_2\}\ .$$

Here, P_1 and P_2 are programs and p is a probability value in $[0, 1]$. Intuitively, this construct behaves as P_1 with probability p and as P_2 with probability $1-p$.

(ii) The ability to *condition* the distribution of program states with respect to an observation. This is done using statements of the form:

$$\mathsf{observe}\ (G)\ ,$$

where G is a Boolean expression over the program variables. The effect of such an instruction is to block all program executions violating G and rescale the probability of the remaining executions so that they sum up to one. In other words, $\mathsf{observe}\ (G)$ transforms the current distribution μ over states into the conditional distribution $\mu|_G$.

To clarify these features consider the two simple sample programs given below:

```
1: {x := 0} [1/2] {x := 1};        1: {x := 0} [1/2] {x := 1};
2: {y := 0} [1/2] {y := -1}        2: {y := 0} [1/2] {y := -1};
                                   3: observe (x + y = 0)
```

The left program flips two fair (and independent) coins and assigns different values to variables x and y depending on the result of the coin flips. This program admits four executions and yields the outcome

$$\Pr[x{=}0, y{=}0] \ = \ \Pr[x{=}0, y{=}{-}1] \ = \ \Pr[x{=}1, y{=}0] \ = \ \Pr[x{=}1, y{=}{-}1] \ = \ \tfrac{1}{4}\ .$$

The program on the right blocks two of these four executions as they violate the observation $x+y$ equals zero in the last line. The probabilities of the remaining two executions are normalized. This leads to the outcome

$$\Pr[x{=}0, y{=}0] \ = \ \Pr[x{=}1, y{=}{-}1] \ = \ \tfrac{1}{2}\ .$$

---

[2] Alternatively, one can use *random assignments* which sample a value from a distribution and assign it to a program variable; see e.g. [8].

Remarks on Conditioning. The `observe` statement is related to the well–known `assert` statement: both statements `observe` (G) and `assert` (G) block all runs violating the Boolean condition G. The crucial difference, however, is that `observe` (G) normalizes the probability of the remaining runs while `assert` (G) does not. This yields a sub–probability distribution of total mass possibly less than one [1].

We also like to point out that an observation may block *all* program runs. In this case the normalization process is not well–defined and the program admits no feasible run. This is similar to the situation that conditional probabilities are ill–defined when conditioning to an event of probability zero. Section 3 sheds more light on this phenomenon. A possible way out is to only allow conditioning at the end of the program, in particular not inside loops. Whereas this view indeed simplifies matters, modern probabilistic programming languages [7,19,21] do not impose this restriction for good reasons. Instead, they allow the use of `observe` statements at any place in a program, e.g. in loops. Section 4 presents two program semantics that adequately handle such (infeasible) programs.

Loops. Let us now consider loops. Consider the following two loopy programs:

1: $i := 0$;	1: $i := 0$;
2: `repeat`	2: `repeat`
3: $\{b := \mathsf{heads}\}\,[p]\,\{b := \mathsf{tails}\}$;	3: $\{b := \mathsf{heads}\}\,[p]\,\{b := \mathsf{tails}\}$;
4: $i := i + 1$	4: $i := i + 1$
5: `until` $(b = \mathsf{heads})$	5: `until` $(b = \mathsf{heads})$;
	6: `observe` $(\mathsf{odd}(i))$

The left program tosses a (possibly biased) coin until it lands heads and tracks the number of necessary trials. It basically simulates a *geometric distribution* with success probability p and upon program termination we have

$$\Pr[i = N] \;=\; (1 - p)^{N-1}\,p \qquad \text{for } N \geq 1 \;.$$

The program on the right is as the left program but models the situation where on termination we observe that the number of trials until the first heads is odd. The set of program executions complying this observation has an overall probability of $\sum_{N \geq 0} (1 - p)^{2N} p = 1/(2-p)$. This follows from considering a geometric series on even indices. Accordingly, the distribution of variable i is now governed by

$$\begin{aligned} \Pr[i = 2N+1] &= (1 - p)^{2N} p (2 - p) \\ \Pr[i = 2N] &= 0 \end{aligned} \qquad \text{for } N \geq 0 \;.$$

As a final remark regarding the previous pair of loopy programs, observe that we allow the probability value of probabilistic choices to remain unspecified. This allows us to deal with *parametric* programs in which the exact values of the probabilities are not known.

Non–determinism. Our programming model also accounts for the possibility of non–determinism. Let $\{P_1\} \ \square \ \{P_2\}$ represent the non–deterministic choice between the programs P_1 and P_2. Non–deterministic choices are resolved by means of a so–called *scheduler* (akin: adversary). On the occurrence of the non–deterministic choice $\{P_1\} \ \square \ \{P_2\}$ during a program run, a scheduler decides whether to execute P_1 or P_2. This choice can in principle depend on the sequence of program states encountered so far in the run. Consider, for instance

$$1\text{: } \{i := 2j\} \ \square \ \{i := 2j+1\};$$
$$2\text{: } \{i := i+1\} \ [1/3] \ \{i := i+2\} \ .$$

It admits the schedulers $\mathcal{L}$ and $\mathcal{R}$, say. Scheduler $\mathcal{L}$ resolves the non–deterministic choice in favor of the assignment $i := 2j$, whereas scheduler $\mathcal{R}$ selects the assignment $i := 2j+1$. Evidently, imposing either the scheduler $\mathcal{L}$ or $\mathcal{R}$ on this program yields a purely probabilistic program.

As in [17], we consider a *demonic* model to determine the probability of an event in the presence of non–determinism. This amounts to resolving all non–deterministic choices in a way that minimizes the probability of the event at hand. In other words, we assume a scheduler that leads to the event occurring with the least probability. For instance, the probability that i is odd in the above program is computed as follows

$$\Pr[\mathsf{odd}(i)] = \min \left\{ \Pr^{\mathcal{L}}[\mathsf{odd}(i)], \ \Pr^{\mathcal{R}}[\mathsf{odd}(i)] \right\} = \min \left\{ \tfrac{1}{3}, \tfrac{2}{3} \right\} = \tfrac{1}{3} \ .$$

By a similar reasoning it follows that the probability that i is even is also $1/3$. This shows that in the presence of non–determinism the law of total probability, namely $\Pr[A] + \Pr[\neg A] = 1$, does not hold.

Observe that our demonic model of non-determinism impacts directly on the termination behavior of programs. This is because in the probabilistic setting, the termination behaviour of a program is given by the probability of establishing true, which—like the probability of any other event—is to be minimized. To clarify this consider the following example. Assume that P is a program which admits a scheduler that leads to a probability of termination zero, while all other schedulers induce a probability of termination that is strictly positive. We will then say that P is non-terminating, or more formally, that it *diverges almost surely*, since according to our demonic model of non-determinism, the probability of establishing true, i.e., termination, will be zero.

3   Semantic Intricacies

In this section, we investigate semantic difficulties that arise in the context of non–deterministic and probabilistic uncertainty in probabilistic programs, in particular in combination with conditioning. We do this by means of examples. Consider as a first example the following two ordinary (i.e. deterministic and non–probabilistic) programs P_{div} (left) and P_{term} (right):

1: **repeat**	1: **repeat**
2: $x := 1$	2: $x := 0$
3: **until** $(x = 0)$	3: **until** $(x = 0)$

While the left program never terminates as the variable x is always set to one, the right program performs only one loop iteration. The right program is said to *certainly terminate*.

Non–deterministic Uncertainty. The first type of uncertainty we take a look at is non–determinism. For that, consider the following program P_{nd}:

1: **repeat**
2:    $\{x := 1\} \ \square \ \{x := 0\}$
3: **until** $(x = 0)$

In each loop iteration, the variable x is set non–deterministically either to 1 or to 0. A natural question is whether this program terminates or not. Obviously, this depends on the resolution of the non–deterministic choice inside the loop body. For the scheduler that chooses the left branch $x := 1$ in each loop iteration, the probability of termination is zero, while for any other scheduler the probability of termination is one. (As P_{nd} contains no probabilistic choice, any event will occur with probability either zero or one). In view of our demonic model of non-determinism, the program presents a certain behavior: non-termination.

Probabilistic Uncertainty. Consider now the following program P_{pr}, which is obtained from the previous program P_{nd} by replacing the non-deterministic choice by a random choice:

1: **repeat**
2:    $\{x := 1\} \ [1/3] \ \{x := 0\}$
3: **until** $(x = 0)$

In each loop iteration, the variable x is set to 1 with probability $1/3$ and to 0 with probability $2/3$. Again we pose the question: does this program terminate? The answer to that requires a differentiated view: there does exist *a single non–terminating program run*, namely the one in which x is set to 1 in each loop iteration. This infinite run, however, has probability $1/3 \cdot 1/3 \cdot 1/3 \cdots = 0$. Thus, the terminating runs have probability $1 - 0 = 1$. In this case, the program is said to terminate *almost surely*. Note that it does not terminate certainly though, as it admits an infinite run.

Combining Non–deterministic and Probabilistic Uncertainty. Let us consider the two notions of uncertainty in a single program P_{nd+pr}:

```
1: repeat
2:    {{x := 1} [8/9] {x := 0}} □ {{x := 1} [1/9] {x := 0}}
3: until (x = 0)
```

In each loop iteration, the variable x is set to 0 with a certain probability, but this probability is chosen non–deterministically to be $1/9$ or $8/9$. Again we pose the question: does this program terminate almost–surely? As a matter of fact, the scheduler cannot prevent this program from terminating almost–surely. In fact the two programs

```
1: repeat                        1: repeat
2:    {x := 1} [1/9] {x := 0}    2:    {x := 1} [8/9] {x := 0}
3: until (x = 0)                 3: until (x = 0)
```

are semantically equivalent in both our semantic views [11,17].

Still it seems natural to ask whether choosing $1/9$ over $8/9$ as the probability of setting x to 0 would not be—so to say—*more demonic* as this would increase the expected time until termination and therefore the right program converges slower. To the best of our knowledge, however, existing semantics for probabilistic programs with non–determinism do not take this convergence rate into account (and neither do our two semantic views).

Observations. Next, we turn towards the second characteristic feature of probabilistic programs—conditioning—and take a look at termination in this context. Consider the following two programs P_{div} (left) and P_{obs} (right):

```
1: repeat                  1: repeat
2:    x := 1               2:    {x := 1} [1/2] {x := 0};
3: until (x = 0)          3:    observe (x = 1)
                          4: until (x = 0)
```

As noted earlier, the left program certainly diverges. For the right program, things are not so clear any more: On the one hand, the only non–terminating run is the one in which in every iteration x is set to 1. This event of setting x infinitely often to 1, however, has probability 0. So the probability of non–termination would be 0. On the other hand, the global effect of the observe statement within the loop is to condition on exactly this event, which occurs with probability 0. Hence, the conditional termination probability is 0 divided by 0, i.e. *undefined*.

Remark 1. Notice that while in this sample program it is immediate to see that the event to which we condition has probability 0, in general it might be highly non–trivial to identify this. Demanding from a "probabilistic programmer" to condition only to events with non–zero probability would thus be just as (if not even more) far–fetched as requiring an "ordinary programmer" to write only terminating programs. Therefore, a rigorous semantics for probabilistic programs with conditioning has to take the possibility of conditioning to zero–probability events into account: To the program on the right such a semantics should assign a dedicated denotation which represents undefined due to conditioning to a zero–probability event.

Conditioning in Presence of Uncertainty. Our final example in this section blurs the situation even further by incorporating both notions of uncertainty and conditioning into the single program P_{all}:

1: `repeat`

2:     $\{x := 1\} \, [1/2] \, \{x := 0\};$

3:     $\{x := 1\} \, \square \, \{\texttt{observe} \, (x = 1)\}$

4: `until` $(x = 0)$

This program first randomly sets x to 1 or 0. Then it either sets x to 1 or conditions to the event that x was set to 1 in the previous probabilistic choice. The latter choice is made non–deterministically and therefore the semantics of the entire program is certainly not clear: If in line 3, the scheduler always chooses $x := 1$, then this results in certain non–termination. If, on the other hand, the scheduler always chooses `observe` $(x = 1)$, then the global effect of the observe statement is a conditioning to this zero–probability event. Which behavior of the scheduler is more demonic? We take the point of view that certain non–termination is a more well–behaved phenomenon than conditioning to a zero–probability event. Therefore a demonic scheduler should prefer the latter.

4   Expectation Transformer and Operational Semantics

This section presents the two semantic views and their relationship. The first perspective is a semantics in terms of weakest pre–expectations, the quantitative analogue of Dijkstra's weakest pre–conditions [4]. The second view is an operational semantics in terms of Markov decision processes (MDPs) [23]. The relationship between the semantics is established by linking weakest pre–expectations to (conditional) rewards in the MDPs associated to the programs.

4.1   Weakest Pre–expectation Semantics

The semantics of Dijkstra's seminal guarded command language [4] has been given in terms of weakest preconditions. It is in fact a predicate transformer semantics, i.e. a total function between two predicates on the state of a program. The predicate transformer $E = \mathsf{wp}(P, F)$ for program P and postcondition F yields the weakest precondition E on the initial state of P ensuring that the execution of P terminates in a final state satisfying F. There is a direct relation with axiomatic semantics: the Hoare triple $\langle E \rangle\, P\, \langle F \rangle$ holds for total correctness if and only if $E \Rightarrow \mathsf{wp}(P, F)$. The weakest *liberal* precondition $\mathsf{wlp}(P, F)$ yields the weakest precondition for which P either does not terminate or establishes F. It does not ensure termination and corresponds to Hoare logic for partial correctness.

Weakest Pre-expectations. Qualitative annotations in predicate calculus are often insufficient for probabilistic programs as they cannot express quantities such as expectations over program variables. To that end, we adopt the app-roach by McIver and Morgan [17] and consider *expectations* over program vari-able valuations. They are the quantitative analogue of predicates and are in fact just random variables (over variable valuations). An *expectation transformer* is a total function between expectations on the state of a program. Stated col-loquially, the expectation transformer $e = \mathsf{wp}(P, f)$ for pGCL–program P and post–expectation f over final states yields the least expected "value" e on P's initial state ensuring that P's execution terminates with a "value" f. That is to say, $e(\sigma) = \mathsf{wp}(P, f)(\sigma)$ represents the expected value of f with respect to the distribution of final states obtained from executing program P in state σ, where σ is a valuation of the program variables. The annotation $\langle e \rangle\, P\, \langle f \rangle$ holds for total correctness if and only if $e \leq \mathsf{wp}(P, f)$, where $\leq$ is to be interpreted in a point–wise manner. The weakest *liberal* pre–expectation $\mathsf{wlp}(P, f)$ yields the least expectation for which P either does not terminate or establishes f. It does not ensure termination and corresponds to partial correctness.

Determining Weakest Pre-expectations. We explain the transformation of expec-tations by means of an example. Consider the program P:

$$\{\{x := 5\} \ \square \ \{x := 2\}\}\ [p]\ \{x := 2\}$$

We would like to find the (least) average value of x produced by this program. This quantity is given by

$$\mathsf{wp}(P, x) = \mathsf{wp}(\{\{x := 5\} \ \square \ \{x := 2\}\}\ [p]\ \{x := 2\}, x)\ .$$

The expectation of the probabilistic choice is given by the weighted average of the expectations of its sub–programs, thus we obtain

$$p \cdot \mathsf{wp}(\{x := 5\} \ \square \ \{x := 2\}, x) + (1 - p) \cdot \mathsf{wp}(x := 2, x)\ .$$

As non–determinism is resolved in a demonic manner, it yields the expectation given by the minimum between the expectations of the sub–programs

$$p \cdot \min\{\mathsf{wp}(x := 5, x), \mathsf{wp}(x := 2, x)\} + (1 - p) \cdot \mathsf{wp}(x := 2, x)\ .$$

In the last step we apply the assignments and evaluate the expression

$$p \cdot \min\{5, 2\} + (1 - p) \cdot 2\ =\ p \cdot 2 + (1 - p) \cdot 2\ =\ 2\ .$$

For loops, the semantics is as usual defined by a least fixed point; in our case, over the domain of expectations with partial order the point–wise ordering $\leq$ on expectations.

Conditioning. Let $\mathsf{wp}(\mathtt{observe}(G), f) = \mathsf{wlp}(\mathtt{observe}(G), f) = [G] \cdot f$, where $[G]$ stands for the characteristic function of the Boolean expression G over the program variables. For probabilistic programs with observations we define a transformer to determine the conditional expectation $\underline{\mathsf{cwp}}(P, f)$. Intuitively, the conditioning takes place on the probability that all observations in the program are successfully passed. The *conditional expectation* of program P with respect to post–expectation f is given as a pair:

$$\underline{\mathsf{cwp}}(P, f) = \big(\mathsf{wp}(P, f), \mathsf{wlp}(P, 1)\big) .$$

The first component gives the expectation of the random variable f, whereas $\mathsf{wlp}(P, 1)$ is the probability that no observation has been violated (this includes non–terminating runs). The pair $\big(\mathsf{wp}(P, f), \mathsf{wlp}(P, 1)\big)$ is to commonly be interpreted as the quotient

$$\frac{\mathsf{wp}(P, f)}{\mathsf{wlp}(P, 1)} .$$

It is possible though that both $\mathsf{wp}(P, f)$ and $\mathsf{wlp}(P, 1)$ evaluate to 0. In that case, the quotient $\frac{0}{0}$ is undefined due to division by zero. The pair $(0, 0)$, however, is well–defined. Let us give an example. Consider the program P from Section 2:

1: $\{x := 0\} \; [1/2] \; \{x := 1\}$;
2: $\{y := 0\} \; [1/2] \; \{y := -1\}$;
3: $\mathtt{observe} \; (x + y = 0)$

Assume we want to compute the conditional expected value of expression x given that observation $x + y = 0$ is passed. This expected value is given by $\underline{\mathsf{cwp}}(P, x)$ and its computation is sketched below. During the computation we use P_{i-j} to denote the fragment of program P from line i to line j. For the first component of $\underline{\mathsf{cwp}}(P, x)$ we have:

$\mathsf{wp}(P, x)$
$= \mathsf{wp}(P_{1-2}, [x + y = 0] \cdot x)$
$= 1/2 \cdot \mathsf{wp}(P_{1-1}; y := 0, [x + y = 0] \cdot x) + 1/2 \cdot \mathsf{wp}(P_{1-1}; y := -1, [x + y = 0] \cdot x)$
$= 1/2 \cdot \mathsf{wp}(P_{1-1}, [x = 0] \cdot x) + 1/2 \cdot \mathsf{wp}(P_{1-1}, [x = 1] \cdot x)$
$= 1/2 \cdot (1/2 \cdot 1 \cdot 0 + 1/2 \cdot 0 \cdot 1) + 1/2 \cdot (1/2 \cdot 0 \cdot 0 + 1/2 \cdot 1 \cdot 1)$
$= 1/4$

For the second component of $\underline{\mathsf{cwp}}(P, x)$ we derive:

$\mathsf{wlp}(P, 1)$
$= \mathsf{wlp}(P_{1-2}, [x + y = 0] \cdot 1)$
$= 1/2 \cdot \mathsf{wlp}(P_{1-1}; y := 0, [x + y = 0]) + 1/2 \cdot \mathsf{wlp}(P_{1-1}; y := -1, [x + y = 0])$
$= 1/2 \cdot \mathsf{wlp}(P_{1-1}, [x = 0]) + 1/2 \cdot \mathsf{wlp}(P_{1-1}, [x = 1])$
$= 1/2 \cdot (1/2 \cdot 1 + 1/2 \cdot 0) + 1/2 \cdot (1/2 \cdot 0 + 1/2 \cdot 1)$
$= 1/2$

Thus the conditional expected value of x is

$$\frac{\mathsf{wp}(P, x)}{\mathsf{wlp}(P, 1)} = \frac{1/4}{1/2} = \frac{1}{2} \ .$$

Revisiting the purely probabilistic example programs of Section 3 (i.e. those not containing any non–deterministic choices), with respect to post–expectation $x+5$ we would obtain the following conditional expectations and according quotients:

P_{div}	$(0, 1)$	$\frac{0}{1} = 0$
P_{term}	$(5, 1)$	$\frac{5}{1} = 5$
P_{pr}	$(5, 1)$	$\frac{5}{1} = 5$
P_{obs}	$(0, 0)$	$\frac{0}{0} = undefined$

In particular notice that P_{div} and P_{obs} diverge due to different reasons and that our semantics discriminates these two programs by assigning different denotations to them.

Remark 2.   Note that the example for the weakest pre–expectation semantics for programs with conditioning does not contain non–determinism. This is deliberate as it is *impossible* to treat non–determinism in a compositional manner [9]. The problem is that determining the conditional expectation in a compositional fashion is not feasible.

4.2   Operational Semantics

MDPs. Markov decision processes (MDPs [23]) serve as a model for probabilistic systems that involve *non–determinism*. An MDP is a state–transition system in which the target of a transition is a discrete probability distribution over states. As in state–transition systems, several transitions may emanate from a state. An MDP thus reduces to an ordinary state–transition system in case all transitions are equipped with a Dirac distribution. In the sample MDP in Figure 1 there is a choice in state s_0 between distributions (or: transitions) α and β. Choosing α results in a *probabilistic choice* of moving either to state s_1 or to state s_2 with probability $1/2$ in each case. Choosing β results in going to s_3 with probability $9/10$ and to s_1 with probability $1/10$. Additionally, in state s_1 a *reward* (also referred to as cost) of 10 is earned; all other states have reward zero, which is omitted from the figure. The *expected reward* of reaching s_1 from state s_0 equals the reward that on average will be earned with respect to the overall probability of reaching state s_1.

These MDPs serve as an operational model for our probabilistic programs. The MDP states are tuples of the form $\langle P, \sigma \rangle$ where P denotes the remaining program to be executed (or equals $\langle sink \rangle$ if the program successfully terminated), and σ is the current valuation of the program variables. Executing a program statement is mimicked by a state change in the MDP. By equipping the MDP states with rewards it is possible to express the expected outcome of a program

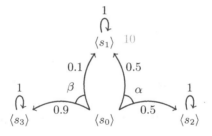

Fig. 1. Sample MDP with four states and a non–deterministic choice between α and β

variable as an expected reward on the MDP. This will become more explicit when discussing the relationship to the weakest pre-expectation semantics at the end of this section. Note that the resulting MDP of a probabilistic program is in general countably infinite (as the variable domains can be infinitely large) and parametric (as probabilistic choices can be parametric).

The Structure of MDPs for Probabilistic Programs. Let us examine the different kinds of *runs* a program can have. First, we have *terminating runs* where—in presence of conditioning—one has to distinguish between runs that satisfy the condition and those that do not. In addition, a program may have *diverging runs*, i.e. runs that do not terminate. Schematically, the MDP of a probabilistic program has the following structure:

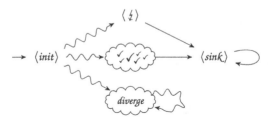

For terminating runs of the program, we use a dedicated $\langle sink \rangle$ state where all terminating runs will end. All diverging runs never reach $\langle sink \rangle$. A program terminates either successfully, i.e. a run passes a ✓–labeled state, or terminates due to violating an observation, i.e. a run passes $\langle \frac{1}{2} \rangle$. Squiggly arrows indicate reachability via possibly multiple paths and states; the clouds indicate that there might be several or even infinitely many states of the particular kind. The ✓– labeled states are the *only ones* where one is allowed to assign positive reward as this corresponds to a desired outcome of the program when subsequently terminating. Note that the sets of paths that eventually reach $\langle \frac{1}{2} \rangle$, eventually reach ✓, or diverge, are pairwise disjoint.

As an example, consider the following program:

$$\{\{x := 5\} \ \square \ \{x := 2\}\} \ [q] \ \{x := 2\};$$
$$\textsf{observe} \ (x > 3)$$

With parametrized probability q, a non–deterministic choice either assigns x with 2 or 5. With probability $1 - q$, x is directly assigned 2, so in this program branch no non–deterministic choice occurs. The event that x exceeds 3 is observed. For the sake of readability, let: $P_1 = \{x := 5\} \,\square\, \{x := 2\}$, $P_2 = x := 2$, $P_3 = \texttt{observe}\ (x > 3)$, and $P_4 = x := 5$. Figure 2 shows the resulting MDP, where σ_I denotes some initial variable valuation for x. Let $\sigma_I[x/y]$ denote the variable valuation that is obtained from σ_I by replacing x by y. Starting in

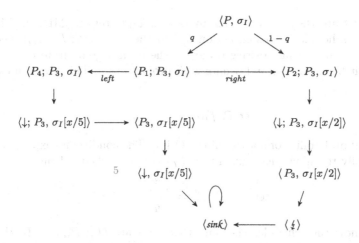

Fig. 2. Reward MDP for the example program

the initial state $\langle P, \sigma_I \rangle$, the probabilistic choice takes place. With probability q, the next state is $\langle P_1; P_3, \sigma_I \rangle$ while with probability $1-q$, the next state is $\langle P_2; P_3, \sigma_I \rangle$. The non–deterministic choice in state $\langle P_1; P_3, \sigma_I \rangle$ is indicated by *left* and *right*. Note that non–deterministic choices yield a choice in the MDP between Dirac distributions.

Conditional Expected Rewards. The *operational semantics* of a probabilistic program P, a program state σ and an expectation (i.e. random variable) f is the reward MDP $\mathfrak{R}^f_\sigma[\![P]\!]$ constructed as described in the paragraph above. Note that in the context of MDPs, the random variable f can also be seen as a *reward function* which adds a positive real–valued reward to certain states of the MDP. In our previous example, the only state with positive reward (5) is $s' := \langle \downarrow, \sigma_I[x/5] \rangle$; all other states have reward zero. In absence of conditioning, we are interested in the *expected reward* to reach a $\langle sink \rangle$–state from the MDP's initial state σ_I:

$$\mathsf{er}(P, f)(\sigma_I) = \mathsf{ExpRew}^{\mathfrak{R}^f_{\sigma_I}[\![P]\!]} (\lozenge\ sink)\,.$$

The right-hand side denotes the sum over all (countably many) paths in the reward MDP $\mathfrak{R}^f_{\sigma_I}[\![P]\!]$ where for each path its likelihood is weighed with its reward. The reward of a path is simply the sum of the rewards of the states it contains.

In the presence of conditioning (i.e. for programs having **observe**–statements), we consider the *conditional expected reward* to reach a $\langle sink \rangle$–state without intermediately passing the $\langle \frac{1}{2} \rangle$–states:

$$\mathsf{cer}(P, f)(\sigma_I) = \frac{\mathsf{ExpRew}^{\mathfrak{R}^f_{\sigma_I} [\![P]\!]} (\Diamond\, sink \cap \neg\Diamond\frac{1}{2})}{\Pr(\neg\Diamond\frac{1}{2})}.$$

Let us illustrate these two notions by our example reward MDP in Figure 2. Consider a scheduler choosing action *left* in the state $\langle P_1; P_3, \sigma_I \rangle$. Then, the only path accumulating positive reward is the path π going from $\langle P, \sigma_I \rangle$ via s' to $\langle sink \rangle$; it has reward 5 and occurs with probability q. This gives an expected reward

$$\mathsf{er}(P, f)(\sigma_I) = 5 \cdot q .$$

The overall probability of not reaching $\langle \frac{1}{2} \rangle$ is q. The conditional expected reward of eventually reaching $\langle sink \rangle$ given that $\langle \frac{1}{2} \rangle$ is not reached is hence

$$\mathsf{cer}(P, f)(\sigma_I) = \frac{5 \cdot q}{q} = 5 .$$

Consider now the scheduler choosing *right* at state $\langle P_1; P_3, \sigma_I \rangle$. In this case, there is no path with positive accumulated reward, yielding an expected reward of 0. The probability of not reaching $\langle \frac{1}{2} \rangle$ is also 0. The conditional expected reward in this case is undefined ($0/0$). Thus, the *right* branch is preferred over the *left* branch by a demonic scheduler, as discussed in Section 3.

4.3 Relating the Two Semantic Views

A key insight is that the operational program semantics in terms of MDPs and the semantics in terms of expectation transformers, as explained in the previous section, correspond in the following sense:

Theorem 1 (Transfer theorem [11]). *For a probabilistic program P without observations, a random variable f, and some initial state σ_I:*

$$\mathsf{wp}(P, f)(\sigma_I) = \mathsf{er}(P, f)(\sigma_I) .$$

Stated in words, this result asserts that the weakest-pre-expectation of program P in initial state σ_I wrt. post-expectation f coincides with the expected reward in the MDP of P where reward f is assigned to successfully terminating states. For probabilistic programs with observations but without non–determinism we can establish a correspondence between the conditional expected reward on the MDP of a program and its conditional pre–expectation:

Theorem 2 (Transfer theorem for conditional expectations [9]). *For a purely probabilistic program P (with observations), a random variable f, and some initial state σ_I, let $\underline{cwp}(P, f) = (g, h)$. Then*

$$\frac{g(\sigma_I)}{h(\sigma_I)} \simeq \mathsf{cer}(P, f)(\sigma_I) \ ,$$

where $x \simeq y$ holds iff either $x = y$ or both sides of the equation are undefined.

For weakest *liberal* pre-expectations, we obtain a similar pair of theorems, where the notions of (conditional) *liberal* expected reward also takes the mere probability of not reaching the target states into account. For further details, the reader is referred to [9–11].

5   Program Transformations

In this section, we use the semantics to show the correctness of a program transformation aimed at removing observations from programs. The program transformation basically allows removing observations from programs through the introduction of a global loop. It is motivated by a well–known technique to simulate a uniform distribution in some interval $[a, b]$ using fair coins [26, Th. 9.2]. The technique is illustrated by a program simulating a six–sided die:

```
1: repeat
2:    {a₀ := 0} [¹/₂] {a₀ := 1};
3:    {a₁ := 0} [¹/₂] {a₁ := 1};
4:    {a₂ := 0} [¹/₂] {a₂ := 1};
5:    i := 4a₀ + 2a₁ + a₀ + 1
6: until (1 ≤ i ≤ 6)
```

The body of the loop simulates a uniform distribution over the interval $[1, 8]$, which is repeatedly sampled (in variable i) until its outcome lies in $[1, 6]$. The effect of the repeated sampling is precisely to condition the distribution of i to $1 \leq i \leq 6$. As a result, $\Pr[i = N] = \frac{1}{6}$ for all $N = 1, \ldots, 6$.

Our program transformation follows the same idea. Given a program P with observations, we repeatedly sample executions from P until the sampled execution satisfies all observations in P. To implement this, we have to take into account three issues. First, we introduce a flag that signals whether all observations along a program execution were satisfied or not. Let variable *flag* be initially **true** and replace every observation **observe** (G) in the original program by the assignment *flag* $:=$ *flag* $\wedge\, G$. In this way, the variable *flag* is true until an observation is violated. Secondly, since a program execution is no longer blocked on violating an observation, we need to modify the program to avoid any possible divergence after an observation has been violated. This is achieved by adapting the loop guards. For instance loop **while** (G) $\{P\}$ is transformed into while $(G \wedge \mathit{flag})$ $\{P\}$, whereas loop repeat $\{P\}$ until (G) is changed into repeat $\{P\}$ until $(G \vee \neg\mathit{flag})$. Finally, observe that we need to keep a permanent copy of the initial program state since every time we sample an execution, the program must start from its original initial state. In general, the transformed program will have the following shape:

1: $s_1, \ldots, s_n := x_1, \ldots, x_n;$

2: repeat

3:     $\mathit{flag} := \text{true};$

4:     $x_1, \ldots, x_n := s_1, \ldots, s_n;$

5:     modified version of original program;

6: until (flag)

Here $x_1, \ldots, x_n$ denote the set of variables that occur in the original program and $s_1, \ldots, s_n$ are auxiliary variables used to store the initial program state; note that if the original program is closed (i.e. independent of its input), Lines 1 and 4 can be omitted. Line 5 includes the modified version of the original program which accounts for the replacement of observations by flag updates and, possibly, the adaptation of loop guards.

We illustrate the program transformation on the left program below:

1: $\{x := 0\}$ $[1/2]$ $\{x := 1\};$ 1: repeat

2: $\{y := 0\}$ $[1/2]$ $\{y := -1\};$ 2: $\mathit{flag} := \text{true};$

3: observe $(x + y = 0)$ 3: $\{x := 0\}$ $[1/2]$ $\{x := 1\};$

     4:     $\{y := 0\}$ $[1/2]$ $\{y := 1\};$

     5:     $\mathit{flag} := \mathit{flag} \wedge (x + y = 0)$

     6: until (flag)

The transformed—observe–free—program is given on the right. Using the operational semantics from Section 4 we establish that the transformation is semantic-preserving:

Theorem 3 (Correctness of the program transformation). *Let P be a probabilistic program and let P' be the result of applying the above transformation to program P. Then for initial state σ_I and reward function f,*

$$\mathsf{cer}(P, f)(\sigma_I) = \mathsf{er}(P', f)(\sigma_I) .$$

In some circumstances it is possible to apply a dual program transformation that replaces program loops with observations. This is applicable when there is no data flow between loop iterations and the samplings across iterations are thus independent and identically distributed. This is the case, e.g. for the earlier program that simulates a six-sided dice. One can show that this program is semantically equivalent to the program

$$
\begin{array}{ll}
1\colon & \{a_0 := 0\}\ [1/2]\ \{a_0 := 1\}; \\
2\colon & \{a_1 := 0\}\ [1/2]\ \{a_1 := 1\}; \\
3\colon & \{a_2 := 0\}\ [1/2]\ \{a_2 := 1\}; \\
4\colon & i := 4a_0 + 2a_1 + a_0 + 1; \\
5\colon & \textsf{observe}\ (1 \le i \le 6)
\end{array}
$$

6 Conclusion

We have presented two views on the semantics of probabilistic programs and showed their relationship for purely probabilistic programs. Whereas the operational semantics can cope with all features—loops, conditioning, non–termination, and non–determinism—the weakest pre–expectation approach cannot be directly applied to handle non–determinism in this setting. We believe that formal semantics, verification, and program analysis has much to offer to improve modern probabilistic programming, and consider this as an interesting and challenging avenue for further research.

References

1. Barthe, G., Kopf, B., Olmedo, F., Béguelin, S.Z.: Probabilistic relational reasoning for differential privacy. ACM Trans. Program. Lang. Syst. **35**(3), 9 (2013)
2. Chakarov, A., Sankaranarayanan, S.: Expectation invariants for probabilistic program loops as fixed points. In: Müller-Olm, M., Seidl, H. (eds.) Static Analysis. LNCS, vol. 8723, pp. 85–100. Springer, Heidelberg (2014)
3. Cock, D.: Verifying probabilistic correctness in Isabelle with pGCL. El. Proc. in Th. Comp. Sc. **102**, 167–178 (2012)
4. Dijkstra, E.W.: A Discipline of Programming. Prentice Hall (1976)
5. Feldman, Y.A., Harel, D.: A probabilistic dynamic logic. In: Proc. of STOC, pp. 181–195. ACM (1982)
6. Gordon, A.D.: An agenda for probabilistic programming: Usable, portable, and ubiquitous (2013). http://research.microsoft.com/en-us/projects/fun
7. Gordon, A.D., Graepel, T., Rolland, N., Russo, C.V., Borgström, J., Guiver, J.: Tabular: a schema-driven probabilistic programming language. In: Proc. of POPL, pp. 321–334. ACM Press (2014)
8. Gordon, A.D., Henzinger, T.A., Nori, A.V., Rajamani, S.K.: Probabilistic programming. In: Proc. of FOSE, pp. 167–181. ACM Press (2014)
9. Gretz, F., Jansen, N., Kaminski, B.L., Katoen, J.P., McIver, A., Olmedo, F.: Conditioning in probabilistic programming. In: Proc. of MFPS, p. 12 (2015)
10. Gretz, F., Jansen, N., Kaminski, B.L., Katoen, J.P., McIver, A., Olmedo, F.: Conditioning in probabilistic programming. CoRR (2015)
11. Gretz, F., Katoen, J.P., McIver, A.: Operational versus weakest pre-expectation semantics for the probabilistic guarded command language. Perform. Eval. **73**, 110–132 (2014)

12. Hur, C.K., Nori, A.V., Rajamani, S.K., Samuel, S.: Slicing probabilistic programs. In: Proc. of PLDI, pp. 133–144. ACM Press (2014)
13. Hurd, J., McIver, A., Morgan, C.: Probabilistic guarded commands mechanized in HOL. Theor. Comput. Sci. **346**(1), 96–112 (2005)
14. Kaminski, B.L., Katoen, J.-P.: On the hardness of almost–sure termination. In: Italiano, G.F., Pighizzini, G., Sannella, D.T. (eds.) MFCS 2015. LNCS, vol. 9234, pp. 307–318. Springer, Heidelberg (2015)
15. Katoen, J.-P., McIver, A.K., Meinicke, L.A., Morgan, C.C.: Linear-invariant generation for probabilistic programs. In: Cousot, R., Martel, M. (eds.) SAS 2010. LNCS, vol. 6337, pp. 390–406. Springer, Heidelberg (2010)
16. Kozen, D.: Semantics of probabilistic programs. J. Comput. Syst. Sci. **22**(3), 328–350 (1981)
17. McIver, A., Morgan, C.: Abstraction, Refinement and Proof for Probabilistic Systems. Springer (2004)
18. Motwani, R., Raghavan, P.: Randomized Algorithms. Cambridge University Press (1995)
19. Nori, A.V., Hur, C.K., Rajamani, S.K., Samuel, S.: R2: An efficient MCMC sampler for probabilistic programs. In: Proc. of AAAI. AAAI Press (July 2014)
20. Olderog, E.R.: Nets, Terms and Formulas: Three Views of Concurrent Processes and their Relationship. Cambridge Tracts in Theoretical Computer Science. Cambridge University Press (1990)
21. Paige, B., Wood, F.: A compilation target for probabilistic programming languages. In: Proc. of ICML. JMLR Proceedings, vol. 32, pp. 1935–1943. JMLR.org. (2014)
22. Pfeffer, A.: Figaro: An object-oriented probabilistic programming language. Technical report, Charles River Analytics (2000)
23. Puterman, M.: Markov Decision Processes: Discrete Stochastic Dynamic Programming. John Wiley and Sons (1994)
24. Sampson, A., Panchekha, P., Mytkowicz, T., McKinley, K.S., Grossman, D., Ceze, L.: Expressing and verifying probabilistic assertions. In: Proc. of PLDI, p. 14. ACM (2014)
25. Sharir, M., Pnueli, A., Hart, S.: Verification of probabilistic programs. SIAM Journal on Computing **13**(2), 292–314 (1984)
26. Shoup, V.: A Computational Introduction to Number Theory and Algebra. Cambridge University Press (2009)

Fairness for Infinitary Control

Jochen Hoenicke[⊠] and Andreas Podelski

Institut für Informatik, Universität Freiburg, Freiburg im Breisgau, Germany
{hoenicke,podelski}@informatik.uni-freiburg.de

Abstract. In 1988, Olderog and Apt developed a fair scheduler for a system with finitely many processes based on the concept of explicit scheduling. In 2010, Hoenicke, Olderog, and Podelski extended the fair scheduler from static to dynamic control. In systems with dynamic control, processes can be created dynamically. Thus, the overall number of processes can be infinite, but the number of created processes is finite at each step of an execution of the system. In this paper we extend the fair scheduler to infinitary control. In systems with infinitary control, the number of created processes can be infinite. The fair scheduler for infinitary control is perhaps interesting for its apparent unfairness: instead of treating all processes equal, the scheduler discriminates each process against finitely many other processes. However, it also privileges each process against infinitely many other processes (in fact, all but finitely many).

1 Introduction

When Ernst-Rüdiger Olderog was even younger than today, a Dutch/German colleague offered him a generous advice: "Do not work on fairness. I have solved it all!" Whether it was in immediate response or not, Olderog and Apt developed an alternative approach to fairness, called explicit fair scheduling [6]. In fact, research on fairness has never stopped. Olderog and Podelski [7] investigated explicit fair scheduling in a setting of *infinitary control*. In this setting, a system is composed of infinitely many parallel processes. The motivation of this setting was to accommodate the setting where processes can be created dynamically. In this setting of *dynamic control*, the overall number of processes can be infinite but the number of created processes is finite at each step of an execution of the system. This contrasts with the setting of *static control* in [6] where the number of processes is arbitrary but fixed. The concept of explicit scheduling is fundamental. It applies not only to the setting of static control but also to the settings of dynamic control and even infinitary control. This motivates the question whether the fair schedulers of [6] can be extended to fair schedulers for dynamic and infinitary control. This paper settles the last remaining case of strong fairness for infinitary control.

As shown in [7], the scheduler for weak fairness from [6] extends without change to the setting of dynamic control (but not to infinitary control). The scheduler for strong fairness from [6] is not sufficient for dynamic control (and

© Springer International Publishing Switzerland 2015
R. Meyer et al. (Eds.): Olderog-Festschrift, LNCS 9360, pp. 33–43, 2015.
DOI: 10.1007/978-3-319-23506-6_5

neither for the general setting of infinitary control). Hoenicke, Olderog, and Podelski [5] developed a scheduler for strong fairness for dynamic control. They also showed that this scheduler is not sufficient for infinitary control. This opened the question whether a scheduler exists for the last case of strong fairness for infinitary control.

The contribution of this paper is to close the final gap and present a scheduler for strong fairness and infinitary control. The fair scheduler for infinitary control is perhaps interesting for its apparent unfairness: instead of treating all processes equal, the scheduler discriminates each process against finitely many other processes. However, it also privileges each process against infinitely many other processes (in fact, all but finitely many). We note that the scheduler applies also to the less general settings of dynamic and even static control. The moral of the story is that *fairness can be obtained by giving up égalité*. From the perspective of a single process, it is irrelevant that the process has less priority than others. It only matters that it has a higher priority than all but finitely many other processes.

Summary of Results. We summarize the results of this paper and compare them with those of [5] and [6] in Table 1. Theorems 1 and 2 are established in the setting of infinitary control, where a possibly infinite number of processes is allowed. The setting of dynamic control is a special case of infinitary control where the number of processes in every step is finite but unbounded. The setting of static control is a special case of dynamic control where the number of processes is finite and the same for every step. So results for infinitary control carry over to the setting of dynamic control and results for dynamic control carry over to the setting of static control. We omit the reference if the result carries over.

Table 1. Schedulers for strong fairness and the extend of their validity to different settings. S_{2015} (defined in this paper), S_{2010} [5], and S_{1988} [6] are the schedulers for strong fairness for infinitary, dynamic, and static control, respectively. Static control is the setting of an arbitrary, but fixed number of processes. Dynamic control is the setting where processes can be created dynamically and the number of processes is finite but unbounded (even in a single execution). Infinitary control refers to the theoretically motivated setting where we have infinitely many processes at the same moment.

	static control	dynamic control	infinitary control
S_{2015} is sound	yes	yes	yes (Theorem 1)
is universal	yes	yes	yes (Theorem 2)
S_{2010} is sound	yes	yes [5]	no [5]
is universal	yes	yes	yes [5]
S_{1988} is sound	yes [6]	no [5]	no
is universal	yes	yes	yes [5]

Roadmap. Presently you are still reading the introduction. In Section 2 we introduce the concepts on which we build in this paper. We generalize Dijkstra's guarded command programs to *infinitary* guarded command programs, i.e., with infinitely many branches in the **do** loops. These programs formalize the setting of *infinitary control* where infinitely many processes can be active at the same moment. In *dynamic control* only finitely many processes can be active at each moment. We then reformulate the classical notion of (strong) fairness for infinitary guarded command programs. Next, we present the notion of explicit scheduling and the specific schedulers for strong fairness from [6] and [5]. In Section 3 we present a new scheduler $\mathbb{S}_{2015}$ that is valid for infinitary control. We prove its correctness and its universality. Section 4 concludes this paper.

2   Basic Concepts

Though the motivation for considering fairness stems from concurrency, it is easier and more elegant to study it in terms of structured nondeterministic programs such as Dijkstra's guarded commands [4]. We follow this approach in this paper. In this section, we carry the classical definitions of fairness from Dijkstra's guarded command language over to an infinitary guarded command language, i.e., with infinitely many branches in **do** loops. It is perhaps a surprise that the definitions carry over directly. We then immediately have the definitions of fairness of programs with dynamically created processes because we will define those formally as a subclass of infinitary guarded command programs.

2.1   Infinite Control

We introduce programs with *infinitary control* by extending Dijkstra's language of guarded command programs [3] with **do** loops that have infinitely many branches. Syntactically, these **do** loops are statements of the form

$$S \equiv \textbf{do} \ []_{i=0}^{\infty} \ B_i \to S_i \ \textbf{od} \tag{1}$$

where for each $i \in \mathbb{N}$ the *component* $B_i \to S_i$ consists of a Boolean expression B_i, its *guard*, and the statement S_i, its *command*. Therefore a component $B_i \to S_i$ is called a *guarded command* and S is called an *infinitary guarded command*.

We define the class of programs with dynamic control as a subclass of programs with infinitary control. At each moment each of the infinitely many processes "exists" (whether it has been created or not). Each process is modeled by a branch in the infinitary **do** loop. However, at each moment, only finitely many processes have been created (or activated or allocated). All others are dormant.

Processes are referred to by natural numbers. The process (with number) i is represented by the guarded command $B_i \to S_i$. To model process creation we use a Boolean expression cr_i for each process i such that this process is considered as being created if cr_i evaluates to true. All other processes are treated as not being created yet. It is an important assumption that a created process can disappear but not reappear, i.e., once the value of the expression cr_i has changed from true to false it cannot go back to true.

We define a structural operational semantics in the sense of Plotkin [8] for infinitary guarded commands. As usual, it is defined in terms of transitions between configurations. A *configuration* K is a pair $<S, \sigma>$ consisting of a statement S that is to be executed and a state σ that assigns a value to each program variable. A *transition* is written as a step $K \rightarrow K'$ between configurations. To express termination we use the empty statement E: a configuration $<E, \sigma>$ denotes termination in the state σ. For a Boolean expression B we write $\sigma \models B$ if B evaluates to true in the state σ. Process i is *created* in a state σ if $\sigma \models cr_i$ and it is *enabled* in state σ if it is created and its guard B_i evaluates to true, formally, $\sigma \models cr_i \wedge B_i$.

For the infinitary **do** loop S as in (1) we have two cases of transitions:

1. $<S, \sigma> \rightarrow <S_i; S, \sigma>$ if $\sigma \models cr_i \wedge B_i$ for each $i \in \mathbb{N}$,

2. $<S, \sigma> \rightarrow <E, \sigma>$ if $\sigma \models \bigwedge_{i=1}^{\infty} \neg (cr_i \wedge B_i)$.

Case 1 states that each *enabled* component $B_i \rightarrow S_i$ of S, i.e., with both the expression cr_i and the guard B_i evaluating to true in the current state σ, can be entered. If more than one component of S is enabled, one of them will be chosen nondeterministically. The successor configuration $<S_i; S, \sigma>$ formalizes the repetition of the **do** loop: once the command S_i is executed the whole loop S has to be executed again. Formally, the transitions of the configuration $<S_i; S, \sigma>$ are determined by the transition rules for the other statements of the guarded command language. For further details see, e.g., [1]. Case 2 states that the **do** loop terminates if none of the components is enabled any more, i.e, if all expressions $cr_i \wedge B_i$ evaluate to false in the state σ.

In this paper we investigate programs with only *one* infinitary **do** loop S of the form (1). This simplifies its definition of fairness and is sufficient for modeling dynamic control. An *execution* of S starting in a state σ_0 is a sequence of transitions

$$K_0 \rightarrow K_1 \rightarrow K_2 \rightarrow \dots, \tag{2}$$

with $K_0 = <S, \sigma_0>$ as the initial configuration, which is either infinite or maximally finite, i.e., the sequence cannot be extended further by some transition.

Consider a program S of the form (1). Then for S having *infinitary control* there is no further requirement on the set of created processes. A program S has *dynamic control* if for every execution (2) of S the set of created processes is finite in every state of a configuration in (2).

A program S has *bounded control* if for every execution (2) there exists some $n \in \mathbb{N}$ such that the number of created processes is bounded by n in every state of a configuration in (2). A program S has *static control* if there is a fixed finite set F of processes such that for every execution (2) the set of created processes is contained in F in every state of a configuration in (2).

Note that we have the following hierarchy: programs with static control are a special case of programs with bounded control, which are a special case of programs with dynamic control, which in turn are a special case of programs with infinitary control.

2.2   Fairness

In this paper we extend the definition of fairness[1] of [6] from programs with static control to programs with process creation and infinitary control. Since fairness can be expressed in terms of created, enabled, and selected processes only, we abstract from all other details in executions and define it on runs.

We now pick an execution as in (2) and define the corresponding run. A transition $K_j \to K_{j+1}$ with $j \in \mathbb{N}$ is a *select transition* if it consists of the selection of an enabled process of S, formally, if $K_j = <S, \sigma>$ and $K_{j+1} = <S_i; S, \sigma>$ with $\sigma \models cr_i \wedge B_i$ for some $i \in \mathbb{N}$, so process i has been *selected* for execution in this transition. We define the *selection* of the transition $K_j \to K_{j+1}$ as the triple (C_j, E_j, i_j), where C_j is the set of all created processes, i.e.,

$$C_j = \{i \in \mathbb{N} \mid \sigma \models cr_i\},$$

and E_j is the subset of all enabled processes, i.e.,

$$E_j = \{i \in C_j \mid \sigma \models B_i\},$$

and i_j is the (index of the) selected process, i.e., $i_j = i$. Obviously, the selected command is among the enabled components. A *run of the execution* (2) is the sequence of all its selections, formally, the sequence

$$(C_{j_0}, E_{j_0}, i_{j_0})(C_{j_1}, E_{j_1}, i_{j_1})\ldots$$

such that $C_{j_0} C_{j_1} \ldots$ is the subsequence of configurations with outgoing select transitions. Computations that do not pass through any select transition yield the empty run. A *run of a program* S is the run of one of its executions.

A run

$$(C_0, E_0, i_0)(C_1, E_1, i_1)(C_2, E_2, i_2)\ldots \tag{3}$$

is called *fair* if it satisfies the condition

$$\forall i \in \mathbb{N} : (\overset{\infty}{\exists} j \in \mathbb{N} : i \in E_j \to \overset{\infty}{\exists} j \in \mathbb{N} : i = i_j).$$

where the quantifier $\overset{\infty}{\exists}$ denotes "there exist infinitely many". By our assumption (see Subsection 2.1), the fact that the process i is infinitely often enabled, formally $\overset{\infty}{\exists} j \in \mathbb{N} : i \in E_j$, implies by $E_j \subseteq C_j$ that process i is created at some moment and stays created forever, formally $\exists j_0 \in \mathbb{N} \; \forall j \geq j_0 : i \in C_j$.

In a fair run, every process i which is enabled infinitely often, is selected infinitely often. Note that every finite run is trivially fair. An *execution* of a program S of the form (1) is *fair* if its run is fair. Thus for fairness only select transitions are relevant; transitions inside the commands S_i of S do not matter. Again, every finite execution is trivially fair. Thus we concentrate on infinite executions throughout this paper.

---

[1] In the literature, this notion of fairness is qualified as *strong fairness* (or *compassion*). For brevity, we simply refer to this notion without the qualifier in this paper.

Although we are not interested in the case where infinitely many processes can be enabled at the same time (continuously or infinitely often) and although this case is perhaps not practically relevant, the definition of fairness still makes sense, i.e., there exist fair executions in this case.

2.3   Explicit Scheduling

We extend the definition of a scheduler from [6] to the setting of infinitary control. In a given state σ the scheduler inputs a set C of created processes and a subset $E \subseteq C$ of enabled processes. It outputs some process $i \in E$ and transitions to a new state σ'. We require that the scheduler is totally defined, i.e., for every scheduler state and every input set E the scheduler will produce an output $i \in E$ and update its scheduler state. Thus a scheduler can never block the execution of a program but only influence its direction. Summarizing, we arrive at the following definition.

Definition 1 ([6]). *A scheduler is a triple* $\mathbb{S} = (\Sigma, \Sigma_0, \delta)$*, where*

- Σ *is a set of* states *with typical element* σ,
- $\Sigma_0 \subseteq \Sigma$ *is the set of* initial *states, and*
- δ *is a* transition relation *of the form*

$$\delta \subseteq \Sigma \times 2^{\mathbb{N}} \times 2^{\mathbb{N}} \times \mathbb{N} \times \Sigma$$

which is total *in the following sense:*

$$\forall \sigma \in \Sigma \ \forall C \in 2^{\mathbb{N}} \ \forall E \in 2^C \setminus \{\emptyset\} \ \exists i \in E \ \exists \sigma' \in \Sigma : (\sigma, C, E, i, \sigma') \in \delta.$$

Thus for every state σ*, every set* C *of created processes, and every nonempty subset* $E \subseteq C$ *of enabled processes there exists a process* $i \in E$ *and the updated state* σ' *such that the tuple* $(\sigma, C, E, i, \sigma')$ *satisfies the transition relation* δ*.*

A run $(C_0, E_0, i_0)(C_1, E_1, i_1)(C_2, E_2, i_2)\ldots$ *is produced by a scheduler* $\mathbb{S}$ *if there exists an infinite sequence* $\sigma_0 \sigma_1 \sigma_2 \ldots \in \Sigma^\omega$ *with* $\sigma_0 \in \Sigma_0$ *such that*

$$(\sigma_j, C_j, E_j, i_j, \sigma_{j+1}) \in \delta$$

holds for all $j \in \mathbb{N}$*. A scheduler* $\mathbb{S}$ *is* sound *if every run that is produced by* $\mathbb{S}$ *is* fair*. A scheduler* $\mathbb{S}$ *is* universal *if every fair run is produced by* $\mathbb{S}$*. A scheduler* $\mathbb{S}$ *is* valid *if it is both sound and universal.*

2.4   The Schedulers $\mathbb{S}_{1988}$ and $\mathbb{S}_{2010}$

The schedulers $\mathbb{S}_{1988}$ and $\mathbb{S}_{2010}$ given in [5,6] use auxiliary integer-valued variables (so called *scheduling variables*), one for each process, to keep track of the relative urgency of each process (relative to the other processes). Making it more urgent is implemented by decrementing its scheduling value. Thus, scheduling values can become negative. The crucial step is the non-deterministic update to a *nonnegative* integer each time after the process has been selected. Then, the process is not necessarily less urgent than all other processes. However, it is definitely less urgent than those that already have a negative scheduling value.

In [5] a scheduler for fairness of programs with dynamic control was proposed. With each process i it associates a *scheduling variable* $z[i]$ representing a priority assigned to that process. A process i has a higher priority than a process j if $z[i] < z[j]$ holds.

Definition 2 ([6],[5]). *The schedulers* $\mathbb{S}_{1988}$ *and* $\mathbb{S}_{2010}$ *are defined as the scheduler* $(\Sigma, \Sigma_0, \delta)$ *where*

- *The states* $\sigma \in \Sigma$ *are given by the values of an infinitary array* z *of type* $\mathbb{N} \to \mathbb{Z}$, *i.e.,* $z[i]$ *is a positive or negative integer for each* $i \in \mathbb{N}$.
- *The initial states in* Σ_0 *are those where each scheduler variable* $z[i]$ *has some nonnegative integer value.*
- *The relation* $(\sigma, C, E, i, \sigma') \in \delta$ *holds for states* $\sigma, \sigma' \in \Sigma$, *a set* C *of created processes, a set* $E \subseteq C$ *of enabled processes, and a process* $i \in E$ *if the value of* $z[i]$ *is minimal in* σ, *i.e., if*

$$z[i] = min\{z[k] \mid k \in E\}$$

holds in σ, *and* σ' *is obtained from* σ *by executing the statement UPDATE$_i$ for* $\mathbb{S}_{1988}$ *resp. S-UPDATE$_i$ for* $\mathbb{S}_{2010}$ *where*

$$UPDATE_i \equiv z[i] := ?;$$
$$\textbf{for all}\ \ j \in E \setminus \{i\}\ \textbf{do}\ z[j] := z[j] - 1\ \textbf{od}.$$
$$S\text{-}UPDATE_i \equiv z[i] := ?;$$
$$\textbf{for all}\ \ j \in C \setminus \{i\}\ \textbf{do}\ z[j] := z[j] - 1\ \textbf{od}.$$

Note that the transition relation δ is total as required by Definition 1. The update of the scheduling variables guarantees that the priorities of all created but not selected processes j (resp. of all created processes for *S-UPDATE*) are increased. The priority of the selected process i, however, is reset to an arbitrary nonnegative number. The idea is that by gradually increasing the priority of processes that are not taken, their activation cannot be refused forever.

We next present the two main results of [5].

- The scheduler $\mathbb{S}_{2010}$ is valid for dynamic control.
- The scheduler $\mathbb{S}_{2010}$ is not valid for infinitary control.

We give the proof from [5] which shows that $\mathbb{S}_{2010}$ is *not sound* for programs with infinitary control. Table 2 shows the initial segment of a run produced by $\mathbb{S}_{2010}$ where *every* process is treated unfairly. More precisely, every process is always enabled but selected *only once*, in the ith selection of the run: $(\mathbb{N}, \mathbb{N}, 0)(\mathbb{N}, \mathbb{N}, 1)(\mathbb{N}, \mathbb{N}, 2)\ldots$ This is possible by choosing the corresponding sequence $\sigma_0 \sigma_1 \sigma_3 \ldots$ of scheduler states as follows:

$$\sigma_j(z[i]) = \begin{cases} i + 1 - j & \text{if } i < j \\ -j & \text{if } i \geq j \end{cases}$$

Table 2. A run of a system with infinitely many processes under the scheduler $\mathbb{S}_{2010}$. The run is not fair (in fact, every process is treated unfairly). Each entry in the table shows the value of the scheduling variable $z[i]$ of the process i in the scheduler state σ_j of the run. A star $*$ after a value in, say, the i-th row and the j-th column indicates that in the state σ_j the process i is scheduled.

process	$\sigma_0(z)$	$\sigma_1(z)$	$\sigma_2(z)$	$\sigma_3(z)$	$\sigma_4(z)$	...
0	0*	0	-1	-2	-3	...
1	0	-1*	0	-1	-2	...
2	0	-1	-2*	0	-1	...
3	0	-1	-2	-3*	0	...
4	0	-1	-2	-3	-4*	...
...	...	...	...	...	...	...

3   The Scheduler $\mathbb{S}_{2015}$ for Infinitary Fairness

The counterexample given in Table 2 leads to the following observation regarding infinitary fairness. Namely, it is not enough to treat all processes equal to be fair. In the counterexample every process is scheduled exactly once and the run is not fair for *every process*. A fair schedule needs to treat the processes in an unequal manner. In particular, one can show that in a fair run some processes are scheduled arbitrarily often before other processes are scheduled for the first time.

We use this idea to change the scheduler into a new fair scheduler for infinitary fairness. It will prefer processes with a small process identifier over processes with a large process identifier. Although this sounds unfair, it is exactly what is needed to treat all processes fair. The intuition is that in the end every process has a "small" process identifier (in the sense that it is smaller than almost all other process identifiers). Thus the changed scheduler will schedule every process more often.

The only change from the scheduler $\mathbb{S}_{2010}$ is the definition of the priority. Instead of choosing the process with the minimal value $z[i]$ we choose the process i with the minimal value $z[i] + i$, i. e., we add the process id to the integer representing the urgency.

Definition 3. *The scheduler $\mathbb{S}_{2015}$ results from $\mathbb{S}_{2010}$ by replacing the relation δ with*

- *The relation $(\sigma, C, E, i, \sigma') \in \delta$ holds for states $\sigma, \sigma' \in \Sigma$, a set C of created processes, a set $E \subseteq C$ of enabled processes, and a process $i \in E$ if the value of $z[i] + i$ is minimal in σ, i.e., if*

$$z[i] + i = min\{z[k] + k \mid k \in E\}$$

holds in σ, and σ' is obtained from σ by executing the following statement:

$S\text{-}UPDATE_i \equiv z[i] := ?;$
$\qquad$ **for all** $j \in C \setminus \{i\}$ **do** $z[j] := z[j] - 1$ **od**.

Theorem 1. *The scheduler* $\mathbb{S}_{2015}$ *is sound for infinitary control.*

Proof. Consider a run

$$(C_0, E_0, i_0)\ldots(C_j, E_j, i_j)\ldots \tag{4}$$

of a program of the form (1) with infinitary control that is produced by $\mathbb{S}_{2015}$ using the sequence $\sigma_0\,\sigma_1\ldots\sigma_j\,\sigma_{j+1}\ldots$ of scheduler states. We claim that (4) is fair.

Suppose the contrary holds. Then there exists some process i that is enabled infinitely often, but from some moment on never selected. Formally, for some $j_0 \geq 0$

$$(\overset{\infty}{\exists} j \in \mathbb{N} : i \in E_j) \wedge (\forall j \geq j_0 : i \neq i_j)$$

holds in (4). Then the variable $z[i]$ of $\mathbb{S}_{2015}$, which gets decremented whenever the process i is not selected, becomes arbitrarily small. Thus, we can choose j_0 large enough so that $z[i] + i < 0$ holds in σ_{j_0}. Consider the set

$$Cr_{i,j} = \{k \in \mathbb{N} \mid k \in C_j \wedge \sigma_j \models z[k] + k \leq z[i] + i\}$$

of all created processes in C_j whose priority is least that of the neglected process i, formally, whose scheduling variable has at most the value of the scheduling variable of i. Since $z[k]$ is either set to a nonnegative number or it is decremented by one, it cannot go below $-j_0$ in j_0 steps. Hence, there are at most j_0 processes k with $z[k] + k < 0$. Therefore, the set Cr_{i,j_0} is finite in σ_{j_0}.

We show that for all $j \geq j_0$:

$$Cr_{i,j+1} \subseteq Cr_{i,j} \quad \text{and} \quad Cr_{i,j+1} \neq Cr_{i,j} \text{ if } i \in E_j. \tag{5}$$

Consider a process p that was not in $Cr_{i,j}$. We show $p \notin Cr_{i,j+1}$ to prove the inclusion. If p was scheduled in step j, then $\sigma_{j+1} \models z[i] + i < 0 \leq z[p] + p$, thus $p \notin Cr_{i,j+1}$.

If process p is newly created in step j we exploit two facts. (1) By the definition of $S\text{-}UPDATE_i$, its scheduling variable $z[p]$ is not decremented as long as p is not created. (2) The process p has not been created before by the assumption that a created process can disappear but not reappear, stated in Subsection 2.1. By (1) and (2), $z[p]$ has still its initial nonnegative value in state σ_{j+1}, thus $\sigma_{j+1} \models z[p] \geq 0$. So $p \notin Cr_{i,j+1}$.

If we take a process p different from the selected process then in the successor state σ_{j+1} the validity of the inequality $z[p] + p \leq z[i] + i$ is preserved (both p and i have their scheduling variable decremented by the definition of $S\text{-}UPDATE_i$).

If process i is enabled in step j, the scheduler needs to select a process p from $Cr_{i,j}$. As seen before, the scheduled process is not in $Cr_{i,j+1}$, thus $Cr_{i,j} \neq Cr_{i,j+1}$. This proves property (5).

By assumption, i is enabled infinitely often. By (5), the set $Cr_{i,j}$ is strictly decreasing infinitely often. This contradicts the fact that Cr_{i,j_0} is finite. $\qquad\square$

Theorem 2. *The scheduler* $\mathbb{S}_{2015}$ *is universal for infinitary control.*

Proof. Consider a fair run

$$(C_0, E_0, i_0)(C_1, E_1, i_1)(C_2, E_2, i_2). \ldots \tag{6}$$

We show that (6) can be produced by $\mathbb{S}_{2015}$ by constructing a sequence $\sigma_0 \ldots \sigma_j \ldots$ of scheduler states satisfying $(\sigma_j, C_j, E_j, i_j, \sigma_{j+1}) \in \delta$ for every $j \in \mathbb{N}$. The first step of the construction determines for every step the maximum process number of every process that was scheduled until this step,

$$p(j) = \max\{i_k \mid k < j\}.$$

The function $p(j)$ is monotone.

Because the run (6) is fair, there must be for every process i and every step j a step $m(i, j)$, such that either (1) $m(i, j) \geq j$ and i is scheduled in step $m(i, j)$ or (2) i becomes disabled forever in step $m(i, j)$. The value $m(i, j)$ is defined as

$$m(i, j) = min \left\{ m \in \mathbb{N} \left| \begin{array}{c} (1) \ (j \leq m \wedge i_m = i) \\ \vee \\ (2) \ (\forall n \geq m : i \notin E_n) \end{array} \right. \right\}.$$

The construction proceeds by assigning appropriate values to the scheduling variables $z[i]$ of $\mathbb{S}_{2015}$. For $i, j \in \mathbb{N}$ we put

$$\sigma_j(z[i]) = max(0, p(m(i, j)) - i) + |\{k \in \mathbb{N} \mid j \leq k < m(i, j) \wedge i \in C_k\}|$$
$$- |\{k \in \mathbb{N} \mid m(i, j) \leq k < j \wedge i \in C_k\}|, \tag{7}$$

In case (1) of the definition of $m(i, j)$, i.e., when i is eventually selected, the value $\sigma_j(z[i])$ is nonnegative. However, in case (2) of the definition of $m(i, j)$, i.e., when i is not enabled any more, the value $\sigma_j(z[i])$ can denote arbitrarily negative values.

This construction of values $\sigma_j(z[i])$ is possible with the assignments in $\mathbb{S}_{2015}$. Initially the value chosen by $\sigma_0(z[i])$ is non-negative since the last set is empty. If i is not scheduled in step j, then $m(i, j) = m(i, j + 1)$. If $j \geq m(i, j)$ the value j is added to the second set in (7) if $i \in C_j$. If $j < m(i, j)$ the value j is removed from the first set in (7) if $i \in C_j$. Thus, the value of $z[i]$ decreases if and only if $i \in C_j$ as demanded by *S-UPDATE*. If i is scheduled in step j, then $m(i, j + 1) \geq j + 1$ (since $i \in E_j$) and $\sigma_{j+1}(z[i])$ is a non-negative number.

In the constructed run the selected process i_j of step j has the scheduling value $z[i_j] + i = p(j)$, since $m(i_j, j) = j$. For all other enabled processes k, we have $m(k, j) > j$. Hence, $z[k] + k > p(m(k, j))$ since the first set in (7) contains k and the second set is empty. Thus, $z[k] + k > p(m(i, j))$. So i is the unique enabled process with the minimum of all scheduling values. $\qquad\square$

4 Conclusion

We have given an explicit scheduler for the setting of infinitary control. This setting encompasses static control and dynamic control. The generality of our setting allows us to shed a new light on the notion of fairness: *fairness is different from égalité*.

We see the potential of explicit scheduling in its use for program analysis [2]. Instead of implementing fairness in the program analyzer, we can apply a generic program analyzer to the program with an explicit fair scheduler. By using a universal scheduler as the one presented in this paper, this approach is oblivious to a particular instance of an operating system.

References

1. Apt, K.-R., Olderog, E.-R.: Verification of Sequential and Concurrent Programs, 2nd edn., Springer (1997)
2. Cousot, P., Cousot, R.: Abstract interpretation: A unified lattice model for static analysis of programs by construction or approximation of fixedpoints. In: POPL, pp. 238–252. ACM (1977)
3. Dijkstra, E.W.: Guarded commands, nondeterminacy and formal derivation of programs. Comm. of the ACM **18**, 453–457 (1975)
4. Francez, N.: Fairness. Springer-Verlag, New York (1986)
5. Hoenicke, J., Olderog, E.-R., Podelski, A.: Fairness for dynamic control. In: Esparza, J., Majumdar, R. (eds.) TACAS 2010. LNCS, vol. 6015, pp. 251–265. Springer, Heidelberg (2010)
6. Olderog, E.R., Apt, K.R.: Fairness in parallel programs, the transformational approach. ACM TOPLAS **10**, 420–455 (1988)
7. Olderog, E.-R., Podelski, A.: Explicit fair scheduling for dynamic control. In: Dams, D., Hannemann, U., Steffen, M. (eds.) Concurrency, Compositionality, and Correctness. LNCS, vol. 5930, pp. 96–117. Springer, Heidelberg (2010)
8. Plotkin, G.: A structural approach to operational semantics. J. of Logic and Algebraic Programming **60–61**, 17–139 (2004)

Evaluation Trees for Proposition Algebra
The Case for Free and Repetition-Proof Valuation Congruence

Jan A. Bergstra and Alban Ponse$^{(\boxtimes)}$

Section Theory of Computer Science, Informatics Institute,
Faculty of Science, University of Amsterdam, Amsterdam, The Netherlands
{j.a.bergstra,a.ponse}@uva.nl
https://staff.fnwi.uva.nl/

Abstract. Proposition algebra is based on Hoare's conditional connective, which is a ternary connective comparable to if-then-else and used in the setting of propositional logic. Conditional statements are provided with a simple semantics that is based on evaluation trees and that characterizes so-called free valuation congruence: two conditional statements are free valuation congruent if, and only if, they have equal evaluation trees. Free valuation congruence is axiomatized by the four basic equational axioms of proposition algebra that define the conditional connective. A valuation congruence that is axiomatized in proposition algebra and that identifies more conditional statements than free valuation congruence is repetition-proof valuation congruence, which we characterize by a simple transformation on evaluation trees.

Keywords: Conditional composition · Evaluation tree · Proposition algebra · Short-circuit evaluation · Short-circuit logic

1 Introduction

In 1985, Hoare's paper *A couple of novelties in the propositional calculus* [12] was published. In this paper the ternary connective $\_ \lhd \_ \rhd \_$ is introduced as the *conditional*.[1] A more common expression for a conditional statement

$$P \lhd Q \rhd R$$

Dedicated to Ernst-Rüdiger Olderog on the occasion of his sixtieth birthday. Jan Bergstra recalls many discussions during various meetings as well as joint work with Ernst-Rüdiger and Jan Willem Klop on readies, failures, and chaos back in 1987. Alban Ponse has pleasant memories of the process of publishing [8], the *Selected Papers from the Workshop on Assertional Methods*, of which Ernst-Rüdiger, who was one of the invited speakers at this workshop (held at CWI in November 1992), is one of the guest editors. An extended version of this paper appeared as report [6].

[1] To be distinguished from Hoare's *conditional* introduced in his 1985 book on CSP [11] and in his well-known 1987 paper *Laws of Programming* [10] for expressions $P \lhd b \rhd Q$ with P and Q denoting programs and b a Boolean expression.

© Springer International Publishing Switzerland 2015
R. Meyer et al. (Eds.): Olderog-Festschrift, LNCS 9360, pp. 44–61, 2015.
DOI: 10.1007/978-3-319-23506-6_6

Table 1. The set CP of equational axioms for free valuation congruence

$$x \triangleleft \mathsf{T} \triangleright y = x \tag{CP1}$$

$$x \triangleleft \mathsf{F} \triangleright y = y \tag{CP2}$$

$$\mathsf{T} \triangleleft x \triangleright \mathsf{F} = x \tag{CP3}$$

$$x \triangleleft (y \triangleleft z \triangleright u) \triangleright v = (x \triangleleft y \triangleright v) \triangleleft z \triangleright (x \triangleleft u \triangleright v) \tag{CP4}$$

is "if Q then P else R", but in order to reason systematically with conditional statements, a notation such as $P \triangleleft Q \triangleright R$ is preferable. In a conditional statement $P \triangleleft Q \triangleright R$, first Q is evaluated, and depending on that evaluation result, then either P or R is evaluated (and the other is not) and determines the final evaluation result. This evaluation strategy is reminiscent of *short-circuit* evaluation.[2] In [12], Hoare proves that propositional logic can be characterized by extending equational logic with eleven axioms on the conditional, some of which employ constants for the truth values *true* and *false*.

In 2011, we introduced *Proposition Algebra* in [4] as a general approach to the study of the conditional: we defined several *valuation congruences* and provided equational axiomatizations of these congruences. The most basic and least identifying valuation congruence is *free* valuation congruence, which is axiomatized by the axioms in Table 1, where we use constants T and F for the truth values *true* and *false*. These axioms stem from [12] and define the conditional as a primitive connective. We use the name CP (for Conditional Propositions) for this set of axioms. Interpreting a conditional statement as an if-then-else expression, axioms (CP1)-(CP3) are natural, and axiom (CP4) (distributivity) can be clarified by case analysis: if z evaluates to *true* and y as well, then x determines the result of evaluation; if z evaluates to *true* and y evaluates to *false*, then v determines the result of evaluation, and so on and so forth. A simple example, taken from [4], is the conditional statement that a pedestrian evaluates before crossing a road with two-way traffic driving on the right:

$$(look\text{-}left\text{-}and\text{-}check \triangleleft look\text{-}right\text{-}and\text{-}check \triangleright \mathsf{F}) \triangleleft look\text{-}left\text{-}and\text{-}check \triangleright \mathsf{F}.$$

This statement requires one, or two, or three atomic evaluations and cannot be simplified to one that requires less.[3]

In Section 2 we characterize free valuation congruence with help of *evaluation trees*, which are simple binary trees proposed by Daan Staudt in [13] (that appeared in 2012). Given a conditional statement, its evaluation tree represents all possible consecutive atomic evaluations followed by the final evaluation result (comparable to a truth table in the case of propositional logic).

---

[2] Short-circuit evaluation denotes the semantics of binary propositional connectives in which the second argument is evaluated only if the first argument does not suffice to determine the value of the expression.

[3] Note that $look\text{-}left\text{-}and\text{-}check \triangleleft (look\text{-}right\text{-}and\text{-}check \triangleleft look\text{-}left\text{-}and\text{-}check \triangleright \mathsf{F}) \triangleright \mathsf{F}$ prescribes by axioms (CP4) and (CP2) the same evaluation.

Two conditional statements are equivalent with respect to free valuation congruence if their evaluation trees are equal. Free valuation congruence identifies less than the equivalence defined by Hoare's axioms in [12]. For example, the atomic proposition a and the conditional statement $\mathsf{T} \lhd a \rhd a$ are not equivalent with respect to free valuation congruence, although they are equivalent with respect to *static* valuation congruence, which is the valuation congruence that characterizes propositional logic.

A valuation congruence that identifies more than free and less than static valuation congruence is *repetition-proof* valuation congruence, which is axiomatized by CP extended with two (schematic) axioms, one of which reads

$$x \lhd a \rhd (y \lhd a \rhd z) = x \lhd a \rhd (z \lhd a \rhd z),$$

and thus expresses that if atomic proposition a evaluates to *false*, a consecutive evaluation of a also evaluates to *false*, so the conditional statement at the y-position will not be evaluated and can be replaced by any other. As an example, $\mathsf{T} \lhd a \rhd a = \mathsf{T} \lhd a \rhd (\mathsf{T} \lhd a \rhd \mathsf{F}) = \mathsf{T} \lhd a \rhd (\mathsf{F} \lhd a \rhd \mathsf{F})$, and the left-hand and right-hand conditional statements are equivalent with respect to repetition-proof valuation congruence, but not with respect to free valuation congruence.

In Section 3 we characterize repetition-proof valuation congruence by defining a transformation on evaluation trees that yields *repetition-proof evaluation trees*: two conditional statements are equivalent with respect to repetition-proof valuation congruence if, and only if, they have equal repetition-proof evaluation trees. Although this transformation on evaluation trees is simple and natural, our proof of the mentioned characterization—which is phrased as a completeness result—is non-trivial and we could not find a proof that is essentially simpler.

In section 4 we discuss the general structure of the proof of this last result, which is based on normalization of conditional statements, and we conclude with a brief digression on *short-circuit logic* and an example on the use of repetition-proof valuation congruence.

The approach followed in this paper also works for most other valuation congruences defined in [4] and the case for repetition-proof valuation congruence is prototypical, as we show in [6].

2   Evaluation Trees for Free Valuation Congruence

Consider the signature $\Sigma_{\mathrm{CP}}(A) = \{\_ \lhd \_ \rhd \_, \mathsf{T}, \mathsf{F}, a \mid a \in A\}$ with constants T and F for the truth values *true* and *false*, respectively, and constants a for atomic propositions, further called *atoms*, from some countable set A. We write

$$C_A$$

for the set of closed terms, or *conditional statements*, over the signature $\Sigma_{\mathrm{CP}}(A)$. Given a conditional statement $P \lhd Q \rhd R$, we refer to Q as its *central condition*.

We define the *dual* P^d of $P \in C_A$ as follows:

$$\mathsf{T}^d = \mathsf{F}, \qquad\qquad\qquad a^d = a \quad (\text{for } a \in A),$$
$$\mathsf{F}^d = \mathsf{T}, \qquad\qquad (P \lhd Q \rhd R)^d = R^d \lhd Q^d \rhd P^d.$$

Observe that CP is a self-dual axiomatization: when defining $x^d = x$ for each variable x, the dual of each axiom is also in CP, and hence

$$\text{CP} \vdash P = Q \quad \Longleftrightarrow \quad \text{CP} \vdash P^d = Q^d.$$

A natural view on conditional statements in C_A involves short-circuited evaluation, similar to how we consider the evaluation of an "if y then x else z" expression. The following definition is taken from [13].

Definition 2.1. *The set $\mathcal{T}_A$ of **evaluation trees** over A **with leaves in** $\{\mathsf{T}, \mathsf{F}\}$ is defined inductively by*

$$\mathsf{T} \in \mathcal{T}_A,$$
$$\mathsf{F} \in \mathcal{T}_A,$$
$$(X \trianglelefteq a \trianglerighteq Y) \in \mathcal{T}_A \text{ for any } X, Y \in \mathcal{T}_A \text{ and } a \in A.$$

*The function $\_ \trianglelefteq a \trianglerighteq \_$ is called **post-conditional composition over** a. In the evaluation tree $X \trianglelefteq a \trianglerighteq Y$, the root is represented by a, the left branch by X and the right branch by Y.*

We refer to trees in $\mathcal{T}_A$ as evaluation trees, or trees for short. Post-conditional composition and its notation stem from [2]. Evaluation trees play a crucial role in the main results of [13]. In order to define our "evaluation tree semantics", we first define an auxiliary function on trees.

Definition 2.2. *Given evaluation trees $Y, Z \in \mathcal{T}_A$, the **leaf replacement** function $[\mathsf{T} \mapsto Y, \mathsf{F} \mapsto Z] : \mathcal{T}_A \to \mathcal{T}_A$, for which post-fix notation*

$$X[\mathsf{T} \mapsto Y, \mathsf{F} \mapsto Z]$$

is adopted, is defined as follows, where $a \in A$:

$$\mathsf{T}[\mathsf{T} \mapsto Y, \mathsf{F} \mapsto Z] = Y,$$
$$\mathsf{F}[\mathsf{T} \mapsto Y, \mathsf{F} \mapsto Z] = Z,$$
$$(X_1 \trianglelefteq a \trianglerighteq X_2)[\mathsf{T} \mapsto Y, \mathsf{F} \mapsto Z] = X_1[\mathsf{T} \mapsto Y, \mathsf{F} \mapsto Z] \trianglelefteq a \trianglerighteq X_2[\mathsf{T} \mapsto Y, \mathsf{F} \mapsto Z].$$

We note that the order in which the replacements of leaves of X is listed is irrelevant and we adopt the convention of not listing identities inside the brackets, e.g., $X[\mathsf{F} \mapsto Z] = X[\mathsf{T} \mapsto \mathsf{T}, \mathsf{F} \mapsto Z]$. Furthermore, repeated leaf replacements satisfy the following equation:

$$\left(X[\mathsf{T} \mapsto Y_1, \mathsf{F} \mapsto Z_1]\right)[\mathsf{T} \mapsto Y_2, \mathsf{F} \mapsto Z_2]$$
$$= X[\mathsf{T} \mapsto Y_1[\mathsf{T} \mapsto Y_2, \mathsf{F} \mapsto Z_2], \ \mathsf{F} \mapsto Z_1[\mathsf{T} \mapsto Y_2, \mathsf{F} \mapsto Z_2]].$$

We now have the terminology and notation to define the interpretation of conditional statements in C_A as evaluation trees by a function se (abbreviating short-circuit evaluation).

Definition 2.3. *The* **short-circuit evaluation function** $se : C_A \to T_A$ *is defined as follows, where* $a \in A$:

$$se(\mathsf{T}) = \mathsf{T},$$
$$se(\mathsf{F}) = \mathsf{F},$$
$$se(a) = \mathsf{T} \trianglelefteq a \trianglerighteq \mathsf{F},$$
$$se(P \triangleleft Q \triangleright R) = se(Q)[\mathsf{T} \mapsto se(P), \mathsf{F} \mapsto se(R)].$$

Example 2.4. *The conditional statement* $a \triangleleft (\mathsf{F} \triangleleft a \triangleright \mathsf{T}) \triangleright \mathsf{F}$ *yields the following evaluation tree:*

$$se(a \triangleleft (\mathsf{F} \triangleleft a \triangleright \mathsf{T}) \triangleright \mathsf{F}) = se(\mathsf{F} \triangleleft a \triangleright \mathsf{T})[\mathsf{T} \mapsto se(a), \mathsf{F} \mapsto se(\mathsf{F})]$$
$$= (\mathsf{F} \trianglelefteq a \trianglerighteq \mathsf{T})[\mathsf{T} \mapsto se(a)]$$
$$= \mathsf{F} \trianglelefteq a \trianglerighteq (\mathsf{T} \trianglelefteq a \trianglerighteq \mathsf{F}).$$

A more pictorial representation of this evaluation tree is the following, where $\trianglelefteq$ *yields a left branch and* $\trianglerighteq$ *a right branch:*

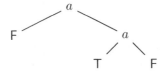

As we can see from the definition on atoms, evaluation continues in the left branch if an atom evaluates to *true* and in the right branch if it evaluates to *false*. We shall often use the constants T and F to denote the result of an evaluation (instead of *true* and *false*).

Definition 2.5. *Let* $P \in C_A$. *An* **evaluation** *of* P *is a pair* (σ, B) *where* $\sigma \in (A\{\mathsf{T}, \mathsf{F}\})^*$ *and* $B \in \{\mathsf{T}, \mathsf{F}\}$, *such that if* $se(P) \in \{\mathsf{T}, \mathsf{F}\}$, *then* $\sigma = \epsilon$ *(the empty string) and* $B = se(P)$, *and otherwise,*

$$\sigma = a_1 B_1 a_2 B_2 \cdots a_n B_n,$$

where $a_1 a_2 \cdots a_n B$ *is a complete path in* $se(P)$ *and*

- *for* $i < n$, *if* a_{i+1} *is a left child of* a_i *then* $B_i = \mathsf{T}$, *and otherwise* $B_i = \mathsf{F}$,
- *if* B *is a left child of* a_n *then* $B_n = \mathsf{T}$, *and otherwise* $B_n = \mathsf{F}$.

We refer to σ *as the* **evaluation path** *and to* B *as the* **evaluation result**.

So, an evaluation of a conditional statement P is a complete path in $se(P)$ (from root to leaf) and contains evaluation values for all occurring atoms. For instance, the evaluation tree $\mathsf{F} \trianglelefteq a \trianglerighteq (\mathsf{T} \trianglelefteq a \trianglerighteq \mathsf{F})$ from Example 2.4 encodes the evaluations $(a\mathsf{T}, \mathsf{F})$, $(a\mathsf{F}a\mathsf{T}, \mathsf{T})$, and $(a\mathsf{F}a\mathsf{F}, \mathsf{F})$. As an aside, we note that this particular evaluation tree encodes all possible evaluations of $\neg a$ && a, where && is the connective that prescribes *short-circuited conjunction* (we return to this connective in Section 4).

In turn, each evaluation tree gives rise to a *unique* conditional statement. For Example 2.4, this is $\mathsf{F} \triangleleft a \triangleright (\mathsf{T} \triangleleft a \triangleright \mathsf{F})$ (note the syntactical correspondence).

Definition 2.6. ***Basic forms over*** A *are defined by the following grammar*

$$t ::= \mathsf{T} \mid \mathsf{F} \mid t \triangleleft a \triangleright t \quad for\ a \in A.$$

We write BF_A *for the set of basic forms over* A. *The* **depth** $d(P)$ *of* $P \in BF_A$ *is defined by* $d(\mathsf{T}) = d(\mathsf{F}) = 0$ *and* $d(Q \triangleleft a \triangleright R) = 1 + \max\{d(Q), d(R)\}$.

The following two lemmas exploit the structure of basic forms and are stepping stones to our first completeness result (Theorem 2.11).

Lemma 2.7. *For each* $P \in C_A$ *there exists* $Q \in BF_A$ *such that* $\mathrm{CP} \vdash P = Q$.

Proof. First we establish an auxiliary result: if P, Q, R are basic forms, then there is a basic form S such that $\mathrm{CP} \vdash P \triangleleft Q \triangleright R = S$. This follows by structural induction on Q.

The lemma's statement follows by structural induction on P. The base cases $P \in \{\mathsf{T}, \mathsf{F}, a \mid a \in A\}$ are trivial, and if $P = P_1 \triangleleft P_2 \triangleright P_3$ there exist by induction basic forms Q_i such that $\mathrm{CP} \vdash P_i = Q_i$, hence $\mathrm{CP} \vdash P_1 \triangleleft P_2 \triangleright P_3 = Q_1 \triangleleft Q_2 \triangleright Q_3$. Now apply the auxiliary result. $\qquad\square$

Lemma 2.8. *For all basic forms* P *and* Q, $se(P) = se(Q)$ *implies* $P = Q$.

Proof. By structural induction on P. The base cases $P \in \{\mathsf{T}, \mathsf{F}\}$ are trivial. If $P = P_1 \triangleleft a \triangleright P_2$, then $Q \notin \{\mathsf{T}, \mathsf{F}\}$ and $Q \neq Q_1 \triangleleft b \triangleright Q_2$ with $b \neq a$, so $Q = Q_1 \triangleleft a \triangleright Q_2$ and $se(P_i) = se(Q_i)$. By induction we find $P_i = Q_i$, and hence $P = Q$. $\qquad\square$

Definition 2.9. ***Free valuation congruence***, *notation* $=_{se}$, *is defined on* C_A *as follows:*

$$P =_{se} Q \iff se(P) = se(Q).$$

Lemma 2.10. *Free valuation congruence is a congruence relation.*

Proof. Let $P, Q, R \in C_A$ and assume $P =_{se} P'$, thus $se(P) = se(P')$. Then $se(P \triangleleft Q \triangleright R) = se(Q)[\mathsf{T} \mapsto se(P), \mathsf{F} \mapsto se(R)] = se(Q)[\mathsf{T} \mapsto se(P'), \mathsf{F} \mapsto se(R)] = se(P' \triangleleft Q \triangleright R)$, and thus $P \triangleleft Q \triangleright R =_{se} P' \triangleleft Q \triangleright R$. The two remaining cases can be proved in a similar way. $\qquad\square$

Theorem 2.11 (Completeness of CP**).** *For all* $P, Q \in C_A$,

$$\mathrm{CP} \vdash P = Q \iff P =_{se} Q.$$

Proof. We first prove $\Rightarrow$. By Lemma 2.10, $=_{se}$ is a congruence relation and it easily follows that all CP-axioms are sound. For example, soundness of axiom (CP4) follows from

$$se(P \triangleleft (Q \triangleleft R \triangleright S) \triangleright U)$$
$$= se(Q \triangleleft R \triangleright S)[\mathsf{T} \mapsto se(P), \mathsf{F} \mapsto se(U)]$$
$$= \big(se(R)[\mathsf{T} \mapsto se(Q), \mathsf{F} \mapsto se(S)]\big)\,[\mathsf{T} \mapsto se(P), \mathsf{F} \mapsto se(U)]$$
$$= se(R)[\mathsf{T} \mapsto se(Q)[\mathsf{T} \mapsto se(P), \mathsf{F} \mapsto se(U)],$$
$$\qquad\quad \mathsf{F} \mapsto se(S)[\mathsf{T} \mapsto se(P), \mathsf{F} \mapsto se(U)]]$$
$$= se(R)[\mathsf{T} \mapsto se(P \triangleleft Q \triangleright U), \mathsf{F} \mapsto se(P \triangleleft S \triangleright U)]$$
$$= se((P \triangleleft Q \triangleright U) \triangleleft R \triangleright (P \triangleleft S \triangleright U)).$$

In order to prove $\Leftarrow$, let $P =_{se} Q$. According to Lemma 2.7 there exist basic forms P' and Q' such that $CP \vdash P = P'$ and $CP \vdash Q = Q'$, so $CP \vdash P' = Q'$. By soundness ($\Rightarrow$) we find $P' =_{se} Q'$, so by Lemma 2.8, $P' = Q'$. Hence, $CP \vdash P = P' = Q' = Q$. $\qquad\square$

A consequence of the above results is that for each $P \in C_A$ there is a *unique* basic form P' with $CP \vdash P = P'$, and that for each basic form, its se-image has exactly the same syntactic structure (replacing $\triangleleft$ by $\trianglelefteq$, and $\triangleright$ by $\trianglerighteq$). In the remainder of this section, we make this precise.

Definition 2.12. *The **basic form function** $bf : C_A \to BF_A$ is defined as follows, where $a \in A$:*

$$bf(\mathsf{T}) = \mathsf{T},$$
$$bf(\mathsf{F}) = \mathsf{F},$$
$$bf(a) = \mathsf{T} \triangleleft a \triangleright \mathsf{F},$$
$$bf(P \triangleleft Q \triangleright R) = bf(Q)[\mathsf{T} \mapsto bf(P), \mathsf{F} \mapsto bf(R)].$$

Given $Q, R \in BF_A$, the auxiliary function $[\mathsf{T} \mapsto Q, \mathsf{F} \mapsto R] : BF_A \to BF_A$ for which post-fix notation $P[\mathsf{T} \mapsto Q, \mathsf{F} \mapsto R]$ is adopted, is defined as follows:

$$\mathsf{T}[\mathsf{T} \mapsto Q, \mathsf{F} \mapsto R] = Q,$$
$$\mathsf{F}[\mathsf{T} \mapsto Q, \mathsf{F} \mapsto R] = R,$$
$$(P_1 \triangleleft a \triangleright P_2)[\mathsf{T} \mapsto Q, \mathsf{F} \mapsto R] = P_1[\mathsf{T} \mapsto Q, \mathsf{F} \mapsto R] \triangleleft a \triangleright P_2[\mathsf{T} \mapsto Q, \mathsf{F} \mapsto R].$$

(The notational overloading with the leaf replacement function on evaluation trees is harmless).

So, for given $Q, R \in BF_A$, the auxiliary function $[\mathsf{T} \mapsto Q, \mathsf{F} \mapsto R]$ applied to $P \in BF_A$ (thus, $P[\mathsf{T} \mapsto Q, \mathsf{F} \mapsto R]$) replaces all T-occurrences in P by Q, and all F-occurrences in P by R. The following two lemmas imply that bf is a normalization function.

Lemma 2.13. *For all $P \in C_A$, $bf(P)$ is a basic form.*

Proof. By structural induction. The base cases are trivial. For the inductive case we find $bf(P \triangleleft Q \triangleright R) = bf(Q)[\mathsf{T} \mapsto bf(P), \mathsf{F} \mapsto bf(R)]$, so by induction, $bf(P)$, $bf(Q)$, and $bf(R)$ are basic forms. Furthermore, replacing all T-occurrences and F-occurrences in $bf(Q)$ by basic forms $bf(P)$ and $bf(R)$, respectively, yields a basic form. $\qquad\square$

Lemma 2.14. *For each basic form P, $bf(P) = P$.*

Proof. By structural induction on P. $\qquad\square$

Definition 2.15. *The binary relation $=_{bf}$ on C_A is defined as follows:*

$$P =_{bf} Q \quad \Longleftrightarrow \quad bf(P) = bf(Q).$$

Lemma 2.16. *The relation $=_{bf}$ is a congruence relation.*

Proof. Let $P, Q, R \in C_A$ and assume $P =_{bf} P'$, thus $bf(P) = bf(P')$. Then $bf(P \triangleleft Q \triangleright R) = bf(Q)[\mathsf{T} \mapsto bf(P), \mathsf{F} \mapsto bf(R)] = bf(Q)[\mathsf{T} \mapsto bf(P'), \mathsf{F} \mapsto bf(R)] = bf(P' \triangleleft Q \triangleright R)$, and thus $P \triangleleft Q \triangleright R =_{bf} P' \triangleleft Q \triangleright R$. The two remaining cases can be proved in a similar way.   □

Before proving that CP is an axiomatization of the relation $=_{bf}$, we show that each instance of the axiom (CP4) satisfies $=_{bf}$.

Lemma 2.17. *For all $P, P_1, P_2, Q_1, Q_2 \in C_A$,*

$$bf(Q_1 \triangleleft (P_1 \triangleleft P \triangleright P_2) \triangleright Q_2) = bf((Q_1 \triangleleft P_1 \triangleright Q_2) \triangleleft P \triangleright (Q_1 \triangleleft P_2 \triangleright Q_2)).$$

Proof. By definition, the lemma's statement is equivalent with

$$\big(bf(P)[\mathsf{T} \mapsto bf(P_1), \mathsf{F} \mapsto bf(P_2)]\big)\,[\mathsf{T} \mapsto bf(Q_1), \mathsf{F} \mapsto bf(Q_2)]$$
$$= bf(P)[\mathsf{T} \mapsto bf(Q_1 \triangleleft P_1 \triangleright Q_2), \mathsf{F} \mapsto bf(Q_1 \triangleleft P_2 \triangleright Q_2)]. \tag{1}$$

By Lemma 2.13, $bf(P)$, $bf(P_i)$, and $bf(Q_i)$ are basic forms. We prove (1) by structural induction on the form that $bf(P)$ can have. If $bf(P) = \mathsf{T}$, then

$$\big(\mathsf{T}[\mathsf{T} \mapsto bf(P_1), \mathsf{F} \mapsto bf(P_2)]\big)\,[\mathsf{T} \mapsto bf(Q_1), \mathsf{F} \mapsto bf(Q_2)]$$
$$= bf(P_1)[\mathsf{T} \mapsto bf(Q_1), \mathsf{F} \mapsto bf(Q_2)]$$

and

$$\mathsf{T}[\mathsf{T} \mapsto bf(Q_1 \triangleleft P_1 \triangleright Q_2), \mathsf{F} \mapsto bf(Q_1 \triangleleft P_2 \triangleright Q_2)]$$
$$= bf(Q_1 \triangleleft P_1 \triangleright Q_2)$$
$$= bf(P_1)[\mathsf{T} \mapsto bf(Q_1), \mathsf{F} \mapsto bf(Q_2)].$$

If $bf(P) = \mathsf{F}$, then equation (1) follows in a similar way.

The inductive case $bf(P) = R_1 \triangleleft a \triangleright R_2$ is trivial (by definition of the last defining clause of the auxiliary functions $[\mathsf{T} \mapsto Q, \mathsf{F} \mapsto R]$ in Definition 2.12).   □

Theorem 2.18. *For all $P, Q \in C_A$, $\mathrm{CP} \vdash P = Q \iff P =_{bf} Q$.*

Proof. We first prove $\Rightarrow$. By Lemma 2.16, $=_{bf}$ is a congruence relation and it easily follows that arbitrary instances of the CP-axioms (CP1)-(CP3) satisfy $=_{bf}$. By Lemma 2.17 it follows that arbitrary instances of axiom (CP4) also satisfy $=_{bf}$.

In order to prove $\Leftarrow$, assume $P =_{bf} Q$. According to Lemma 2.7, there exist basic forms P' and Q' such that $\mathrm{CP} \vdash P = P'$ and $\mathrm{CP} \vdash Q = Q'$, so $\mathrm{CP} \vdash P' = Q'$. By $\Rightarrow$ it follows that $P' =_{bf} Q'$, which implies by Lemma 2.14 that $P' = Q'$. Hence, $\mathrm{CP} \vdash P = P' = Q' = Q$.   □

Corollary 2.19. *For all $P \in C_A$, $P =_{bf} bf(P)$ and $P =_{se} bf(P)$.*

Proof. By Lemma 2.13 and Lemma 2.14, $bf(P) = bf(bf(P))$, thus $P =_{bf} bf(P)$. By Theorem 2.18, $\mathrm{CP} \vdash P = bf(P)$, and by Theorem 2.11, $P =_{se} bf(P)$.   □

3   Evaluation Trees for Repetition-proof Valuation Congruence

In [4] we defined *repetition-proof* CP as the extension of the axiom set CP with the following two axiom schemes, where a ranges over A:

$$(x \lhd a \rhd y) \lhd a \rhd z = (x \lhd a \rhd x) \lhd a \rhd z, \tag{CPrp1}$$

$$x \lhd a \rhd (y \lhd a \rhd z) = x \lhd a \rhd (z \lhd a \rhd z). \tag{CPrp2}$$

We write $\mathrm{CP}_{rp}(A)$ for this extension. These axiom schemes characterize that for each atom a, a consecutive evaluation of a yields the same result, so in both cases the conditional statement at the y-position will not be evaluated and can be replaced by any other. Note that (CPrp1) and (CPrp2) are each others dual.

We define a proper subset of basic forms with the property that each conditional statement can be proved equal to such a basic form.

Definition 3.1. ***Rp-basic forms*** *are inductively defined:*

- T *and* F *are rp-basic forms, and*
- $P_1 \lhd a \rhd P_2$ *is an rp-basic form if* P_1 *and* P_2 *are rp-basic forms, and if* P_i *is not equal to* T *or* F*, then either the central condition in* P_i *is different from* a*, or* P_i *is of the form* $Q_i \lhd a \rhd Q_i$.

It will turn out useful to define a function that transforms conditional statements into rp-basic forms and that is comparable to the function bf.

Definition 3.2. *The **rp-basic form function** $rpbf : C_A \to C_A$ is defined by*

$$rpbf(P) = rpf(bf(P)).$$

The auxiliary function $rpf : BF_A \to BF_A$ is defined as follows:

$$rpf(\mathsf{T}) = \mathsf{T},$$
$$rpf(\mathsf{F}) = \mathsf{F},$$
$$rpf(P \lhd a \rhd Q) = rpf(f_a(P)) \lhd a \rhd rpf(g_a(Q)).$$

For $a \in A$, the auxiliary functions $f_a : BF_A \to BF_A$ and $g_a : BF_A \to BF_A$ are defined by

$$f_a(\mathsf{T}) = \mathsf{T},$$
$$f_a(\mathsf{F}) = \mathsf{F},$$
$$f_a(P \lhd b \rhd Q) = \begin{cases} f_a(P) \lhd a \rhd f_a(P) & \text{if } b = a, \\ P \lhd b \rhd Q & \text{otherwise,} \end{cases}$$

and

$$g_a(\mathsf{T}) = \mathsf{T},$$
$$g_a(\mathsf{F}) = \mathsf{F},$$
$$g_a(P \lhd b \rhd Q) = \begin{cases} g_a(Q) \lhd a \rhd g_a(Q) & \text{if } b = a, \\ P \lhd b \rhd Q & \text{otherwise.} \end{cases}$$

Thus, *rpbf* maps a conditional statement P to $bf(P)$ and then transforms $bf(P)$ according to the auxiliary functions rpf, f_a, and g_a.

Lemma 3.3. *For all* $a \in A$ *and* $P \in BF_A$, $g_a(f_a(P)) = f_a(f_a(P)) = f_a(P)$ *and* $f_a(g_a(P)) = g_a(g_a(P)) = g_a(P)$.

Proof. By structural induction on P. The base cases $P \in \{\mathsf{T}, \mathsf{F}\}$ are trivial. For the inductive case $P = Q \triangleleft b \triangleright R$ we have to distinguish the cases $b = a$ and $b \neq a$. If $b = a$, then

$$
\begin{aligned}
g_a(f_a(Q \triangleleft a \triangleright R)) &= g_a(f_a(Q)) \triangleleft a \triangleright g_a(f_a(Q)) \\
&= f_a(Q) \triangleleft a \triangleright f_a(Q) && \text{by IH} \\
&= f_a(Q \triangleleft a \triangleright R),
\end{aligned}
$$

and $f_a(f_a(Q \triangleleft a \triangleright R)) = f_a(Q \triangleleft a \triangleright R)$ follows in a similar way. If $b \neq a$, then $f_a(P) = g_a(P) = P$, and hence $g_a(f_a(P)) = f_a(f_a(P)) = f_a(P)$.

The second pair of equalities can be derived in a similar way. $\qquad\square$

In order to prove that for all $P \in C_A$, $rpbf(P)$ is an rp-basic form, we use the following auxiliary lemma.

Lemma 3.4. *For all* $a \in A$ *and* $P \in BF_A$, $d(P) \geq d(f_a(P))$ *and* $d(P) \geq d(g_a(P))$.

Proof. Fix some $a \in A$. We prove these inequalities by structural induction on P. The base cases $P \in \{\mathsf{T}, \mathsf{F}\}$ are trivial. For the inductive case $P = Q \triangleleft b \triangleright R$ we have to distinguish the cases $b = a$ and $b \neq a$. If $b = a$, then

$$
\begin{aligned}
d(Q \triangleleft a \triangleright R) &= 1 + \max\{d(Q), d(R)\} \\
&\geq 1 + d(Q) \\
&\geq 1 + d(f_a(Q)) && \text{by IH} \\
&= d(f_a(Q) \triangleleft a \triangleright f_a(Q)) \\
&= d(f_a(Q \triangleleft a \triangleright R)),
\end{aligned}
$$

and $d(Q \triangleleft a \triangleright R) \geq d(g_a(Q \triangleleft a \triangleright R))$ follows in a similar way.

If $b \neq a$, then $f_a(P) = g_a(P) = P$, and hence $d(P) \geq d(f_a(P))$ and $d(P) \geq d(g_a(P))$. $\qquad\square$

Lemma 3.5. *For all* $P \in C_A$, $rpbf(P)$ *is an rp-basic form.*

Proof. We first prove an auxiliary result:

$$\text{For all } P \in BF_A, \ rpf(P) \text{ is an rp-basic form.} \tag{2}$$

This follows by induction on the depth $d(P)$ of P. If $d(P) = 0$, then $P \in \{\mathsf{T}, \mathsf{F}\}$, and hence $rpf(P) = P$ is an rp-basic form. For the inductive case $d(P) = n + 1$ it must be the case that $P = Q \triangleleft a \triangleright R$. We find

$$rpf(Q \triangleleft a \triangleright R) = rpf(f_a(Q)) \triangleleft a \triangleright rpf(g_a(R)),$$

which is an rp-basic form because

- by Lemma 3.4, $f_a(Q)$ and $g_a(R)$ are basic forms with depth smaller than or equal to n, so by the induction hypothesis, $rpf(f_a(Q))$ and $rpf(g_a(R))$ are rp-basic forms,
- $rpf(f_a(Q))$ and $rpf(g_a(R))$ both satisfy the following property: if the central condition (if present) is a, then the outer arguments are equal. We show this first for $rpf(f_a(Q))$ by a case distinction on the form of Q:
  1. If $Q \in \{\mathsf{T},\mathsf{F}\}$, then $rpf(f_a(Q)) = Q$, so there is nothing to prove.
  2. If $Q = Q_1 \triangleleft a \triangleright Q_2$, then $f_a(Q) = f_a(Q_1) \triangleleft a \triangleright f_a(Q_1)$ and thus by Lemma 3.3, $rpf(f_a(Q)) = rpf(f_a(Q_1)) \triangleleft a \triangleright rpf(f_a(Q_1))$.
  3. If $Q = Q_1 \triangleleft b \triangleright Q_2$ with $b \neq a$, then $f_a(Q) = Q_1 \triangleleft b \triangleright Q_2$ and thus $rpf(f_a(Q)) = rpf(f_b(Q_1)) \triangleleft b \triangleright rpf(g_b(Q_2))$, so there is nothing to prove.

  The fact that $rpf(g_a(R))$ satisfies this property follows in a similar way.

This finishes the proof of auxiliary result (2).

The lemma's statement now follows by structural induction: the base cases (comprising a single atom a) are again trivial, and for the inductive case,

$$rpbf(P \triangleleft Q \triangleright R) = rpf(bf(P \triangleleft Q \triangleright R)) = rpf(S)$$

for some basic form S by Lemma 2.13, and by auxiliary result (2), $rpf(S)$ is an rp-basic form. □

The following, rather technical result is used in Proposition 3.7 and Lemma 3.8.

Lemma 3.6. *If $Q \triangleleft a \triangleright R$ is an rp-basic form, then $Q = rpf(Q) = rpf(f_a(Q))$ and $R = rpf(R) = rpf(g_a(R))$.*

Proof. We first prove an auxiliary result:

If $Q \triangleleft a \triangleright R$ is an rp-basic form, then $f_a(Q) = g_a(Q)$ and $f_a(R) = g_a(R)$. (3)

We prove both equalities by simultaneous induction on the structure of Q and R. The base case, thus $Q, R \in \{\mathsf{T},\mathsf{F}\}$, is trivial. If $Q = Q_1 \triangleleft a \triangleright Q_1$ and $R = R_1 \triangleleft a \triangleright R_1$, then Q and R are rp-basic forms with central condition a, so

$$
\begin{aligned}
f_a(Q) &= f_a(Q_1) \triangleleft a \triangleright f_a(Q_1) \\
&= g_a(Q_1) \triangleleft a \triangleright g_a(Q_1) \qquad\qquad \text{by IH} \\
&= g_a(Q),
\end{aligned}
$$

and the equality for R follows in a similar way. If $Q = Q_1 \triangleleft a \triangleright Q_1$ and $R \neq R_1 \triangleleft a \triangleright R_1$, then $f_a(R) = g_a(R) = R$, and the result follows as above. All remaining cases follow in a similar way, which finishes the proof of (3).

We now prove the lemma's statement by simultaneous induction on the structure of Q and R. The base case, thus $Q, R \in \{\mathsf{T},\mathsf{F}\}$, is again trivial. If $Q = Q_1 \triangleleft a \triangleright Q_1$ and $R = R_1 \triangleleft a \triangleright R_1$, then by auxiliary result (3),

$$rpf(Q) = rpf(f_a(Q_1)) \triangleleft a \triangleright rpf(f_a(Q_1)),$$

and by induction, $Q_1 = rpf(Q_1) = rpf(f_a(Q_1))$. Hence, $rpf(Q) = Q_1 \triangleleft a \triangleright Q_1$, and

$$
\begin{aligned}
rpf(f_a(Q)) &= rpf(f_a(f_a(Q_1))) \triangleleft a \triangleright rpf(g_a(f_a(Q_1))) \\
&= rpf(f_a(Q_1)) \triangleleft a \triangleright rpf(f_a(Q_1)) && \text{by Lemma 3.3} \\
&= Q_1 \triangleleft a \triangleright Q_1,
\end{aligned}
$$

and the equalities for R follow in a similar way.

If $Q = Q_1 \triangleleft a \triangleright Q_1$ and $R \neq R_1 \triangleleft a \triangleright R_1$, the lemma's equalities follow in a similar way, although a bit simpler because $g_a(R) = f_a(R) = R$.

For all remaining cases, the lemma's equalities follow in a similar way. $\square$

Proposition 3.7 (*rpbf* **is a normalization function**)**.** *For all* $P \in C_A$, $rpbf(P)$ *is an rp-basic form, and for each rp-basic form* P, $rpbf(P) = P$.

Proof. The first statement is Lemma 3.5. For the second statement, it suffices by Lemma 2.14 to prove that for each rp-basic form P, $rpf(P) = P$. This follows by case distinction on P. The cases $P \in \{\mathsf{T}, \mathsf{F}\}$ follow immediately, and otherwise $P = P_1 \triangleleft a \triangleright P_2$, and thus $rpf(P) = rpf(f_a(P_1)) \triangleleft a \triangleright rpf(g_a(P_2))$. By Lemma 3.6, $rpf(f_a(P_1)) = P_1$ and $rpf(g_a(P_2)) = P_2$, hence $rpf(P) = P$. $\square$

Lemma 3.8. *For all* $P \in BF_A$, $\mathrm{CP}_{rp}(A) \vdash P = rpf(P)$.

Proof. We apply structural induction on P. The base cases $P \in \{\mathsf{T}, \mathsf{F}\}$ are trivial. Assume $P = P_1 \triangleleft a \triangleright P_2$. By induction $\mathrm{CP}_{rp}(A) \vdash P_i = rpf(P_i)$. We proceed by a case distinction on the form that P_1 and P_2 can have:

1. If $P_i \in \{\mathsf{T}, \mathsf{F}, Q_i \triangleleft b_i \triangleright Q_i'\}$ with $b_i \neq a$, then $f_a(P_1) = P_1$ and $g_a(P_2) = P_2$, and hence $rpf(P) = rpf(P_1) \triangleleft a \triangleright rpf(P_2)$, and thus $\mathrm{CP}_{rp}(A) \vdash P = rpf(P)$.
2. If $P_1 = R_1 \triangleleft a \triangleright R_2$ and $P_2 \in \{\mathsf{T}, \mathsf{F}, Q' \triangleleft b \triangleright Q''\}$ with $b \neq a$, then $g_a(P_2) = P_2$ and by auxiliary result (2) in the proof of Lemma 3.5, $rpf(R_1)$ and $rpf(P_2)$ are rp-basic forms. We derive

$$
\begin{aligned}
\mathrm{CP}_{rp}(A) \vdash P &= (R_1 \triangleleft a \triangleright R_2) \triangleleft a \triangleright P_2 \\
&= (R_1 \triangleleft a \triangleright R_1) \triangleleft a \triangleright P_2 && \text{by (CPrp1)} \\
&= (rpf(R_1) \triangleleft a \triangleright rpf(R_1)) \triangleleft a \triangleright rpf(P_2) && \text{by IH} \\
&= (rpf(f_a(R_1)) \triangleleft a \triangleright rpf(f_a(R_1))) \triangleleft a \triangleright rpf(g_a(P_2)) && \text{by Lemma 3.6} \\
&= rpf(f_a(R_1 \triangleleft a \triangleright R_2)) \triangleleft a \triangleright rpf(g_a(P_2)) \\
&= rpf((R_1 \triangleleft a \triangleright R_2) \triangleleft a \triangleright P_2) \\
&= rpf(P).
\end{aligned}
$$

3. If $P_1 \in \{\mathsf{T}, \mathsf{F}, Q' \triangleleft b \triangleright Q''\}$ with $b \neq a$ and $P_2 = S_1 \triangleleft a \triangleright S_2$, we can proceed as in the previous case, but now using axiom scheme (CPrp2) and the identity $f_a(P_1) = P_1$, and the fact that $rpf(P_1)$ and $rpf(S_2)$ are rp-basic forms.
4. If $P_1 = R_1 \triangleleft a \triangleright R_2$ and $P_2 = S_1 \triangleleft a \triangleright S_2$, we can proceed as in two previous cases, now using both (CPrp1) and (CPrp2), and the fact that $rpf(R_1)$ and $rpf(S_2)$ are rp-basic forms.

$\square$

Theorem 3.9. *For all* $P \in C_A$, $\mathrm{CP}_{rp}(A) \vdash P = rpbf(P)$.

Proof. By Theorem 2.18 and Corollary 2.19 we find $\mathrm{CP}_{rp}(A) \vdash P = bf(P)$. By Lemma 3.8, $\mathrm{CP}_{rp}(A) \vdash bf(P) = rpf(bf(P))$, and $rpf(bf(P)) = rpbf(P)$.   □

Definition 3.10. *The binary relation* $=_{rpbf}$ *on* C_A *is defined as follows:*

$$P =_{rpbf} Q \iff rpbf(P) = rpbf(Q).$$

Theorem 3.11. *For all* $P, Q \in C_A$, $\mathrm{CP}_{rp}(A) \vdash P = Q \iff P =_{rpbf} Q$.

Proof. Assume $\mathrm{CP}_{rp}(A) \vdash P = Q$. By Theorem 3.9, $\mathrm{CP}_{rp}(A) \vdash rpbf(P) = rpbf(Q)$. In [4] the following two statements are proved (Theorem 6.3 and an auxiliary result in its proof), where $=_{rpf}$ is a binary relation on C_A:

1. For all $P, Q \in C_A$, $\mathrm{CP}_{rp}(A) \vdash P = Q \iff P =_{rpf} Q$.
2. For all rp-basic forms P and Q, $P =_{rpf} Q \Rightarrow P = Q$.

By Lemma 3.5 these statements imply $rpbf(P) = rpbf(Q)$, that is, $P =_{rpbf} Q$.
    Assume $P =_{rpbf} Q$. By Lemma 2.14, $bf(rpbf(P)) = bf(rpbf(Q))$. By Theorem 2.18, $\mathrm{CP} \vdash rpbf(P) = rpbf(Q)$. By Theorem 3.9, $\mathrm{CP}_{rp}(A) \vdash P = Q$.   □

So, the relation $=_{rpbf}$ is axiomatized by $\mathrm{CP}_{rp}(A)$ and is thus a congruence. With this observation in mind, we define a transformation on evaluation trees that mimics the function $rpbf$ and prove that equality of two such transformed trees characterizes the congruence that is axiomatized by $\mathrm{CP}_{rp}(A)$.

Definition 3.12. *The unary **repetition-proof evaluation function***

$$rpse : C_A \to \mathcal{T}_A$$

*yields **repetition-proof evaluation trees** and is defined by*

$$rpse(P) = rp(se(P)).$$

The auxiliary function $rp : \mathcal{T}_A \to \mathcal{T}_A$ *is defined as follows* $(a \in A)$:

$$rp(\mathsf{T}) = \mathsf{T},$$
$$rp(\mathsf{F}) = \mathsf{F},$$
$$rp(X \trianglelefteq a \trianglerighteq Y) = rp(F_a(X)) \trianglelefteq a \trianglerighteq rp(G_a(Y)).$$

For $a \in A$, *the auxiliary functions* $F_a : \mathcal{T}_A \to \mathcal{T}_A$ *and* $G_a : \mathcal{T}_A \to \mathcal{T}_A$ *are defined by*

$$F_a(\mathsf{T}) = \mathsf{T},$$
$$F_a(\mathsf{F}) = \mathsf{F},$$
$$F_a(X \trianglelefteq b \trianglerighteq Y) = \begin{cases} F_a(X) \trianglelefteq a \trianglerighteq F_a(X) & \text{if } b = a, \\ X \trianglelefteq b \trianglerighteq Y & \text{otherwise,} \end{cases}$$

and

$$G_a(\mathsf{T}) = \mathsf{T},$$
$$G_a(\mathsf{F}) = \mathsf{F},$$
$$G_a(X \trianglelefteq b \trianglerighteq Y) = \begin{cases} G_a(Y) \trianglelefteq a \trianglerighteq G_a(Y) & \text{if } b = a, \\ X \trianglelefteq b \trianglerighteq Y & \text{otherwise.} \end{cases}$$

Example 3.13. *Let* $P = a \triangleleft (\mathsf{F} \triangleleft a \triangleright \mathsf{T}) \triangleright \mathsf{F}$. *We depict* $se(P)$ *(as in Example 2.4) and the repetition-proof evaluation tree* $rpse(P) = \mathsf{F} \trianglelefteq a \trianglerighteq (\mathsf{F} \trianglelefteq a \trianglerighteq \mathsf{F})$:

The similarities between *rpse* and the function *rpbf* can be exploited:

Lemma 3.14. *For all* $a \in A$ *and* $X \in \mathcal{T}_A$, $G_a(F_a(X)) = F_a(F_a(X)) = F_a(X)$ *and* $F_a(G_a(X)) = G_a(G_a(X)) = G_a(X)$.

Proof. By structural induction on X (cf. the proof of Lemma 3.3). □

We use the following lemma in the proof of our final completeness result.

Lemma 3.15. *For all* $P \in BF_A$, $rp(se(P)) = se(rpf(P))$.

Proof. We first prove an auxiliary result:

For all $P \in BF_A$ and for all $a \in A$, $rp(F_a(se(P))) = se(rpf(f_a(P)))$
and $rp(G_a(se(P))) = se(rpf(g_a(P)))$. $\qquad(4)$

We prove the first equality of (4) by structural induction on P. The base cases $P \in \{\mathsf{T}, \mathsf{F}\}$ are trivial. For the inductive case $P = Q \triangleleft a \triangleright R$, let $b \in A$. We have to distinguish the cases $b = a$ and $b \neq a$. If $b = a$, then

$$
\begin{aligned}
&rp(F_a(se(Q \triangleleft a \triangleright R))) \\
&\quad = rp(F_a(se(Q) \trianglelefteq a \trianglerighteq se(R))) \\
&\quad = rp(F_a(se(Q)) \trianglelefteq a \trianglerighteq F_a(se(Q))) \\
&\quad = rp(F_a(F_a(se(Q)))) \trianglelefteq a \trianglerighteq rp(G_a(F_a(se(Q)))) \\
&\quad = rp(F_a(se(Q))) \trianglelefteq a \trianglerighteq rp(F_a(se(Q))) &&\text{by Lemma 3.14} \\
&\quad = se(rpf(f_a(Q))) \trianglelefteq a \trianglerighteq se(rpf(f_a(Q))) &&\text{by IH} \\
&\quad = se(rpf(f_a(Q)) \triangleleft a \triangleright rpf(f_a(Q))) \\
&\quad = se(rpf(f_a(f_a(Q))) \triangleleft a \triangleright rpf(g_a(f_a(Q)))) &&\text{by Lemma 3.3} \\
&\quad = se(rpf(f_a(Q \triangleleft a \triangleright f_a(Q)))) \\
&\quad = se(rpf(f_a(Q \triangleleft a \triangleright R))).
\end{aligned}
$$

If $b \neq a$, then

$$
\begin{aligned}
rp(F_b(se(Q \triangleleft a \triangleright R))) &= rp(F_b(se(Q) \trianglelefteq a \trianglerighteq se(R))) \\
&= rp(se(Q) \trianglelefteq a \trianglerighteq se(R)) \\
&= rp(F_a(se(Q))) \trianglelefteq a \trianglerighteq rp(G_a(se(R))) \\
&= se(rpf(f_a(Q))) \trianglelefteq a \trianglerighteq se(rpf(g_a(R))) \qquad \text{by IH} \\
&= se(rpf(f_a(Q)) \triangleleft a \triangleright rpf(g_a(R))) \\
&= se(rpf(Q \triangleleft a \triangleright R)) \\
&= se(rpf(f_b(Q \triangleleft a \triangleright R))).
\end{aligned}
$$

The second equality can be proved in a similar way, and this finishes the proof of (4).

The lemma's statement now follows by a case distinction on P. The cases $P \in \{T, F\}$ follow immediately, and otherwise $P = Q \triangleleft a \triangleright R$, and thus

$$
\begin{aligned}
rp(se(Q \triangleleft a \triangleright R)) &= rp(se(Q) \trianglelefteq a \trianglerighteq se(R)) \\
&= rp(F_a(se(Q))) \trianglelefteq a \trianglerighteq rp(G_a(se(R))) \\
&= se(rpf(f_a(Q))) \trianglelefteq a \trianglerighteq se(rpf(g_a(R))) \qquad \text{by (4)} \\
&= se(rpf(f_a(Q)) \triangleleft a \triangleright rpf(g_a(R))) \\
&= se(rpf(Q \triangleleft a \triangleright R)).
\end{aligned}
$$

$\square$

Finally, we relate conditional statements by means of their repetition-proof evaluation trees.

Definition 3.16. *Repetition-proof valuation congruence,* *notation* $=_{rpse}$, *is defined on* C_A *as follows:*

$$
P =_{rpse} Q \iff rpse(P) = rpse(Q).
$$

The following characterization result immediately implies that $=_{rpse}$ is a congruence relation on C_A (and hence justifies calling it a congruence).

Proposition 3.17. *For all* $P, Q \in C_A$, $P =_{rpse} Q \iff P =_{rpbf} Q$.

Proof. In order to prove $\Rightarrow$, assume $rpse(P) = rpse(Q)$, thus $rp(se(P)) = rp(se(Q))$. By Corollary 2.19,

$$
rp(se(bf(P))) = rp(se(bf(Q))),
$$

so by Lemma 3.15, $se(rpf(bf(P))) = se(rpf(bf(Q)))$. By Lemma 2.8 and auxiliary result (2) (see the proof of Lemma 3.5), it follows that $rpf(bf(P)) = rpf(bf(Q))$, that is, $P =_{rpbf} Q$.

In order to prove $\Leftarrow$, assume $P =_{rpbf} Q$, thus $rpf(bf(P)) = rpf(bf(Q))$. Then $se(rpf(bf(P))) = se(rpf(bf(Q)))$ and thus by Lemma 3.15,

$$
rp(se(bf(P))) = rp(se(bf(Q))).
$$

By Corollary 2.19, $se(bf(P)) = se(P)$ and $se(bf(Q)) = se(Q)$, so $rp(se(P)) = rp(se(Q))$, that is, $P =_{rpse} Q$.   □

We end this section with a last completeness result.

Theorem 3.18 (Completeness of $\text{CP}_{rp}(A)$). *For all $P, Q \in C_A$,*

$$\text{CP}_{rp}(A) \vdash P = Q \iff P =_{rpse} Q.$$

Proof. Combine Theorem 3.11 and Proposition 3.17.   □

4   Conclusions

In [4] we introduced proposition algebra using Hoare's conditional $x \triangleleft y \triangleright z$ and the constants T and F. We defined a number of varieties of so-called *valuation algebras* in order to capture different semantics for the evaluation of conditional statements, and provided axiomatizations for the resulting valuation congruences. In [3,5] we introduced an alternative valuation semantics for proposition algebra in the form of *Hoare-McCarthy algebras* (HMA's) that is more elegant than the semantical framework provided in [4]: HMA-based semantics has the advantage that one can define a valuation congruence without first defining the valuation *equivalence* it is contained in.

In this paper, we use Staudt's evaluation trees [13] to define free valuation congruence as the relation $=_{se}$ (see Section 2) and this appears to be a relatively simple and stand-alone exercise, resulting in a semantics that is elegant and much simpler than HMA-based semantics [3,5] and the semantics defined in [4]. By Theorem 2.11, $=_{se}$ coincides with "free valuation congruence as defined in [4]" because both relations are axiomatized by CP (see [4, Thm.4.4andThm.6.2]). The advantage of "evaluation tree semantics" is that for a given conditional statement P, the evaluation tree $se(P)$ determines all relevant atomic evaluations, and $P =_{se} Q$ is determined by evaluation trees that contain no more atoms than those that occur in P and Q; this is comparable to how truth tables can be used in the setting of propositional logic.

In Section 3 we define repetition-proof valuation congruence $=_{rpse}$ on C_A by $P =_{rpse} Q$ if, and only if, $rpse(P) = rpse(Q)$, where $rpse(P) = rp(se(P))$ and rp is a transformation function on evaluation trees. It is obvious that this transformation is "natural", given the axiom schemes (CPrp1) and (CPrp2) that are characteristic for $\text{CP}_{rp}(A)$. The equivalence on C_A that we want to prove is

$$\text{CP}_{rp}(A) \vdash P = Q \iff P =_{rpse} Q, \tag{5}$$

by which $=_{rpse}$ coincides with "repetition-proof valuation congruence as defined in [4]" because both are axiomatized by $\text{CP}_{rp}(A)$ (see [4, Thm.6.3]). However, equivalence (5) implies that $=_{rpse}$ is a *congruence* relation on C_A and we could not find a direct proof of this fact. We chose to simulate the transformation $rpse$ by the transformation $rpbf$ on conditional statements and to prove that the resulting equivalence relation $=_{rpbf}$ is a congruence axiomatized by $\text{CP}_{rp}(A)$.

This is Theorem 3.11, the proof of which depends on [4, Thm.6.3]) *and* on Theorem 3.9, that is,

$$\text{For all } P \in C_A, \text{ CP}_{rp}(A) \vdash P = rpbf(P).$$

In order to prove equivalence (5) (which is Theorem 3.18), it is thus sufficient to prove that $=_{rpbf}$ and $=_{rpse}$ coincide, and this is Proposition 3.17.

In [6] we define evaluation trees for most of the other valuation congruences defined in [4] by transformations on *se*-images that are also "natural", and this also results in elegant "evaluation tree semantics" for each of these congruences.

We conclude with a brief digression on *short-circuit logic*, which we defined in [7] (see [5] for a quick introduction), and an example on the use of $\text{CP}_{rp}(A)$. Familiar binary connectives that occur in the context of imperative programming and that prescribe short-circuit evaluation, such as && (in C called "logical AND"), are often defined in the following way:

$$P \text{ \&\& } Q =_{\text{def}} \text{ if } P \text{ then } Q \text{ else } \textit{false},$$

independent of the precise syntax of P and Q, hence, $P \text{ \&\& } Q =_{\text{def}} Q \triangleleft P \triangleright \mathsf{F}$. It easily follows that && is associative (cf. Footnote 3). In a similarly way, negation can be defined by $\neg P =_{\text{def}} \mathsf{F} \triangleleft P \triangleright \mathsf{T}$. In [7] we focus on this question:

Question 4.1. *Which are the logical laws that characterize short-circuit evaluation of binary propositional connectives?*

A first approach to this question is to adopt the conditional as an auxiliary operator, as is done in [5,7], and to answer Question 4.1 using definitions of the binary propositional connectives as above and the axiomatization for the valuation congruence of interest in proposition algebra (or, if "mixed conditional statements" are at stake, axiomatizations for the appropriate valuation congruences). An alternative and more direct approach to Question 4.1 is to establish axiomatizations for short-circuited binary connectives in which the conditional is *not* used. For free valuation congruence, an equational axiomatization of short-circuited binary propositional connectives is provided by Staudt in [13], where $se(P \text{ \&\& } Q) =_{\text{def}} se(P)[\mathsf{T} \mapsto se(Q)]$ and $se(\neg P) =_{\text{def}} se(P)[\mathsf{T} \mapsto \mathsf{F}, \mathsf{F} \mapsto \mathsf{T}]$ (and where the function se is also defined for short-circuited disjunction), and the associated completeness proof is based on decomposition properties of such evaluation trees. For repetition-proof valuation congruence it is an open question whether a finite, equational axiomatization of the short-circuited binary propositional connectives exists, and an investigation of repetition-proof evaluation trees defined by such connectives might be of interest in this respect. We end with an example on the use of $\text{CP}_{rp}(A)$ that is based on [7, Ex.4].

Example 4.2. *Let A be a set of atoms of the form $(e{=}{=}e')$ and $(\mathsf{n}{=}e)$ with n some initialized program variable and e, e' arithmetical expressions over the integers that may contain n. Assume that $(e{=}{=}e')$ evaluates to true if e and e' represent the same value, and $(\mathsf{n}{=}e)$ always evaluates to true with the effect that* *e's value is assigned to* n. *Then these atoms satisfy the axioms of* $CP_{rp}(A)$.[4] *Notice that if* n *has initial value 0 or 1,* ((n=n+1) && (n=n+1)) && (n==2) *and* (n=n+1) && (n==2) *evaluate to different results, so the atom* (n=n+1) *does not satisfy the law* a && $a = a$, *by which this example is typical for the repetition-proof characteristic of* $CP_{rp}(A)$.

We acknowledge the helpful comments of two anonymous reviewers.

References

1. Bergstra, J.A., Bethke, I., Rodenburg, P.H.: A propositional logic with 4 values: true, false, divergent and meaningless. Journal of Applied Non-Classical Logics **5**(2), 199–218 (1995)
2. Bergstra, J.A., Loots, M.E.: Program algebra for sequential code. Journal of Logic and Algebraic Programming **51**(2), 125–156 (2002)
3. Bergstra, J.A., Ponse, A.: On Hoare-McCarthy algebras [cs.LO] (2010). http://arxiv.org/abs/1012.5059
4. Bergstra, J.A., Ponse, A.: Proposition algebra. ACM Transactions on Computational Logic **12**(3), Article 21, 36 pages (2011)
5. Bergstra, J.A., Ponse, A.: Proposition algebra and short-circuit logic. In: Arbab, F., Sirjani, M. (eds.) FSEN 2011. LNCS, vol. 7141, pp. 15–31. Springer, Heidelberg (2012)
6. Bergstra, J.A., Ponse A.: Evaluation trees for proposition algebra. arXiv:1504.08321v2 [cs.LO] (2015)
7. Bergstra, J.A., Ponse, A., Staudt, D.J.C.: Short-circuit logic. arXiv:1010.3674v4 [cs.LO, math.LO] (version v1: October 2010) (2013)
8. de Boer, F.S., de Vries, F.-J., Olderog, E.-R., Ponse, A. (guest editors): Selected papers from the Workshop on Assertional Methods. Formal Aspects of Computing, 6(1 Supplement; Special issue) (1994)
9. Harel, D.: Dynamic logic. In: Gabbay, D., Günthner, F. (eds.) Handbook of Philosophical Logic, vol. II, pp. 497–604. Reidel Publishing Company (1984)
10. Hayes, I.J., He Jifeng, Hoare, C.A.R., Morgan, C.C., Roscoe, A.W., Sanders, J.W., Sorensen, I.H., Spivey, J.M., Sufrin B.A.: Laws of programming. Communications of the ACM **3**(8), 672–686 (1987)
11. Hoare, C.A.R.: Communicating Sequential Processes. Prentice Hall International (1985)
12. Hoare, C.A.R.: A couple of novelties in the propositional calculus. Zeitschrift für Mathematische Logik und Grundlagen der Mathematik **31**(2), 173–178 (1985); Republished. In: Hoare, C.A.R., Jones, C.B. (eds.) Essays in Computing Science. Series in Computer Science, pp. 325–331. Prentice Hall International (1989)
13. Staudt, D.J.C.: Completeness for two left-sequential logics. MSc. thesis Logic, University of Amsterdam (May 2012). arXiv:1206.1936v1 [cs.LO] (2012)
14. Wortel, L.: Side effects in steering fragments. MSc. thesis Logic, University of Amsterdam (September 2011). arXiv:1109.2222v1 [cs.LO] (2011)

---

[4] Of course, not all equations that are valid in the setting of Example 4.2 follow from $CP_{rp}(A)$, e.g., $CP_{rp}(A) \nvdash$ (0==0) = T. We note that a particular consequence of $CP_{rp}(A)$ in the setting of short-circuit logic is $(\neg a$ && $a)$ && $x = \neg a$ && a (cf. Example 3.13), and that Example 4.2 is related to the work of Wortel [14], where an instance of *Propositional Dynamic Logic* [9] is investigated in which assignments can be turned into tests; the assumption that such tests always evaluate to *true* is natural because the assumption that assignments always succeed is natural.

Process Algebra

On Applicative Similarity, Sequentiality, and Full Abstraction

Raphaëlle Crubillé[1], Ugo Dal Lago[2], Davide Sangiorgi[2(✉)], and Valeria Vignudelli[2]

[1] ENS-Lyon, Lyon, France
raphaelle.crubille@ens-lyon.fr
[2] Universitá di Bologna and INRIA, Bologna, Italy
{ugo.dallago,davide.sangiorgi2,valeria.vignudelli2}@unibo.it

Abstract. We study how applicative bisimilarity behaves when instantiated on a call-by-value probabilistic λ-calculus, endowed with Plotkin's parallel disjunction operator. We prove that congruence and coincidence with the corresponding context relation hold for both bisimilarity and similarity, the latter known to be impossible in sequential languages.

Keywords: Probabilistic lambda calculus · Bisimulation · Coinduction · Sequentiality

1 Introduction

The work in this paper is part of a general effort in trying to transport techniques and concepts for program correctness and verification that have been introduced and successfully applied to ordinary (first-order) concurrency (CCS, CSP, Petri Nets), following pioneering work by Bergstra, Hoare, Milner, Olderog, and others, onto formalisms with higher-order features, in which the values exchanged or manipulated may include pieces of code. Specifically, we focus on the prototypical higher-order language, the λ-calculus, enriched with a probabilistic choice, and use coinductive methods and logics to understand and characterise behavioural equivalences.

Probabilistic models are more and more pervasive. Examples of application areas in which they have proved to be useful include natural language processing [16], robotics [23], computer vision [3], and machine learning [19]. Sometimes, being able to "flip a fair coin" while computing is a *necessity* rather than an alternative, like in cryptography (where, e.g., secure public key encryption schemes are bound to be probabilistic [11]): randomness is not only a modeling tool, but a capability algorithms can exploit.

The specification of probabilistic models and algorithms can be made easier by the design of programming languages. And indeed, various probabilistic programming languages have been introduced in the last years, from abstract

---

The authors are partially supported by the ANR project 12IS02001 PACE.

R. Meyer et al. (Eds.): Olderog-Festschrift, LNCS 9360, pp. 65–82, 2015.
DOI: 10.1007/978-3-319-23506-6_7

ones [13,18,22] to more concrete ones [12,20]. A common scheme consists in endowing deterministic languages with one or more primitives for probabilistic choice, like binary probabilistic choice or primitives for distributions. Many of them, as a matter of fact, are designed around the λ-calculus or one of its incarnations, like Scheme. This, in turn, has stimulated foundational research about probabilistic λ-calculi, and in particular about the nature of program equivalence in a probabilistic setting. This has already started to produce some interesting results in the realm of denotational semantics, where adequacy and full-abstraction results have recently appeared [8,10].

Operational techniques for program equivalence, and in particular coinductive methodologies, have the advantage of not requiring a too complicated mathematical machinery. Various notions of bisimilarity have been proved adequate and, in some cases, fully abstract, for deterministic and nondeterministic computation [1,15,17]. A recent paper [6] generalizes Abramsky's applicative bisimulation [1] to a call-by-*name*, untyped λ-calculus endowed with binary, fair, probabilistic choice [7]. Probabilistic applicative bisimulation is shown to be a congruence, thus included in context equivalence. Completeness, however, fails, but can be recovered if call-by-*value* evaluation is considered, as shown in [4]. This can appear surprising, given that in nondeterministic λ-calculi, both when call-by-name *and* call-by-value evaluation are considered, applicative bisimilarity is a congruence, but *finer* than context equivalence [15]. But there is another, even less expected result: the aforementioned correspondence does not hold anymore if we consider applicative *simulation* and the contextual *preorder*.

The reason why this happens can be understood if one looks at the testing-based characterization of similarity and bisimilarity from the literature [9,24]: the class of tests characterizing *bi*similarity is simple enough to allow any test to be implementable by a program context. This is impossible for tests characterizing similarity, which include not only conjunction (which can be implemented as copying) but also disjunction, an operator that seems to require the underlying language to be parallel.

In this paper, we show that, indeed, the presence of Plotkin's disjunction [2,21] turns applicative similarity into a relation which coincides with the context preorder. This is done by checking that the proof of precongruence for applicative bisimilarity [4,6] continues to hold, and by showing how tests involving conjunction and disjunction can be implemented by contexts. This somehow completes the picture about how applicative (bi)similarity behaves in a probabilistic scenario.

2   Programs and Their Operational Semantics

In this section, we present the syntax and operational semantics of $\Lambda_{\oplus or}$, the language on which we define applicative bisimulation. $\Lambda_{\oplus or}$ is a λ-calculus endowed with probabilistic choice and parallel disjunction operators.

The terms of $\Lambda_{\oplus or}$ are built up from variables, using the usual constructs of λ-calculus, binary choice and parallel disjunction. In the following, $Var = \{x, y, \ldots\}$ is a countable set of variables

Definition 1. *The terms of $\Lambda_{\oplus or}$ are expressions generated by the following grammar:*

$$M, N, L ::= x \mid \lambda x.M \mid M \oplus N \mid MN \mid [M \parallel N] \rightarrowtail L$$

where $x \in Var$.

In what follows, we consider terms of $\Lambda_{\oplus or}$ as α-equivalence classes of syntax trees. We let $FV(M)$ denote the set of free variables of the term M. A term M is closed if $FV(M) = \emptyset$. Given a set $\overline{x}$ of variables, $\Lambda_{\oplus or}(\overline{x})$ is the set of terms M such that $FV(M) \subseteq \overline{x}$. We write $\Lambda_{\oplus or}$ for $\Lambda_{\oplus or}(\emptyset)$. The (capture-avoiding) substitution of N for the free occurrences of x in M is denoted by $M[N/x]$.

The constructs of the λ-calculus have their usual meanings. The construct $M \oplus N$ is a binary choice operator, to be interpreted probabilistically, as in $\Lambda_{\oplus}$ [7]. The construct $[M \parallel N] \rightarrowtail L$ corresponds to the so-called parallel disjunction operator: if the evaluation of M or N terminates, then the behaviour of $[M \parallel N] \rightarrowtail L$ is the same as the behaviour of L, otherwise this term does not terminate. Since we are in a probabilistic calculus, this means that $[M \parallel N] \rightarrowtail L$ converges to L with a probability that is equal to the probability that either M or N converge. (This formulation of parallel disjunction is equivalent to the binary one, without the third term.)

Example 1. Relevant examples of terms are $\Omega = (\lambda x.xx)(\lambda x.xx)$, and $I = \lambda x.x$: the first one always diverges, while the second always converges (to itself). In between, one can find terms such as $I \oplus \Omega$, and $I \oplus (I \oplus \Omega)$, converging with probability one half and three quarters, respectively.

2.1 Operational Semantics

Because of the probabilistic nature of choice in $\Lambda_{\oplus or}$, a program doesn't evaluate to a value, but to a probability distribution on values. Therefore, we need the following notions to define an evaluation relation.

Definition 2. *Values are terms of the form $V ::= \lambda x.M$. We will call $\mathcal{V}_{\oplus or}$ the set of values. A value distribution is a function $\mathscr{D} : \mathcal{V}_{\oplus or} \rightarrow [0, 1]$, such that $\sum_{V \in \mathcal{V}_{\oplus or}} \mathscr{D}(V) \leq 1$. Given a value distribution $\mathscr{D}$, we let $\mathsf{S}(\mathscr{D})$ denote the set of those values V such that $\mathscr{D}(V) > 0$. Given a set X of values, $\mathscr{D}(X)$ is the sum of the probabilities assigned to every element of X, i.e., $\mathscr{D}(X) = \sum_{V \in X} \mathscr{D}(V)$. Moreover, we define $\sum \mathscr{D} = \sum_V \mathscr{D}(V)$, which corresponds to the total weight of the distribution $\mathscr{D}$. A value distribution $\mathscr{D}$ is finite whenever $\mathsf{S}(\mathscr{D})$ has finite cardinality. If V is a value, we write $\{V^1\}$ for the value distribution $\mathscr{D}$ such that $\mathscr{D}(W) = 1$ if $W = V$ and $\mathscr{D}(V) = 0$ otherwise. We'll note $\mathscr{D} \leq \mathscr{E}$ for the pointwise preorder on value distributions.*

We first define an *approximation* semantics, which attributes *finite* probability distributions to terms, and only later define the actual semantics, which is the least upper bound of all distributions obtained through the approximation semantics. Big-step semantics is given by means of a binary relation $\Downarrow$ between

$$\frac{}{M \Downarrow \emptyset} \, b_e \qquad \frac{}{V \Downarrow \{V^1\}} \, b_v \qquad \frac{M \Downarrow \mathscr{D} \qquad N \Downarrow \mathscr{E}}{M \oplus N \Downarrow \frac{1}{2}\mathscr{D} + \frac{1}{2}\mathscr{E}} \, b_s$$

$$\frac{M \Downarrow \mathscr{K} \qquad N \Downarrow \mathscr{F} \qquad \{P[V/x] \Downarrow \mathscr{E}_{P,V}\}_{\lambda x.P \in S(\mathscr{K}),\, V \in S(\mathscr{F})}}{MN \Downarrow \sum_{V \in S(\mathscr{F})} \mathscr{F}(V) \cdot \left(\sum_{\lambda x.P \in S(\mathscr{K})} \mathscr{K}(\lambda x.P) \cdot \mathscr{E}_{P,V}\right)} \, b_a$$

$$\frac{M \Downarrow \mathscr{D} \qquad N \Downarrow \mathscr{E} \qquad L \Downarrow \mathscr{F}}{[M \parallel N] \rightarrowtail L \Downarrow \left(\sum \mathscr{D} + \sum \mathscr{E} - \left(\sum \mathscr{D} \cdot \sum \mathscr{E}\right)\right) \cdot \mathscr{F}} \, b_{or}$$

Fig. 1. Evaluation

closed terms and value distributions, which is defined by the set of rules from Figure 1. This evaluation relation is the natural extension to $\Lambda_{\oplus or}$ of the evaluation relation given in [7] for the untyped probabilistic λ-calculus. Since the calculus has a call-by-value evaluation strategy, function arguments are evaluated before being passed to functions.

Lemma 1. *For every term M, if $M \Downarrow \mathscr{D}$, and $M \Downarrow \mathscr{E}$, then there exists a distribution $\mathscr{F}$ such that $M \Downarrow \mathscr{F}$ with $\mathscr{D} \leq \mathscr{F}$, and $\mathscr{E} \leq \mathscr{F}$.*

Proof. The proof is by induction on the structure of derivations for $M \Downarrow \mathscr{D}$. We only consider two cases, since the others are the same as in [7]:

- If the derivation for $M \Downarrow \mathscr{D}$ is: $\frac{}{M \Downarrow \emptyset} \, b_e$: Then it is enough to take $\mathscr{F} = \mathscr{E}$, and since $\emptyset \leq \mathscr{E}$ and $\mathscr{E} \leq \mathscr{E}$, the result holds.
- If the derivation for $M \Downarrow \mathscr{D}$ is of the form:

$$\frac{P \Downarrow \mathscr{G} \qquad N \Downarrow \mathscr{H} \qquad L \Downarrow \mathscr{I}}{M = [P \parallel N] \rightarrowtail L \Downarrow \mathscr{D} = \left(\sum \mathscr{G} + \sum \mathscr{H} - \left(\sum \mathscr{G} \cdot \sum \mathscr{H}\right)\right) \cdot \mathscr{I}} \, b_{or}$$

Since $M = [P \parallel N] \rightarrowtail L$, there are only two possible structures for the derivation of $M \Downarrow \mathscr{E}$: either $\mathscr{E} = \emptyset$ and the result holds by $\mathscr{F} = \mathscr{D}$, or the structure of $M \Downarrow \mathscr{E}$ is the following:

$$\frac{P \Downarrow \mathscr{G}_2 \qquad N \Downarrow \mathscr{H}_2 \qquad L \Downarrow \mathscr{I}_2}{M = [P \parallel N] \rightarrowtail L \Downarrow \mathscr{E} = \left(\sum \mathscr{G}_2 + \sum \mathscr{H}_2 - \left(\sum \mathscr{G}_2 \cdot \sum \mathscr{H}_2\right)\right) \cdot \mathscr{I}_2} \, b_{or}$$

By applying the induction hypothesis, we obtain that there exist $\mathscr{J}, \mathscr{K}, \mathscr{L}$ value distributions such that $P \Downarrow \mathscr{J}$, $N \Downarrow \mathscr{K}$, $L \Downarrow \mathscr{L}$, and, moreover, $\mathscr{G}, \mathscr{G}_2 \leq \mathscr{J}$, $\mathscr{H}, \mathscr{H}_2 \leq \mathscr{K}$, and $\mathscr{I}, \mathscr{I}_2 \leq \mathscr{L}$. We define $\mathscr{F} = \left(\sum \mathscr{J} + \sum \mathscr{K} - \left(\sum \mathscr{J} \cdot \sum \mathscr{K}\right)\right) \cdot \mathscr{L}$, and we have that $M \Downarrow \mathscr{F}$. We must show that $\mathscr{D} \leq \mathscr{F}$ and $\mathscr{E} \leq \mathscr{F}$. Let $f : [0,1] \times [0,1] \to [0,1]$ be the function defined by $f(x,y) = x + y - x \cdot y$. The result follows from the fact that f is an increasing function, which holds since its two partial derivatives are positive. $\square$

Definition 3. *For any closed term M, we define the* big-steps semantics $[\![M]\!]$ *of M as* $\sup_{M \Downarrow \mathscr{D}} \mathscr{D}$.

Since distributions form an ω-complete partial order, and for every M the set of those distributions $\mathscr{D}$ such that $M \Downarrow \mathscr{D}$ is a countable directed set (by Lemma 1), this definition is well-posed, and associates a unique value distribution to every term.

2.2   The Contextual Preorder

The general idea of the contextual preorder is the following: a term M is smaller than a term N if the probability of convergence of any program L where M occurs is less than or equal to the probability of convergence of the program obtained by replacing M by N in L. The notion of context allows us to formalize this idea.

Definition 4. *A context C of $\Lambda_{\oplus or}$ is a syntax tree with a unique hole:*

$$C ::= [\cdot] \mid \lambda x.C \mid CM \mid MC \mid C \oplus M \mid M \oplus C$$

$$\mid [C \parallel M] \rightarrowtail N \mid [M \parallel C] \rightarrowtail N \mid [M \parallel N] \rightarrowtail C.$$

We let $\mathscr{C}$ denote the set of all contexts.

Definition 5. *Terms $M, N \in \Lambda_{\oplus or}(\overline{x})$ are put in relation by the contextual preorder $(M \leq N)$ if for every context C of $\Lambda_{\oplus or}$ such that $C[M]$ and $C[N]$ are closed terms, it holds that $\sum [\![C[M]]\!] \leq \sum [\![C[N]]\!]$. M, N are contextually equivalent $(M = N)$ if $M \leq N$, and $N \leq M$.*

Note that the contextual preorder is directly defined on open terms, by requiring contexts to bind the free variables of terms. It is easy to verify that the contextual preorder is indeed a preorder, and analogously for equivalence.

Example 2. To see how things differ when we consider the contextual preorder in $\Lambda_\oplus$ and in $\Lambda_{\oplus or}$, consider the following terms of $\Lambda_\oplus$:

$$M = \lambda y.(\Omega \oplus I) \qquad N = (\lambda y.\Omega) \oplus (\lambda y.I).$$

where Ω and I are defined as in Example 1. We let $\leq_\oplus$ and $=_\oplus$ respectively denote the contextual preorder and equivalence for the language $\Lambda_\oplus$, i.e., the relations restricted to terms and contexts without the parallel disjunction construct. In [4] it is proved that $M \leq_\oplus N$. The converse does not hold, since if we take the $\Lambda_\oplus$ context

$$C = (\lambda x.(xI)(xI))[\cdot]$$

we have that in $C[M]$ the term $\lambda y.(\Omega \oplus I)$ is copied with probability one, while in $C[N]$ both term $\lambda y.\Omega$ and term $\lambda y.I$ are copied with probability one half. Hence, $C[M]$ converges with probability one quarter (i.e., the probability that $\Omega \oplus I$ converges two times in a row) while $C[N]$ has probability one half of diverging (i.e., one half times the probability that Ω diverges two times in a row) and one half of converging (i.e., one half times the probability that I converges two times in a row). In $\Lambda_{\oplus or}$ we still have that $N \nleq M$, since the contexts of $\Lambda_\oplus$ are contexts of $\Lambda_{\oplus or}$ as well, but we also have that $M \nleq N$. Consider the context

$$C = (\lambda x. [(xI) \parallel (xI)] \rightarrowtail I)[\cdot]$$

If we put term M in context C then $\lambda y.(\Omega \oplus I)$ is copied, which has probability one half of converging when applied to I. Hence, by summing the probabilities of convergence of the two copies of $(\lambda y.(\Omega \oplus I))I$ and subtracting the probability that they both converge, we obtain that $[\![C[M]]\!] = \frac{3}{4} \cdot \{I^1\}$. Term $C[N]$ only converges with probability one half, since with one half probability we have the parallel disjunction of two terms that never converge and with one half probability we have the parallel disjunction of two terms that always converge. Hence, both in $\Lambda_\oplus$ and in $\Lambda_{\oplus or}$ terms M, N are not contextually equivalent, but it is only in $\Lambda_{\oplus or}$ that neither M is below N nor N is below M in the contextual preorder. We will see in the following section that this corresponds to what happens when we consider the simulation preorder.

3   Applicative Simulation

In this section we introduce the notions of probabilistic applicative simulation and bisimulation for $\Lambda_{\oplus or}$. Then we define probabilistic simulation and bisimulation on labelled Markov chains (LMCs, which also appear as Reactive Probabilistic Labelled Transition Systems in the literature). Bisimilarity on this class of structures was defined in [14]. We show how to define a labelled Markov chain representing terms of $\Lambda_{\oplus or}$ and their evaluation. Two states in the labelled Markov chain corresponding to terms M, N are in the simulation preorder (respectively, bisimilar) if and only if terms M, N are in the applicative simulation preorder (respectively: applicative bisimilar). Recall that, given a relation $\mathcal{R} \subseteq X \times Y$ and a set $Z \subseteq X$, $\mathcal{R}(Z) = \{y | \exists x \in Z \text{ such that } x\mathcal{R}y\}$.

Definition 6. *A relation $\mathcal{R} \subseteq \Lambda_{\oplus or} \times \Lambda_{\oplus or}$ is a probabilistic applicative simulation if $M\mathcal{R}N$ implies:*
- *for all $X \subseteq \mathcal{V}_{\oplus or}$, $[\![M]\!](X) \leq [\![N]\!](\mathcal{R}(X))$*
- *if $M = \lambda x.L$ and $N = \lambda x.P$ then $L[V/x]\mathcal{R}P[V/x]$ for all $V \in \mathcal{V}_{\oplus or}$.*

A relation $\mathcal{R}$ is a probabilistic applicative bisimulation if both $\mathcal{R}$ and $\mathcal{R}^{-1}$ are probabilistic applicative simulations. We say that M is simulated by N ($M \precsim_a N$) if there exists a probabilistic applicative simulation $\mathcal{R}$ such that $M\mathcal{R}N$. Terms M, N are bisimilar ($M \sim_a N$) if there exists a probabilistic applicative bisimulation $\mathcal{R}$ such that $M\mathcal{R}N$.

Definition 7. *A labelled Markov chain (LMC) is a triple $\mathcal{M} = (\mathcal{S}, \mathcal{L}, \mathcal{P})$, where $\mathcal{S}$ is a countable set of states, $\mathcal{L}$ is a set of labels, and $\mathcal{P}$ is a transition probability matrix, i.e., a function $\mathcal{P} : \mathcal{S} \times \mathcal{L} \times \mathcal{S} \to \mathbb{R}$ such that for every state $s \in \mathcal{S}$ and for every label $l \in \mathcal{L}$, $\sum_{u \in \mathcal{S}} \mathcal{P}(s, l, u) \leq 1$.*

Definition 8. *Let $(\mathcal{S}, \mathcal{L}, \mathcal{P})$ be a labelled Markov chain. A probabilistic simulation is a relation $\mathcal{R}$ on $\mathcal{S}$ such that $(s, t) \in \mathcal{R}$ implies that for every $X \subseteq \mathcal{S}$ and for every $l \in \mathcal{L}$, $\mathcal{P}(s, l, X) \leq \mathcal{P}(t, l, \mathcal{R}(X))$. A probabilistic bisimulation is a relation $\mathcal{R}$ on $\mathcal{S}$ such that both $\mathcal{R}$ and $\mathcal{R}^{-1}$ are probabilistic simulation relations. We say that s is simulated by t ($s \precsim t$) if there exists a probabilistic simulation $\mathcal{R}$ such that $s\mathcal{R}t$. States s, t are bisimilar ($s \sim t$) if there exists a probabilistic bisimulation $\mathcal{R}$ such that $s\mathcal{R}t$.*

Labelled Markov chains allow for external nondeterminism (every state can reach different probability distributions, depending on the chosen label) but they do not allow for internal nondeterminism (given a state and a label there is only one associated probability distribution). This is the reason why bisimilarity coincides with simulation equivalence on labelled Markov chains, i.e., $\sim = \precsim \cap \precsim^{-1}$.

Lemma 2. *For any labelled Markov chain $(\mathcal{S}, \mathcal{L}, \mathcal{P})$:*
1. *relations $\precsim$ and $\sim$ are the largest simulation and the largest bisimulation on $\mathcal{S}$, respectively;*
2. *relation $\precsim$ is a preorder and relation $\sim$ is an equivalence.*

Proof. Let us examine the two points separately:
1. Simulations and bisimulations are closed under union, hence the results follows.
2. The identity relation is a simulation, hence $\precsim$ is reflexive. Given two simulation relations $\mathcal{R}_1, \mathcal{R}_2$, relation $\mathcal{R}_1; \mathcal{R}_2 = \{(s, t)|s\mathcal{R}_1 u\mathcal{R}_2 t$ for some $u\}$ is a simulation. Hence, $\precsim$ is transitive as well. By definition, relation $\sim$ is symmetric, which implies that it is an equivalence. $\qquad\square$

We will now define a labelled Markov chain that has among its states all terms of $\Lambda_{\oplus or}$ and that models the evaluation of these terms.

Definition 9. *The labelled Markov chain $\mathcal{M}_{\oplus or} = (\mathcal{S}_{\oplus or}, \mathcal{L}_{\oplus or}, \mathcal{P}_{\oplus or})$ is given by:*
- *A set of states $\mathcal{S}_{\oplus or} = \{\Lambda_{\oplus or}\} \uplus \{\hat{\mathcal{V}}_{\oplus or}\}$, where terms and values are taken modulo α-equivalence and $\hat{\mathcal{V}}_{\oplus or} = \{\hat{V}|V \in \mathcal{V}_{\oplus or}\}$ is a set containing copies of the values in $\Lambda_{\oplus or}$ decorated with^. We call these values distinguished values.*
- *A set of labels $\mathcal{L}_{\oplus or} = \mathcal{V}_{\oplus or} \uplus \{eval\}$, where, again, terms are taken modulo α-equivalence.*
- *A transition probability matrix $\mathcal{P}_{\oplus or}$ such that:*
  - *for every $M \in \Lambda_{\oplus or}$ and for every $\hat{V} \in \hat{\mathcal{V}}_{\oplus or}$, $\mathcal{P}_{\oplus or}(M, eval, \hat{V}) = [\![M]\!](V)$ and $\mathcal{P}_{\oplus or}(M, eval, M') = 0$ for all $M' \in \Lambda_{\oplus or}$.*
  - *for every $\lambda\hat{x}.M \in \hat{\mathcal{V}}_{\oplus or}$ and for every $V \in \mathcal{V}_{\oplus or}$, $\mathcal{P}_{\oplus or}(\lambda\hat{x}.M, V, M[V/x]) = 1$ and $\mathcal{P}_{\oplus or}(\lambda\hat{x}.M, V, M') = 0$ for all $M' \in \Lambda_{\oplus or}$ such that $M' \neq M[V/x]$.*

Please observe that if $V \in \mathcal{V}_{\oplus or}$, then both V and $\hat{V}$ are states of the Markov chain $\mathcal{M}_{\oplus or}$. A similar labelled Markov chain is defined in [5] for a call-by-name untyped probabilistic λ-calculus $\Lambda_{\oplus}$, and for a call-by-value typed probabilistic version of PCF in [4]. Actions in $\mathcal{V}_{\oplus or}$ and action *eval* respectively represent the application of a term to a value and the evaluation of a term.

Following [9], given a state and an action we allow the sum of the probabilities of reaching other states in the labelled Markov chain to be smaller than 1, modelling divergence this way. The definition of simulation implies that whenever M is simulated by N we have that $\sum[\![M]\!] \leq \sum[\![N]\!]$. Analogously, if M is bisimilar to N, then $\sum[\![M]\!] = \sum[\![N]\!]$.

An applicative simulation $\mathcal{R}$ on terms of $\Lambda_{\oplus or}$ can be easily seen as a simulation relation $\mathcal{R}'$ on states of $\mathcal{M}_{\oplus or}$, obtained by adding to relation $\mathcal{R}$ the pairs $\{(\hat{V}, \hat{W}) | V\mathcal{R}W\}$. Analogously, a simulation relation on $\mathcal{M}_{\oplus or}$ corresponds to an applicative simulation for $\Lambda_{\oplus or}$.

Theorem 1. *On terms of $\Lambda_{\oplus or}$, $\precsim_a = \precsim$ and $\sim_a = \sim$.*

In what follows, we will mainly use the definitions of simulation and bisimulation for the labelled Markov chain $\mathcal{M}_{\oplus or}$. By Lemma 2, $\precsim$ coincides with the simulation preorder defined in [4], which requires simulations to be preorders themselves. For instance, I and II are (applicative) bisimilar since

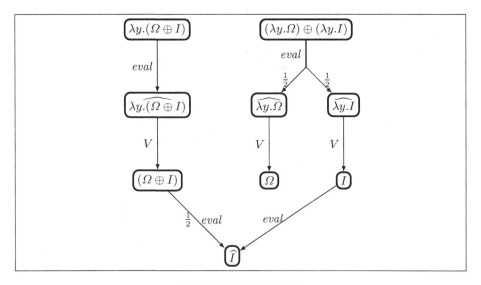

Fig. 2. LMC for M, N.

$\mathcal{R} = \{(I, (II))\} \cup \mathcal{ID} \cup \{(\hat{V}, \hat{V}) | V \in \mathcal{V}_{\oplus or}\}$, where $\mathcal{ID}$ is the identity relation on $\Lambda_{\oplus or}$, is a bisimulation on $\mathcal{M}_{\oplus or}$. Consider now the terms M and N defined in Example 2 and represented in Figure 2 as states in $\mathcal{M}_{\oplus or}$. Term M is not simulated by N: if a simulation $\mathcal{R}$ relates them, then it must also relate term $(\Omega \oplus I)$ to both term Ω and term I. However, $(\Omega \oplus I)$ can perform *eval* and reach I with probability one half, while Ω has zero probability of becoming a value, which means that $\mathcal{R}$ cannot be a simulation relation. In the other direction, we have that N cannot be simulated by M either. If $\mathcal{R}$ is simulation such that $N \mathcal{R} M$ then it must relate term I to term $(\Omega \oplus I)$, but the former has probability one of convergence and the latter has probability one half of convergence.

4   The Simulation Preorder is a Precongruence

The extension $\precsim_\circ$ of the applicative simulation preorder to open terms is defined by considering all closing substitutions, i.e., for all $M, N \in \Lambda_{\oplus or}(x_1, \ldots, x_n)$, we have $M \precsim_\circ N$ if

$$M[V_1, \ldots, V_n/x_1, \ldots, x_n] \precsim_\circ N[V_1, \ldots, V_n/x_1, \ldots, x_n], \text{ for all } V_1, \ldots, V_n \in \mathcal{V}_{\oplus or}.$$

Here we show that $\precsim_\circ$ is a precongruence, i.e., closed with respect to the operators of $\Lambda_{\oplus or}$.

It is here convenient to work with generalizations of relations called $\Lambda_{\oplus or}$-relations, i.e. sets of triples in the form $(\overline{x}, M, N)$, where $M, N \in \Lambda_{\oplus or}(\overline{x})$. Given a relation $\mathcal{R}$ on open terms, if $M \mathcal{R} N$ and $M, N \in \Lambda_{\oplus or}(\overline{x})$ then the triple $(\overline{x}, M, N)$ is in the corresponding $\Lambda_{\oplus or}$-relation. We denote this by $\overline{x} \vdash M \mathcal{R} N$. We extend the usual notions of symmetry, reflexivity and transitivity to $\Lambda_{\oplus or}$-relations as expected.

Definition 10. *A $\Lambda_{\oplus or}$-relation $\mathcal{R}$ is compatible if and only if the following conditions hold:*
(Com1) $\forall \overline{x}, \forall x \in \overline{x}, \overline{x} \vdash x \mathcal{R} x$;
(Com2) $\forall \overline{x}, \forall x \notin \overline{x}, \forall M, N, \overline{x} \cup \{x\} \vdash M \mathcal{R} N \implies \overline{x} \vdash \lambda x.M \mathcal{R} \lambda x.N;$
(Com3) $\forall \overline{x}, \forall M, N, P, Q, \overline{x} \vdash M \mathcal{R} N \wedge \overline{x} \vdash P \mathcal{R} Q \implies \overline{x} \vdash MP \mathcal{R} NQ;$
(Com4) $\forall \overline{x}, \forall M, N, P, Q, \overline{x} \vdash M \mathcal{R} N \wedge \overline{x} \vdash P \mathcal{R} Q \implies \overline{x} \vdash M \oplus P \mathcal{R} N \oplus Q;$
(Com5) $\forall \overline{x}, \forall M, N, P, Q, T, \overline{x} \vdash M \mathcal{R} N \wedge \overline{x} \vdash P \mathcal{R} Q \implies \overline{x} \vdash [M \parallel P] \rightarrowtail$
    $T \mathcal{R} [N \parallel Q] \rightarrowtail T;$

It follows from these properties that a compatible relation is reflexive, since this holds by *(Com1)* on variables, and it is preserved by the other operators by *(Com2)-(Com5)*:

Proposition 1. *If a relation is compatible, then it is reflexive.*

4.1   Howe's Method

The main idea of Howe's method consists in defining an auxiliary relation $\precsim_\circ^H$ such that it is easy to see that it is compatible, and then prove that $\precsim_\circ = \precsim_\circ^H$.

Definition 11. *Let $\mathcal{R}$ be a relation. We define inductively the relation $\mathcal{R}^H$ by the rules in Figure 3.*

We are now going to show that if the relation $\mathcal{R}$ we start from satisfies minimal requirements, namely that it is reflexive and transitive, then $\mathcal{R}^H$ is guaranteed to be compatible and to contain $\mathcal{R}$. This is a direct consequence of the following results, whose proofs are standard inductions:

$$\frac{\overline{x} \cup \{x\} \vdash x \; \mathcal{R} \; M}{\overline{x} \cup \{x\} \vdash x \; \mathcal{R}^H \; M} \qquad \frac{\overline{x} \cup \{x\} \vdash M \; \mathcal{R}^H \; N \qquad \overline{x} \vdash \lambda x.N \; \mathcal{R} \; L}{\overline{x} \vdash \lambda x.M \; \mathcal{R}^H \; L}$$

$$\frac{\overline{x} \vdash M \; \mathcal{R}^H \; N \qquad \overline{x} \vdash L \; \mathcal{R}^H \; P \qquad \overline{x} \vdash NP \; \mathcal{R} \; R}{\overline{x} \vdash ML \; \mathcal{R}^H \; R}$$

$$\frac{\overline{x} \vdash M \; \mathcal{R}^H \; N \qquad \overline{x} \vdash L \; \mathcal{R}^H \; P \qquad \overline{x} \vdash N \oplus P \; \mathcal{R} \; R}{\overline{x} \vdash M \oplus L \; \mathcal{R}^H \; R}$$

$$\frac{\overline{x} \vdash M \; \mathcal{R}^H \; N \qquad \overline{x} \vdash L \; \mathcal{R}^H \; P \qquad \overline{x} \vdash [N \parallel P] \rightarrowtail T \; \mathcal{R} \; R}{\overline{x} \vdash [M \parallel L] \rightarrowtail T \; \mathcal{R}^H \; R}$$

Fig. 3. Howe's Construction

- Let $\mathcal{R}$ be a reflexive relation. Then $\mathcal{R}^H$ is compatible.
- Let $\mathcal{R}$ be transitive. Then:

$$\left(\overline{x} \vdash M \; \mathcal{R}^H \; N\right) \wedge \left(\overline{x} \vdash N \; \mathcal{R} \; L\right) \Rightarrow \left(\overline{x} \vdash M \; \mathcal{R}^H \; L\right) \qquad (1)$$

- If $\mathcal{R}$ is reflexive, then $\overline{x} \vdash M \; \mathcal{R} \; N$ implies $\overline{x} \vdash M \; \mathcal{R}^H \; N$.

We can now apply Howe's construction to $\precsim_\circ$, since it is clearly reflexive and transitive. The properties above then tell us that $\precsim_\circ^H$ is compatible and that $\precsim_\circ \subseteq \precsim_\circ^H$. What we are left with, then, is proving that $\precsim_\circ^H$ is also a simulation.[1]

Lemma 3. $\precsim_\circ^H$ is value-substitutive: for all terms M, N and values V, W such that $x \vdash M \precsim_\circ^H N$ and $\emptyset \vdash V \precsim_\circ^H W$, it holds that $\emptyset \vdash M[V/x] \precsim_\circ^H N[W/x]$

Proof. By induction on the derivation of $x \vdash M \precsim_\circ^H N$.

We also need an auxiliary, technical, lemma about probability assignments:

Definition 12. $\mathbb{P} = \left(\{p_i\}_{1 \le i \le n}, \{r_I\}_{I \subseteq \{1,\ldots,n\}}\right)$ is said to be a probability assignment if for every $I \subseteq \{1,..,n\}$, it holds that $\sum_{i \in I} p_i \le \sum_{J \cap I \ne \emptyset} r_J$.

Lemma 4 (Disentangling Sets). Let $P = \left(\{p_i\}_{1 \le i \le n}, \{r_I\}_{I \subseteq \{1,\ldots,n\}}\right)$ be a probability assignment. Then for every non-empty $I \subseteq \{1,\ldots,n\}$, and for every $k \in I$, there is an $s_{k,I} \in [0,1]$ satisfying the following conditions:

- for every I, it holds that $\sum_{k \in I} s_{k,I} \le 1$;
- for every $k \in 1,\ldots,n$, it holds that $p_k \le \sum_{\{I | k \in I\}} s_{k,I} \cdot r_I$.

The proof is an application of the Max-Flow Min-Cut Theorem, see e.g., [4,6].

Given a set of set of open terms X, let $\lambda x.X = \{\lambda x.M | M \in X\}$.

---

[1] In the proof of congruence for the probabilistic call-by-value λ-calculus presented in [4], the transitive closure of $\precsim_\circ^H$ is considered, since the definition of simulation required the relation to be preorder, which implies that the transitivity of $\precsim_\circ^H$ is needed. Since we relaxed the definition of simulation, this is not anymore necessary.

Lemma 5 (Key Lemma). *For all terms M, N, if $\emptyset \vdash M \precsim_\circ^H N$, then for every $\lambda x.X \subseteq V_{\oplus or}$ it holds that $[\![M]\!]\,(\lambda x.X) \leq [\![N]\!]\,(\precsim_\circ\,(\lambda x.\,\precsim_\circ^H(X)))$.*

Proof. We show that the inequality holds for every approximation of the semantics of M, which implies the result since the semantics is the supremum of the approximations. In particular, we prove by induction on the structure of the derivation of $M \Downarrow \mathscr{D}$ that, for any M, N, if $M \Downarrow \mathscr{D}$ and $\emptyset \vdash M \precsim_\circ^H N$, then for every $\lambda x.X \subseteq V_{\oplus or}$ it holds that $\mathscr{D}\,(\lambda x.X) \leq [\![N]\!]\,(\precsim_\circ\,(\lambda x.\,\precsim_\circ^H(X)))$. We consider separately every possible rule which can be applied at the bottom of the derivation:

- If the rule is $\dfrac{}{M \Downarrow \emptyset}\,bv$ then $\mathscr{D} = \emptyset$, and for all set of values $\lambda x.X$, $\mathscr{D}(\lambda x.X) = 0$, and it concludes the proof.
- If M is a value $V = \lambda x.L$ and the last rule of the derivation is $\dfrac{}{V \Downarrow \{V^1\}}\,bv$ then $\mathscr{D} = \{V^1\}$ is the Dirac distribution for V and, by the definition of Howe's lifting, $\left(\emptyset \vdash \lambda x.L \precsim_\circ^H N\right)$ was derived by the following rule:

$$\frac{x \vdash L \precsim_\circ^H P \qquad \emptyset \vdash \lambda x.P \precsim_\circ N}{\emptyset \vdash \lambda x.L \precsim_\circ^H N}$$

It follows from the definition of simulation and from $(\emptyset \vdash \lambda x.P \precsim_\circ N)$ that $1 = [\![N]\!](\precsim_\circ\{\lambda x.P\})$. Let $\lambda x.X \subseteq V_{\oplus or}$. If $\lambda x.L \notin \lambda x.X$ then $\mathscr{D}(\lambda x.X) = 0$ and the thesis holds. Otherwise, $\mathscr{D}(\lambda x.X) = \mathscr{D}(\lambda x.L) = 1 = [\![N]\!](\precsim_\circ\{\lambda x.P\})$. It follows from $L \precsim_\circ^H P$ and from $\lambda x.L \in \lambda x.X$ that $\lambda x.P \in \lambda x.(\precsim_\circ^H X)$; hence, $[\![N]\!](\precsim_\circ\{\lambda x.P\}) \leq [\![N]\!](\precsim_\circ \lambda x.(\precsim_\circ^H X))$.
- If the derivation of $M \Downarrow \mathscr{D}$ is of the following form:

$$\frac{M_1 \Downarrow \mathscr{K} \qquad M_2 \Downarrow \mathscr{F} \qquad \{P[V/x] \Downarrow \mathscr{E}_{P,V}\}_{\lambda x.P \in S(\mathscr{K}), V \in S(\mathscr{F})}}{M_1 M_2 \Downarrow \sum_{V \in S(\mathscr{F})} \mathscr{F}(V)\left(\sum_{\lambda x.P \in S(\mathscr{K})} \mathscr{K}(\lambda x.P).\mathscr{E}_{P,V}\right)}$$

Then $M = M_1 M_2$ and we have that the last rule used in the derivation of $\emptyset \vdash M \precsim_\circ^H N$ is:

$$\frac{\emptyset \vdash M_1 \precsim_\circ^H M_1' \qquad \emptyset \vdash M_2 \precsim_\circ^H M_2' \qquad \emptyset \vdash M_1' M_2' \precsim_\circ N}{\emptyset \vdash M_1 M_2 \precsim_\circ^H N}$$

Let $S(\mathscr{K}) = \{\lambda x.P_1, \ldots, \lambda x.P_n\}$ and $K_i = \precsim_\circ\{\lambda x.L \mid x \vdash P_i \precsim_\circ^H L\}$ and, symmetrically, $S(\mathscr{F}) = \{V_1, \ldots, V_l\}$ and $X_k = \precsim_\circ\{\lambda x.L \mid V_k = \lambda x.M'$ and $x \vdash M' \precsim_\circ^H L\}$. Then by the inductive hypothesis on $M_1 \Downarrow \mathscr{K}$ and $M_2 \Downarrow \mathscr{F}$ we have that $\mathscr{K}\left(\bigcup_{i \in I}\{\lambda x.P_i\}\right) \leq [\![M_1']\!](\bigcup_{i \in I} K_i)$ for every $I \subseteq \{1, .., n\}$ and $\mathscr{F}(\bigcup_{k \in I}\{V_k\}) \leq [\![M_2']\!]\left(\bigcup_{k \in I} X_k\right)$ for every $I \subseteq \{1, .., l\}$.

Lemma 4 allows us to derive that for all $U \in \bigcup_{1 \leq i \leq n} K_i$ there exist probability values $r_1^U, \ldots, r_n^U$ and for all $W \in \bigcup_{1 \leq k \leq l} X_k$ there exist probability values $s_1^W, .., s_l^W$ such that:

$$[\![M_1']\!](U) \geq \sum_{1 \leq i \leq n} r_i^U \qquad [\![M_2']\!](W) \geq \sum_{1 \leq k \leq l} s_k^W \qquad \forall U \in \bigcup_{1 \leq i \leq n} K_i, W \in \bigcup_{1 \leq k \leq l} X_k$$

$$\mathscr{K}(\lambda x.P_i) \leq \sum_{U \in K_i} r_i^U \qquad \mathscr{F}(V_k) \leq \sum_{W \in X_k} s_k^W \qquad \forall 1 \leq i \leq n, 1 \leq k \leq l$$

Hence, for every value $Z \in \mathcal{V}_{\oplus or}$, we have that:

$$\mathcal{D}(Z) = \sum_{1 \leq k \leq l} \mathcal{F}(V_k) \cdot \sum_{1 \leq i \leq n} \mathcal{K}(\lambda x.P_i) \cdot \mathcal{E}_{P_i, V_k}(Z)$$

$$\leq \sum_{1 \leq k \leq l} \sum_{W \in X_k} s_k^W \cdot \sum_{1 \leq i \leq n} \sum_{U \in K_i} r_i^U \cdot \mathcal{E}_{P_i, V_k}(Z)$$

If $U = \lambda x.U' \in K_i$ then there exists S such that:

$$(2) \quad \emptyset \vdash \lambda x.S \precsim_\circ U \qquad (3) \quad x \vdash P_i \precsim_\circ^H S$$

By (2), $\emptyset \vdash S[W/x] \precsim_\circ U'[W/x]$. By (3) and by Lemma 3, for $W \in X_k$ we have that $\emptyset \vdash P_i[V_k/x] \precsim_\circ^H S[W/x]$. It follows from (1) that $\emptyset \vdash P_i[V_k/x] \precsim_\circ^H U'[W/x]$. Hence, by the induction hypothesis applied to $P_i[V_k/x]$ we have $\mathcal{E}_{P_i, V_k}(\lambda x.X) \leq [\![U'[W/x]]\!](\precsim_\circ \lambda x.(\precsim_\circ^H X))$. Therefore,

$$\mathcal{D}(\lambda x.X) \leq \sum_{1 \leq k \leq l} \sum_{W \in X_k} s_k^W \cdot \sum_{1 \leq i \leq n} \sum_{U \in K_i} r_i^U \cdot \mathcal{E}_{P_i, V_k}(\lambda x.X)$$

$$\leq \sum_{W \in \bigcup_{1 \leq k \leq l} X_k} \sum_{U \in \bigcup_{1 \leq i \leq n} K_i} \left(\sum_{\{k | W \in X_k\}} s_k^W \right) \cdot \left(\sum_{\{i | U \in K_i\}} r_i^U \right) [\![L_{U,W}]\!](\precsim_\circ \lambda x.(\precsim_\circ^H X))$$

$$\leq \sum_{W \in \bigcup_{1 \leq k \leq l} X_k} \sum_{U \in \bigcup_{1 \leq i \leq n} K_i} [\![M_2']\!](W) \cdot [\![M_1']\!](U) \cdot [\![L_{U,W}]\!](\precsim_\circ \lambda x.(\precsim_\circ^H X))$$

$$\leq [\![M_1' M_2']\!](\precsim_\circ \lambda x.(\precsim_\circ^H X))$$

where $L_{U,W} = U'[W/x]$ for any U such that $U = \lambda x.U'$.

- If $M \Downarrow \mathcal{D}$ is derived by:

$$\frac{M_1 \Downarrow \mathcal{D}_1 \qquad M_2 \Downarrow \mathcal{D}_2}{M_1 \oplus M_2 \Downarrow \frac{1}{2}\mathcal{D}_1 + \frac{1}{2}\mathcal{D}_2}$$

then $\emptyset \vdash M \precsim_\circ^H N$ is derived by:

$$\frac{\emptyset \vdash M_1 \precsim_\circ^H N_1 \qquad \emptyset \vdash M_2 \precsim_\circ^H N_2 \qquad \emptyset \vdash N_1 \oplus N_2 \precsim_\circ N}{\emptyset \vdash M_1 \oplus M_2 \precsim_\circ^H N}$$

By the inductive hypothesis, for $i \in \{1, 2\}$ we have that for any $\lambda x.X \subseteq \mathcal{V}_{\oplus or}$,

$$\mathcal{D}_i(\lambda x.X) \leq [\![N_i]\!](\precsim_\circ \lambda x.(\precsim_\circ^H X))$$

Hence, the result follows from:

$$\frac{1}{2} \cdot \mathcal{D}_1(\lambda x.X) + \frac{1}{2} \cdot \mathcal{D}_2(\lambda x.X) \leq \frac{1}{2} \cdot [\![N_1]\!](\precsim_\circ \lambda x.(\precsim_\circ^H X)) + \frac{1}{2} \cdot [\![N_2]\!](\precsim_\circ \lambda x.(\precsim_\circ^H X))$$

- If the last rule applied in the derivation of $M \Downarrow \mathcal{D}$ is of the following form:

$$\frac{M_1 \Downarrow \mathcal{D}_1 \qquad M_2 \Downarrow \mathcal{D}_2}{[M_1 \parallel M_2] \rightarrowtail T \Downarrow (\sum \mathcal{D}_1 + \sum \mathcal{D}_2 - \sum \mathcal{D}_1 \cdot \sum \mathcal{D}_2) \cdot \{T^1\}}$$

then $M = [M_1 \parallel M_2] \rightarrowtail T$ and $\emptyset \vdash M \precsim_o^H N$ is derived by:

$$\frac{\emptyset \vdash M_1 \precsim_o^H N_1 \qquad \emptyset \vdash M_2 \precsim_o^H N_2 \qquad \emptyset \vdash [N_1 \parallel N_2] \rightarrowtail T \precsim_o N}{\emptyset \vdash [M_1 \parallel M_2] \rightarrowtail T \precsim_o^H N}$$

By inductive hypothesis on $M_1 \Downarrow \mathscr{D}_1$ we have that for any $\lambda x.X \subseteq \mathcal{V}_{\oplus or}$, $\mathscr{D}_1(\lambda x.X) \leq [\![N_1]\!](\precsim_o \lambda x.(\precsim_o^H X))$. Hence, for $\lambda x.X = \mathsf{S}(\mathscr{D}_1)$ we have that:

$$\sum \mathscr{D}_1 = \mathscr{D}_1(\lambda x.X) \leq [\![N_1]\!](\precsim_o \lambda x.(\precsim_o^H X)) \leq [\![N_1]\!](\mathsf{S}([\![N_1]\!])) = \sum [\![N_1]\!]$$

and, symmetrically, by the inductive hypothesis on $M_2 \Downarrow \mathscr{D}_2$ we have $\sum \mathscr{D}_2 \leq \sum [\![N_2]\!]$. Therefore,

$$\sum \mathscr{D}_1 + \sum \mathscr{D}_2 - \sum \mathscr{D}_1 \cdot \sum \mathscr{D}_2 \leq \sum [\![N_1]\!] + \sum [\![N_2]\!] - \sum [\![N_1]\!] \cdot \sum [\![N_2]\!]$$

Let $\lambda x.X \subseteq \mathcal{V}_{\oplus or}$. If $T \notin \lambda x.X$ then $\mathscr{D} = 0$ and the result follows. Otherwise, it follows from $T = \lambda x.T' \in \precsim_o \lambda x.(\precsim_o^H \{T'\})$ (since both $\precsim_o$ and $\precsim_o^H$ are reflexive) that

$$\begin{aligned}\mathscr{D}(\lambda x.X) = \mathscr{D}(\lambda x.T') &= \sum \mathscr{D}_1 + \sum \mathscr{D}_2 - \sum \mathscr{D}_1 \cdot \sum \mathscr{D}_2 \\ &\leq \sum [\![N_1]\!] + \sum [\![N_2]\!] - \sum [\![N_1]\!] \cdot \sum [\![N_2]\!] \\ &= [\![N]\!](\lambda x.T') = [\![N]\!](\precsim_o \lambda x.(\precsim_o^H X))\end{aligned}$$

$\square$

A consequence of the Key Lemma, then, is that relation $\precsim_o^H$ on closed terms is an applicative simulation, thus included in the largest one, namely $\precsim$. Hence, if M, N are open terms and $x_1, \ldots, x_n \vdash M \precsim_o^H N$ then it follows from Lemma 3 that for all $V_1, \ldots, V_n, W_1, \ldots, W_n$ such that $\emptyset \vdash V_i \precsim_o^H W_i$ we have that $\emptyset \vdash M[V_1, \ldots, V_n/x_1, \ldots, x_n] \precsim_o^H N[W_1, \ldots, W_n/x_1, \ldots, x_n]$, which implies (by the reflexivity of $\precsim_o^H$ and by $\precsim_o^H \subseteq \precsim_o$ on closed terms) that for all $V_1, \ldots, V_n$ we have that $\emptyset \vdash M[V_1, \ldots, V_n/x_1, \ldots, x_n] \precsim_o N[V_1, \ldots, V_n/x_1, \ldots, x_n]$, i.e., $M \precsim_o N$. Since $\precsim_o$ is itself included in $\precsim_o^H$, we obtain that $\precsim_o = \precsim_o^H$. Hence, it follows from the transitivity of $\precsim_o$ and from the fact that $\precsim_o^H$ is compatible that:

Theorem 2 (Congruence). $\precsim_o$ *is a precongruence* .

The congruence of $\precsim_o$ allows us to prove that it is a sound with respect to the contextual preorder.

Theorem 3 (Soundness). *If* $M \precsim_o N$ *then* $M \leq N$.

Proof. Let $M \precsim_o N$. Using Theorem 2, it can be easily proved by induction on C that for any context C it holds that $C[M] \precsim_o C[N]$. If $C[M] \precsim_o C[N]$ then $\sum [\![C[M]]\!] \leq \sum [\![C[M]]\!]$, which implies the result. $\square$

5   Full Abstraction

In [24], both bisimilarity and similarity on labelled Markov chains are characterised by a language of test, refining the testing characterization of bisimilarity presented in [14]. This characterisation is used in [4] to show that the bisimilarity relation on terms is fully abstract with respect to the contextual equivalence. The language of tests used to characterize bisimulation is the following:

Definition 13. *Let $\mathcal{M} = (\mathcal{S}, \mathcal{L}, \mathcal{P})$ be a LMC. The test language $\mathscr{T}_0(\mathcal{M})$ is given by the grammar $t ::= \omega \mid a \cdot t \mid \langle t, t \rangle$, where $a \in \mathcal{L}$.*

This language represents tests in the following sense: for any t in the test language $\mathscr{T}_0(\mathcal{M})$, and for any s state of $\mathcal{M}$, we can define the probability $\Pr(s, t)$ that the test t succeeds when executed on s.

The full-abstraction result in [4] is based on the fact that, when we consider the particular Markov chain used to define a bisimulation relation on terms, any of these tests can actually be simulated by a context. However, the characterisation of the simulation preorder requires to add disjunctive tests:

Definition 14. *Let $\mathcal{M} = (\mathcal{S}, \mathcal{L}, \mathcal{P})$ be a LMC. The test language $\mathscr{T}_1(\mathcal{M})$ is given by the grammar $t ::= \omega \mid a \cdot t \mid \langle t, t \rangle \mid t \vee t$, where $a \in \mathcal{L}$.*

We are now going to define the success probability of a test. The success probability of ω is 1 no matter what state we are starting from. The success probability of a disjunctive test corresponds to the probability that at least one of the two tests is successful.

Definition 15. *Let $\mathcal{M} = (\mathcal{S}, \mathcal{L}, \mathcal{P})$ be a LMC. For all $s \in \mathcal{S}$, and $t \in \mathscr{T}_1(\mathcal{M})$, we define:*

$$Pr(s, \omega) = 1; \qquad Pr(s, t \vee u) = Pr(s, t) + Pr(s, u) - Pr(s, t) \cdot Pr(s, u)$$

$$Pr(s, \langle t, u \rangle) = Pr(s, t) \cdot Pr(s, u); \qquad Pr(s, a \cdot t) = \sum_{s' \in \mathcal{S}} \mathcal{P}(s, a, s') \cdot Pr(s', t).$$

The following theorem characterises bisimilarity and the simulation preorder on labelled Markov chains by means of sets of tests.

Theorem 4 ([24]). *Let $\mathcal{M} = (\mathcal{S}, \mathcal{L}, \mathcal{P})$ be a LMC and let $s, s' \in \mathcal{S}$. Then:*
- *$s \sim s'$ if and only if for every $t \in \mathscr{T}_0(\mathcal{M})$ it holds that: $Pr(s, t) = Pr(s', t)$*
- *$s \precsim s'$ if and only if for every $t \in \mathscr{T}_1(\mathcal{M})$ it holds that $Pr(s, t) \leq Pr(s', t)$*

Example 3. Consider the two terms $M = \lambda x.(I \oplus \Omega)$ and $N = (\lambda x.I) \oplus (\lambda x.\Omega)$ from Example 2. We already know that, since they do not verify $M \precsim N$, there exists a test $t \in \mathscr{T}_1(\mathcal{M}_{\oplus or})$ whose success probability when executed on M is strictly greater that its success probability when executed on N. We can actually explicitly give such a test: let $t = eval \cdot (I \cdot eval \cdot \omega \vee I \cdot eval \cdot \omega)$ Then it holds that:

$$\Pr(\lambda x.(I \oplus \Omega), t) = \frac{3}{4}; \qquad \Pr((\lambda x.I) \oplus (\lambda x.\Omega), t) = \frac{1}{2}.$$

5.1   From Tests to Contexts

It is shown in [4] that simulation is not fully abstract for $\mathsf{PCFL}_\oplus$ with respect to the contextual preorder: a direct consequence is that disjunctive tests cannot be simulated by contexts. In other terms, it is not possible to write a program that has access to two sub-programs, and terminates with a probability equal to the probability that at least one of its sub-programs terminates. The proof of [4] is based on an encoding from $\mathscr{T}_0(\mathcal{M}_\oplus)$ to the set of contexts. We are going to extend it into two encodings from $\mathscr{T}_1(\mathcal{M}_{\oplus or})$ to the set of contexts of $\Lambda_{\oplus or}$: one encoding expresses the action of tests on states of the form M, and the other one on states of the form $\hat{V}$. The intuitive idea behind Θ^{val} and Θ^{term} is the following: if we take a test t, its success probability starting from the state M is the same as the convergence probability of the context $\Theta^{term}(t)$ filled by M, and similarly, its success probability starting from the state $\hat{V}$ is the same as the convergence probability of the context $\Theta^{term}(t)$ filled by V.

Definition 16. *Let $\Theta^{val} : \mathscr{T}_1(\mathcal{M}_{\oplus or}) \to \mathscr{C}$ and $\Theta^{term} : \mathscr{T}_1(\mathcal{M}_{\oplus or}) \to \mathscr{C}$ be defined by:*

$$\Theta^{term}(\omega) = \lambda x.[\cdot]; \qquad\qquad \Theta^{val}(\omega) = \lambda x.[\cdot];$$
$$\Theta^{term}(V \cdot t) = \Omega[\cdot]; \qquad\qquad \Theta^{val}(V \cdot t) = \Theta^{term}(t)[([\cdot]V)];$$
$$\Theta^{term}(eval \cdot t) = \lambda x.(\Theta^{val}(t)[x])[\cdot]; \qquad \Theta^{val}(eval \cdot t) = \Omega[\cdot];$$
$$\Theta^{term}(t \vee u) = g(\Theta^{term}(t), \Theta^{term}(u)); \qquad \Theta^{val}(t \vee u) = g(\Theta^{val}(t), \Theta^{val}(u));$$
$$\Theta^{term}(\langle t, u \rangle) = f(\Theta^{term}(t), \Theta^{term}(u)); \qquad \Theta^{val}(\langle t, u \rangle) = f(\Theta^{val}(t), \Theta^{val}(u));$$

where $f, g : \mathscr{C} \times \mathscr{C} \to \mathscr{C}$ are defined by:

$$f(C, D) = (\lambda x.(\lambda y, z.I)(C[xI])(D[xI]))(\lambda x.[\cdot]);$$
$$g(C, D) = (\lambda x.([C[xI] \parallel D[xI]] \rightarrowtail I)(\lambda x.[\cdot]).$$

The apparently complicated structure of f and g comes from the fact that we cannot construct contexts with several holes. However, since our language has copying capability, we can emulate contexts with several holes by means of contexts with only one hole. Intuitively, we could say that $g(C, D)$ would correspond to a multihole context $[C \parallel D] \rightarrowtail I$. Please observe that the encoding of the fragment of $\mathscr{T}_1(\mathcal{M}_{\oplus or})$ corresponding to $\mathscr{T}_0(\mathcal{M}_{\oplus or})$ does not use parallel disjunction, i.e., the image of $\mathscr{T}_0(\mathcal{M}_{\oplus or})$ by the encoding is a subset of $\Lambda_\oplus$. We can now apply this encoding to the test we defined in Example 3.

Example 4. Recall the test $t = eval \cdot (I \cdot eval \cdot \omega \vee I \cdot eval \cdot \omega)$ defined in Example 3. We can apply the embedding to this particular test:

$$\Theta^{term}(t) = (\lambda x.(\lambda z.[(\lambda y.(\lambda w.y))zII \parallel (\lambda y.(\lambda w.y))zII] \rightarrowtail I)(\lambda y.x))[\cdot].$$

We can see that if we consider the terms $M = \lambda x.(I \oplus \Omega)$ and $N = (\lambda x.I) \oplus (\lambda x.\Omega)$ defined in Example 2, the context $\Theta^{term}(t)$ simulates the test t with respect to M and N:

$$\Pr(M, t) = \sum \llbracket \Theta^{term}(t)[M] \rrbracket; \qquad\qquad \Pr(N, t) = \sum \llbracket \Theta^{term}(t)[N] \rrbracket.$$

Theorem 5. *Let t be a test in $\mathscr{T}_1(\mathcal{M}_{\oplus or})$. Then for every M closed term and every V closed value it holds that:*

$$Pr(M,t) = \sum [\![\Theta^{term}(t)[M]]\!]; \qquad Pr(\hat{V},t) = \sum [\![\Theta^{val}(t)[V]]\!].$$

Proof. We are going to show the thesis by induction on the structure of t.

- If $t = \omega$, then for every closed term M, and every closed value V, $Pr(M,\omega) = Pr(\hat{V},\omega) = 1$, and we have defined $\Theta^{term}(\omega) = \Theta^{val}(\omega) = \lambda x.[\cdot]$. Since $\Theta^{term}(\omega)[M]$ and $\Theta^{val}(\omega)[V]$ are values, the weight of their semantics is 1, and so the result holds.

- If $t = \langle u_1, u_2 \rangle$, we can directly adapt the construction proposed in [4] to the untyped case. By the inductive hypothesis, for all $1 \le i \le 2$ it holds that for every closed term M and every closed value V,

$$\Pr(M,u_i) = \sum [\![\Theta^{term}(u_i)[M]]\!]; \Pr(\hat{V},u_i) = \sum [\![\Theta^{val}(u_i)[V]]\!].$$

The overall effect of f is to copy the content of the hole into the holes of the two contexts C and D. For any closed term M, we can express the convergence probability of $f(C,D)[M]$ as a function of the convergence probability of $C[M]$ and $D[M]$:

$$\sum [\![f(C,D)[M]]\!] = \left(\sum [\![C[(\lambda x.M)I]]\!] \right) \cdot \left(\sum [\![D[(\lambda x.M)I]]\!] \right)$$
$$= \left(\sum [\![C[M]]\!] \right) \cdot \left(\sum [\![D[M]]\!] \right)$$

Please recall that we have defined:

$$\Theta^{term}(\langle u_1, u_2 \rangle) = f(\Theta^{term}(u_1), \Theta^{term}(u_2))$$
$$\Theta^{val}(\langle u_1, u_2 \rangle) = f(\Theta^{val}(u_1), \Theta^{val}(u_2))$$

We have that, for any closed term M, and any closed value V:

$$\sum [\![\Theta^{term}(\langle u_1, u_2 \rangle)[M]]\!] = \Pr(M,u_1) \cdot \Pr(M,u_2) = \Pr(M, \langle u_1, u_2 \rangle)$$
$$\sum [\![\Theta^{val}(\langle u_1, u_2 \rangle)[V]]\!] = \Pr(\hat{V},u_1) \cdot \Pr(\hat{V},u_2) = \Pr(\hat{V}, \langle u_1, u_2 \rangle)$$

- Now the case $t = u_1 \vee u_2$. By the inductive hypothesis, for all $1 \le i \le 2$ it holds that for every closed term M and every closed value V,

$$\Pr(M,u_i) = \sum [\![\Theta^{term}(u_i)[M]]\!] \qquad \Pr(\hat{V},u_i) = \sum [\![\Theta^{val}(u_i)[V]]\!].$$

The definition of g allows us to show:

$$\sum [\![g(C,D)[M]]\!] = \sum [\![C[M]]\!] + \sum [\![D[M]]\!] - \sum [\![C[M]]\!] \cdot \sum [\![D[M]]\!]$$

and now it is straightforward to see that:

$$\sum [\![\Theta^{term}(u_1 \vee u_2)[M]]\!] = \Pr(M, u_1 \vee u_2);$$
$$\sum [\![\Theta^{val}(u_1 \vee u_2)[V]]\!] = \Pr(\hat{V}, u_1 \vee u_2).$$

- If $t = a \cdot u$, there are two different kinds of actions:
  - when $a = eval$, we first consider $\Theta^{val}(t)$: since the *eval* action is relevant only for states of $\mathcal{M}_{\oplus or}$ which are terms (and not distinguished values), we want that $\Theta^{val}(t)[V]$ always diverges. Since $\Theta^{val}(t) = \Omega[\cdot]$ and since $[\![\Omega]\!] = \emptyset$, we have that for any closed value V, $[\![\Theta^{val}(t)[V]]\!] = \emptyset$.
    Now, we consider $\Theta^{term}(t)$. By the inductive hypothesis, we know that:

$$\Pr(\hat{V}, u) = \sum [\![\Theta^{val}(u)[V]]\!].$$

Please recall that we have defined: $\Theta^{term}(a \cdot u) = \lambda x.(\Theta^{val}(u)[x])[\cdot]$. Let be M a closed term. Then it holds that:

$$\sum [\![\Theta^{term}(a \cdot u)[M]]\!] = \sum_V [\![M]\!](V) \cdot \sum [\![\Theta^{val}(u)[V]]\!]$$

$$= \sum_V [\![M]\!](V) \cdot \Pr(\hat{V}, u)$$

$$= \sum_{e \in \mathcal{S}_{\oplus or}} \mathcal{P}_{\oplus or}(M, eval, e) \cdot \Pr(e, u) = \Pr(M, u)$$

- When $a = V$, with $V \in \mathcal{V}_{\oplus or}$, we consider first $\Theta^{term}(V \cdot u)$. It has been designed to be a context which diverges whatever its argument is, and so we indeed have: $\Pr(M, V \cdot u) = 0 = \sum [\![\Theta^{term}(V \cdot u)[M]]\!]$. Then we consider $\Theta^{val}(t)$. Recall that we have defined: $\Theta^{val}(V \cdot u) = \Theta^{term}(u)[[\cdot]V]$. Let $W = \lambda x.M$ be a closed value:

$$\sum [\![\Theta^{val}(V \cdot u)[W]]\!] = \sum [\![\Theta^{term}(u)[WV]]\!]$$

$$= \Pr(WV, u)$$

$$= \Pr(M[x/V], u) \qquad \text{since } [\![WV]\!] = [\![M[x/V]]\!]$$

$$= \Pr(W, V \cdot u).$$

$\square$

Theorem 6. $\precsim$ *is fully abstract with respect to the contextual preorder.*

Proof. We already know that $\precsim$ is sound, that is $\precsim \subseteq \leq$. Hence, what is left to show is that $\leq \subseteq \precsim$, which follows from Theorem 5. Let M and N be two closed terms such that $M \leq N$. We want to show that $M \precsim N$. The testing characterisation of simulation allows us to say that it is sufficient to show that, for every test $t \in \mathcal{T}_1(\mathcal{M}_{\oplus or})$, $\Pr(M, t) \leq \Pr(N, t)$, which in turn is a consequence of Theorem 5, since every test t of $\mathcal{T}_1(\mathcal{M}_{\oplus or})$ can be simulated by a context of $\Lambda_{\oplus or}$.

References

1. Abramsky, S.: The lazy λ-Calculus. In: Turner, D. (ed.) Research Topics in Functional Programming, pp. 65–117. Addison Wesley (1990)

2. Abramsky, S., Ong, C.-H.L.: Full abstraction in the lazy lambda calculus. Inf. Comput. **105**(2), 159–267 (1993)
3. Comaniciu, D., Ramesh, V., Meer, P.: Kernel-based object tracking. IEEE Trans. on Pattern Analysis and Machine Intelligence **25**(5), 564–577 (2003)
4. Crubillé, R., Dal Lago, U.: On probabilistic applicative bisimulation and call-by-value λ-calculi (long version). CoRR, abs/1401.3766 (2014)
5. Dal Lago, U., Sangiorgi, D., Alberti, M.: On coinductive equivalences for higher-order probabilistic functional programs (long version). CoRR, abs/1311.1722 (2013)
6. Dal Lago, U., Sangiorgi, D., Alberti, M.: On coinductive equivalences for higher-order probabilistic functional programs. In: POPL, pp. 297–308 (2014)
7. Dal Lago, U., Zorzi, M.: Probabilistic operational semantics for the lambda calculus. RAIRO - Theor. Inf. and Applic. **46**(3), 413–450 (2012)
8. Danos, V., Harmer, R.: Probabilistic game semantics. ACM Trans. Comput. Log. **3**(3), 359–382 (2002)
9. Desharnais, J., Edalat, A., Panangaden, P.: Bisimulation for labelled Markov processes. Inf. Comput. **179**(2), 163–193 (2002)
10. Ehrhard, T., Tasson, C., Pagani, M.: Probabilistic coherence spaces are fully abstract for probabilistic PCF. In: POPL, pp 309–320 (2014)
11. Goldwasser, S., Micali, S.: Probabilistic encryption. J. Comput. Syst. Sci. **28**(2), 270–299 (1984)
12. Goodman, N.D.: The principles and practice of probabilistic programming. In: POPL, pp. 399–402 (2013)
13. Jones, C., Plotkin, G.D.: A probabilistic powerdomain of evaluations. In: LICS, pp. 186–195 (1989)
14. Larsen, K.G., Skou, A.: Bisimulation through probabilistic testing. Inf. Comput. **94**(1), 1–28 (1991)
15. Lassen, S.B.: Relational Reasoning about Functions and Nondeterminism. PhD thesis, University of Aarhus (1998)
16. Manning, C.D., Schütze, H.: Foundations of statistical natural language processing, vol. 999. MIT Press (1999)
17. Ong, C.-H.L.: Non-determinism in a functional setting. In: LICS, pp. 275–286 (1993)
18. Park, S., Pfenning, F., Thrun, S.: A probabilistic language based on sampling functions. ACM Trans. Program. Lang. Syst. **31**(1) (2008)
19. Pearl, J.: Probabilistic reasoning in intelligent systems: networks of plausible inference. Morgan Kaufmann (1988)
20. Pfeffer, A.: IBAL: A probabilistic rational programming language. In: IJCAI, pp. 733–740. Morgan Kaufmann (2001)
21. Plotkin, G.D.: LCF considered as a programming language. Theor. Comput. Sci. **5**(3), 223–255 (1977)
22. Ramsey, N., Pfeffer, A.: Stochastic lambda calculus and monads of probability distributions. In: POPL, pp. 154–165 (2002)
23. Thrun, S.: Robotic mapping: A survey. Exploring Artificial Intelligence in the New Millennium, pp. 1–35 (2002)
24. van Breugel, F., Mislove, M.W., Ouaknine, J., Worrell, J.: Domain theory, testing and simulation for labelled markov processes. Theor. Comput. Sci. **333**(1–2), 171–197 (2005)

Causality, Behavioural Equivalences, and the Security of Cyberphysical Systems

Sibylle Fröschle [✉]

OFFIS and University of Oldenburg, 26121 Oldenburg, Germany
froeschle@informatik.uni-oldenburg.de

Abstract. The large cyberphysical systems that are currently being developed such as Car2X come with sophisticated security architectures that involve a complex interplay of security protocols and security APIs. Although formal methods for security protocols have achieved a mature stage there are still many challenges left. One is to improve the verification of equivalence-based security properties. A second challenge is the compositionality problem: how can the security of a composition of security protocols and APIs be derived from the security of its components. It seems intuitively clear that foundational results on causal equivalences and process calculi could help in this situation. In this talk we first identify four ways to exploit causality in security verification. In particular, this will lead us to review results on causal equivalences. Finally, we discuss how such results could help us to tackle the two challenges.

1 Motivation

Cyberphysical systems such as Car2X are potentially vulnerable against attacks that could have a drastic impact on the safety as well as the privacy of their users. Therefore such systems must be protected by a sophisticated security architecture. Take Car2X as an example. Based on a threat and risk analysis, the ETSI[1] standards advocate a security architecture that includes authenticated Car2X communication by digital signatures, cryptographic keys and credentials management, privacy enabling technologies by pseudonyms, and in-car software and hardware security based on hardware security modules (HSMs) [11,12].

Two mechanisms are central within such a security architecture: *security protocols* and *security APIs*. A security protocol specifies an exchange of cryptographic messages between two or several principals, intended to achieve security objectives such as authentication, key establishment, or confidentiality of data. A security API (Application Programming Interface) is the software interface to a security services layer. At the lowest level this will typically be the API to an HSM that stores and uses sensitive cryptographic keys.

This work is partially supported by the *Niedersächsisches Vorab* of the Volkswagen Foundation and the Ministry of Science and Culture of Lower Saxony as part of the *Interdisciplinary Research Center on Critical Systems Engineering for Socio-Technical Systems*.

[1] European Telecommunications Standards Institute: http://www.etsi.org.

© Springer International Publishing Switzerland 2015
R. Meyer et al. (Eds.): Olderog-Festschrift, LNCS 9360, pp. 83–98, 2015.
DOI: 10.1007/978-3-319-23506-6_8

The security properties that are central for the verification of security protocols and APIs fall into two categories: one is that of *reachability-based* (or *trace-based*) properties, the second is the class of *equivalence-based* properties. Traditional properties such as authentication and syntactic secrecy are trace-based properties. They express properties of protocol runs: if A and B exchange a secret s and during no run of the protocol the attacker can obtain the value s then the protocol satisfies syntactic secrecy of s. In contrast, privacy-type properties such as untraceability, vote secrecy, or anonymity have to be expressed in terms of indistinguishability: if an attacker has no way to distinguish the process in which A votes 'yes' from the process in which she votes 'no' then vote secrecy is satisfied. Formally, indistinguishability is expressed in terms of a notion of behavioural equivalence.

The verification of security protocols has reached a mature state with many tools available that can automatically check whether a security property is satisfied (up to certain assumptions or abstractions) (e.g. [46]). However, there are still many challenges left. Decidability and complexity results as well as automatic tools mainly target reachability-based properties so far. Hence, one challenge is to improve the foundations and verification techniques for equivalence-based properties. Only few results on the applied equivalences are known (c.f. [6]). The tool ProVerif does support the verification of privacy-type properties but it does so by an ad hoc encoding of the situation when processes differ only in their choice of some terms [4].

A second challenge concerns compositionality. Most of the formal methods and automatic tools are only capable of checking one protocol at a time. It is folklore that as long as two protocols are disjoint in that they do not share any data their composition is secure iff each protocol is secure in isolation. However, as exemplified by real security architectures such as that of Car2X this situation is far from reality. A stack of different protocols and APIs is necessary exactly because there are different interconnected phases such as key management on a server, key establishment between a server and a principal, and exchange of confidential information between them. Another problem is that an attacker might deliberately induce that protocols share a key they are not supposed to share, or this might be induced by users who use the same passwords in different situations. For a summary of works that already address the compositionality problem see [2,7].

In this talk we explore how causality as a general theme, and causal equivalences from 'pure' concurrency theory can help us in the verification of complex security architectures. We proceed as follows. In Section 2 we review four ways to exploit causality in security verification, and note that causality has mainly been applied to reachability-based verification problems. We identify that the fourth way to exploit causality could be very relevant for equivalence-based properties. The idea is to apply and lift a positive trend for causal equivalences from concurrency theory to equivalence-based verification. Thereby motivated, in Section 3 we take a closer look at these causal equivalences. We give an overview of known decidability and complexity results, and identify some open problems relevant for their application. Finally, in Section 4 we discuss what it takes to lift this trend into an applied setting, and close with some general remarks.

2 Four Ways to Exploit Causality

2.1 Modelling

1. Modelling: "A causal model says more than a million transitions."

Transition systems are the natural model when it comes to automatic state space explorations. But when it comes to modelling or analysing a system by hand then models that faithfully represent the causal structure of the system are usually the model of choice. The reason for this success of causal models is twofold: they avoid the state explosion problem by modelling concurrency explicitly; and, they typically come with an intuitive graphical notation. So just as "a picture says more than a thousand words" "a causal model says more than a million transitions".

In security verification this is exemplified by the *strand space model*, which is *the* causal model for security protocols. The strand space model was introduced by Thayer, Herzog, and Guttman in their paper "Strand Spaces: Why is a Security Protocol Correct?" [13] as a special-purpose model that allows one to develop correctness proofs by hand. To use the author's own words, it is "distinguished from other work on protocol verification by the simplicity of the model and the ease of producing intelligible and reliable proofs of protocol correctness even without automated support".

2.2 Verification

2. Verification: "Refute that an attack exists by tracing all possible causal constellations to a contradiction."

Everybody has a notion that some event A is a cause of another event B. And that if event A hadn't happened then event B wouldn't have happened either. This translates into a natural proof principle of backwards analysis. Say we wish to show that a bad event B cannot happen. To the contrary assume that B has occurred and analyse what must have happened beforehand. If we can lead all possible causal constellations to a contradiction then we can conclude that our assumption was wrong and B can indeed *not* happen.

This is the proof method that is originally associated with the strand space model [13]. The original manual proof method also sparked off several generations of semi-automated tools which work by backwards search. Among the earliest is the semi-automated tool Athena by Song [47], which translates the backwards reasoning style into a backwards search algorithm. The Athena tool has in turn influenced many of the more recent ones: Cremer's tool Scyther [8], which has recently been used to analyse the large Internet protocol IPSec by using supercomputer power [9], and the successor tool Tamarin [46].

The proof principle has also been employed in the verification of security APIs. The strand space method has inspired our backwards reasoning approach to security API verification [18,23], including case studies of the standard PKCS#11. Moreover, an approach for automated backward analysis for a class of PKCS# 11 configurations as been presented in [35]: by translation into the Tamarin prover. Finally, the strand space method has also been used to tackle the compositionality problem: Among the recent proof methods by hand there is Lowe et al.'s approach to verify layered security architectures including a case study of TLS [33].

2.3   Decidability and Complexity I

3. Decidability and Complexity I: "Exploit causal structure to reduce a search space of attacks to a decidable search space of 'well-structured' attacks."

If we work with a model that faithfully represents the causal structure of a system, we may be able to exploit this extra structure to obtain results we may not be able to formulate and prove otherwise. An explicit representation of causal dependencies gives us concepts at hand such as the causal shape or the causal depth of attacks. Moreover, if we can show that there is an attack iff there is one with a particularly 'good and regular' causal structure then we may be able to reduce our search for attacks to a search for more manageable 'well-structured attacks', and thereby obtain new decidability and complexity results.

We have used this principle to prove that REACHABILITY is decidable in NEXPTIME for protocols with disequality constraints and bounded message size [17]; and more recently, that LEAKINESS is decidable for well-founded protocols [19]. (Leakiness is a type of secrecy that does not admit temporary secrets.)

Well-founded protocols strictly contain a group of protocols that impose conditions that make encrypted messages context-explicit [5,37,43–45]. The idea is that such protocols merely satisfy the prudent engineering practice recommended by Abadi and Needham [1]. For example, a 'light' way to achieve context-explicitness is to tag protocols by introducing a constant into each encryption, and thereby to uniquely identify encrypted subterms occurring in the protocol specification. The decidability of well-founded protocols confirms that even under this static notion of context-explicitness security protocols lose their ability to encode Turing complete models, even without bounding message size or the number of nonces. The key to the result was to introduce a notion of *honest causality*, which captures that honest information is propagated from one event to another: there is a causal chain which contains a backbone of messages and control flow transitions that could not have been manipulated by the intruder. One can then show that the depth of honest causality is bounded for well-founded protocols.

Context-explicit protocols also enjoy good properties wrt the compositionality problem. In particular, in [7] Ciobâcă and Cortier obtain: any attack trace on the composition of two differently tagged protocols can be transformed into an attack against one of the protocols. Hence, the security of a composed protocol can be derived from the security of each component protocol. The result is obtained for a variant of the applied pi-calculus that covers both parallel and sequential composition. It therefore includes the case where one protocol uses a sub-protocol for an initial key exchange phase. This result is extended in [2], and also investigated for equivalence-based properties. A composition result is obtained for the case of key-exchange protocols wrt *diff-equivalence*.

2.4   Decidability and Complexity II

4. Decidability and Complexity II: "Use causal counterparts of equivalences and logics rather than interleaving ones to obtain good decidability and complexity."

Typical for the previous subsection is that the verification problem is still in the realm of interleaving semantics, i.e. can be formulated without relating to causality, but we use causal concepts to obtain the proof. The fourth slogan suggests to exploit causality more directly by taking the verification problem itself into the causal setting. The hope is that then good structural properties of the systems directly translate into good decidability and complexity properties, e.g. by admitting a divide and conquer approach.

While reachability-based security properties naturally fall into the previous category equivalence-based properties seem ideally suited to this approach. Indeed, there are only few decidability results for equivalence checking in the context of security protocols, and most of them have concentrated on how to handle messages rather than the computational power induced by the composition operators. In contrast, in "pure" concurrency theory there is a body of work, which has investigated the decidability and complexity of equivalence-checking for both classical and causal equivalences. Thereby motivated let us next take a closer look at the standard causal equivalences and their computational power compared to their interleaving counterparts.

3   Equivalences

3.1   Three Causal Equivalences

Equivalences for concurrency have mainly been studied in the classical setting where the behaviour of a concurrent system is captured in terms of transitions labelled by atomic actions rather than sending and receiving of terms. The various behavioural equivalences can be classified according to two main distinctions: one is *linear-time* versus *branching-time*; the second is *interleaving* versus *causality*.

In the linear-time view the behaviour of a system is understood in terms of its set of possible runs. If concurrency is abstracted away by nondeterministic interleaving a system run will simply be modelled as a totally ordered sequence of labelled transitions. Thus, the coarsest behavioural equivalence is *trace equivalence*: two systems are *trace equivalent* iff their sets of runs are equivalent up to isomorphism (i.e. as sequences of actions). In the causal approach a system run is more faithfully modelled by a partially ordered set of labelled events. Since events with the same label may occur concurrently, technically we are dealing with *partially ordered multisets* of actions, or *pomsets* as coined by Pratt. The causal counterpart of trace equivalence is then represented by *pomset trace equivalence*: two systems are *pomset trace equivalent* iff their sets of labelled partial order runs are equivalent up to isomorphism (i.e. as pomsets of actions).

In branching-time notions of conflict, choice, or branching that naturally arise during a computation are modelled faithfully. The behaviour of a system is understood in terms of an 'unfolding' that reflects such information, and thereby shows how the system can unfold into many different possible futures. In the interleaving view a system naturally unfolds into a tree, or to be precise into a *synchronization tree* [39]. Attempts to capture what it means to distinguish branching in an *observational* way have culminated in the notion of *bisimulation equivalence* (short: *bisimilarity*) [39,42]. It is best explained in terms of a game between two players, Spoiler and Duplicator, on the two systems to be compared: Spoiler chooses a transition of one of the systems, and in response, Duplicator must choose a transition of the other system such that the labels are matching. The game then continues at the resulting pair of processes. The game continues like this forever, in which case Duplicator wins, or until either Spoiler or Duplicator is unable to move, in which case the other participant wins. Two systems are bisimilar iff Duplicator has a winning strategy in this game.

What are the unfolding structures of the causal approach? It turns out there are two different ways of capturing causality. One way is to stay within a tree-shaped view of the world but keep pointers that indicate when one transition is causally dependent on a previous transition of the same branch. The unfolding structure is then a *causal tree* [10]. The corresponding behavioural equivalence is *history preserving bisimilarity* (short: *hp-b*) [50]. It refines the bisimulation game as follows. Game positions now keep track of the history of the game. Technically, the histories are pairs (r_1, r_2, f) where r_1 is a partial order run of the first system, r_2 is a partial order run of the second system, and f is a pomset isomorphism between them. In her move Duplicator must now respond such that this pair of runs grows pomset isomorphic, i.e. $f \cup \{t_1, t_2\}$ must remain a pomset isomorphism.

The second way of capturing causality while keeping branching information departs from a tree-shaped structure but unfolds a system into an *event structure* [40]. In its most basic form an event structure is a set of events with a partial order that models causal dependence, and a symmetric and irreflexive relation added on that captures when two events are in conflict. Several axioms must hold to implement natural intuitions of this interpretation. They satisfy a basic principle of concurrency: whenever two independent events can occur consecutively they can also occur in the opposite order. The corresponding behavioural equivalence is *hereditary history preserving bisimilarity* (short: *hhp-b*) [3,31]. It further refines the bisimulation game by giving Spoiler the option of a backtrack move: Spoiler may choose a transition in one of the runs that is maximal in the partial order, and backtrack it. Duplicator must respond by backtracking the transition in the other run that is related to Spoiler's transition in f. The game continues at the resulting histories. Note how the backtrack move reflects that history can be traced back in different ways just as independent transitions can be shuffled in their order.

3.2   Finite-State Results

For finite-state systems the decidability and complexity of the discussed equivalences are well-understood. Fig. 1 gives an overview for finite 1-safe Petri nets. Checking trace equivalence on finite-state transition systems is similar to checking language equivalence on finite automata, and turns out to be PSPACE-complete [34]. Bisimilarity on finite-state transition systems is PTIME-complete [34]; it can be solved in polynomial-time by partition-refinement algorithms [34,41]. Based on these classical results Jategaonkar and Meyer have obtained the following results for finite 1-safe Petri nets: trace equivalence is EXPSPACE-complete, and bisimilarity is DEXPTIME-complete respectively [30]. The blow-up in complexity reflects that the transition system induced by a finite 1-safe Petri net is in general exponentially larger than the size of the net.

	finite 1-safe Petri nets	
trace equivalence	EXPSPACE-complete	[30] using [34]
pomset trace equivalence	EXPSPACE-complete	[30]
bisimilarity	DEXPTIME-complete	[30] using [34]
hp-b	DEXPTIME-complete	[30]
hhp-b	undecidable	[32]

Fig. 1. Finite-state results

Hp-b and pomset trace equivalence behave similarly to their interleaving counterparts. Checking hp-b is DEXPTIME-complete [30]. It can be decided analogously to bisimilarity using the following insight: it is not necessary to keep the entire history to capture hp-b, but to see whether pomsets grow isomorphic it is sufficient to record only those events that can act as maximal causes. Moreover, this essential fragment of history can be captured in a finite way: by the *ordered markings* of [51], or the *growth-sites* of [30] respectively. The same insight leads to EXPSPACE-decidability of pomset trace equivalence [30].

In contrast, hhp-b turns out undecidable for finite 1-safe Petri nets [32]. The root cause of the higher power lies in the different way of capturing causality: by allowing Spoiler to backtrack the game is taken to the event-based unfolding level, where the relationship of transitions concerning concurrency and conflict is globally captured. A key insight is to use the following gadget that is inspired by a similar tool in [38]: A tiling system T to be played on the $\omega \times \omega$ grid can be universally encoded by a finite 1-safe Petri net $N(T)$ such that the building of a domino snake can be faithfully mimicked by a special pattern of forwards and backtrack moves in the unfolding structure of $N(T)$. Hence, on their event-based unfolding level, systems such as finite 1-safe Petri nets are strong enough to encode tiling problems, and hence the computations of Turing machines, in a relatively straightforward sense.

All we have established so far is this: *in the finite-state world the causal equivalences are at least as hard as their interleaving counterparts.* So is our suggestion to use causal equivalences for their better composition properties a futile endeavour? Indeed, we have only reviewed here results on the full class of finite 1-safe Petri nets. There is a trend that suggests that as soon as we look at system classes that have good composition properties, and hence, a 'tame' interplay between causality, concurrency and conflict then hhp-b and to a degree also hp-b are better behaved than classical bisimilarity. A survey of results on subclasses of finite 1-safe Petri nets and open problems can be found in [16]. In the following, we will investigate this trend for infinite-state classes generated by process calculi.

3.3   A Hierarchy of Causal Processes

In the interleaving setting equivalence checking has been investigated along a hierarchy of process behaviours that can be captured in terms of rewrite rules. This *Process Rewrite Systems (PRS)* hierarchy is inspired by the Chomsky hierarchy of formal languages but the PRS grammars are interpreted as generators of infinite-state transition systems rather than languages. For borderline investigations of *causal* equivalences we consider the process algebras of the PRS-hierarchy as generators of infinite-state 1-safe Petri nets (or other causal models such as asynchronous transition systems).

Fig. 2 gives an overview of the PRS classes, expanded by the classes *Simple BPP* and *Simple PA* to be explained below. The root of the hierarchy comprises all *finite-state transition systems (FS)*. At the next level there are two extensions that can be seen as two interpretations of context-free grammars: *Basic Process Algebra (BPA)* extends FS by a sequential composition operator while *Basic Parallel Processes (BPP)* integrate a parallel composition operator. The class *Process Algebra (PA)* generalizes BPA and BPP by admitting both parallel and sequential composition. *PDA* is the class of *pushdown processes*, the processes described by pushdown automata, while on the right side we have *Petri nets (PN)*. The process classes on the left are not interesting here: since they do not integrate any parallel operator, the causal equivalences will coincide with their interleaving counterparts. Note that while in the classical interpretation the infinite-state classes contain all finite-state transition systems, under causal semantics up to PN they are incomparable with finite-state 1-safe Petri nets. This is so because BPP and PA restrict the interplay between concurrency and conflict due to the discipline of the grammars.

Given a set *Act* of atomic *actions*, usually denoted by $a, b, \ldots$, and a set *Var* of process *variables*, ranged over by $X, Y, \ldots$, the grammars for *FS, BPP, or PA process expressions* over *Act* and *Var* are defined as follows:

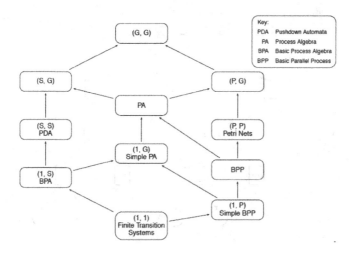

Fig. 2. The causal PRS-hierarchy

$$
\begin{aligned}
FS: \quad & F ::= \mathbf{0} \mid X \mid a.F \mid F + F \\
BPP: \quad & E ::= \mathbf{0} \mid X \mid a.E \mid E + E \mid E \parallel E \\
PA: \quad & P ::= \mathbf{0} \mid X \mid a.P \mid P + P \mid P \parallel P \mid P \cdot P
\end{aligned}
$$

where $\mathbf{0}$ denotes the empty process, X stands for a process variable, and $a.\_$, $\_ + \_$, $\_ \parallel \_$, $\_ \cdot \_$ denote the operations of *action prefix* (for each $a \in Act$), *nondeterministic choice*, *parallel composition*, and *sequential composition* respectively. BPP processes are defined as pairs (E, Δ) where Δ is a finite family of (possibly recursive) defining equations $X_i \stackrel{\text{def}}{=} E_i$. As usual we require that each occurrence of a variable in E_i is guarded, i.e. within the scope of an action prefix. This analogous for FS and PA.

The PRS grammars give rise to BPP, and respectively PA, in normal form. While in the interleaving world they represent the entire process classes, under causal semantics they only describe the subclasses *Simple BPP*, and *Simple PA* respectively. They are defined by the following grammars:

$$
\begin{aligned}
SBPP: \quad & E ::= X \mid S_E \mid E \parallel E \\
SPA: \quad & P ::= X \mid S_P \mid P \parallel P \mid P \cdot P
\end{aligned}
$$

where S_E stands for an *initially sequential SBPP expression* given by the following grammar:

$$
S_E ::= \mathbf{0} \mid a.E \mid S_E + S_E
$$

and analogously for S_P. Thus, SBPP restrict the mixture of choice and parallel composition: general summation is replaced by *guarded summation*. In particular, this excludes processes such as $(P_1 \parallel P_2) + P_3$. This is similar for SPA.

Causal semantics for BPP processes have been provided in terms of net unfoldings, e.g. in [14], and equivalently in terms of event structures via their

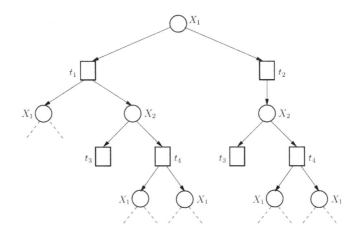

Fig. 3. The unfolding of SBPP $\mathcal{E}$.

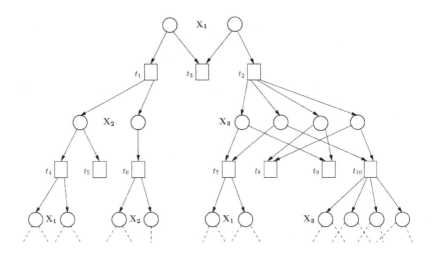

Fig. 4. The unfolding of BPP $\mathcal{E}$.

syntax-tree unfoldings [20]. Under such semantics BPP and SBPP have a tree-like structure. We provide two examples as an illustration.

Example 1. Fig. 3 gives the net unfolding of the SBPP $\mathcal{E} = (\Delta, X_1)$, where
$\Delta = \{X_1 \stackrel{\text{def}}{=} t_1.(X_1 \parallel X_2) + t_2.X_2; \ X_2 \stackrel{\text{def}}{=} t_3.\mathbf{0} + t_4.(X_1 \parallel X_1)\}$.

	SBPP = BPP?	BPP	PA	PN
$=_{tr}$	yes	undecidable [25]	undecidable	undecidable [24,28]
$=_{pom}$	yes [49]	decidable [49]	?	undecidable
$\sim$	yes	PSPACE-complete [29,48]	PSPACE-hard	undecidable [28]
$\sim_{hp}$	no	P [20,36]	?	undecidable
$\sim_{hhp}$	no	P [20,21]	?	undecidable

Fig. 5. Summary

Example 2. Fig. 4 demonstrates the unfolding of the BPP $\mathcal{E} = (\Delta, X_1)$, where $\Delta = \{X_1 \overset{\text{def}}{=} (t_1.X_2 \parallel t_2.X_3) + t_3.\mathbf{0}; X_2 \overset{\text{def}}{=} (t_4.X_1 + t_5.\mathbf{0}) \parallel t_6.X_2; X_3 \overset{\text{def}}{=} (t_7.X_1 \parallel t_8.\mathbf{0}) + (t_9.\mathbf{0} \parallel t_{10}.X_3)\}$.

PA and SPA have not been equipped with causal semantics yet. But one would expect that an appropriately defined unfolding semantics would display a regular structure of fork and join of 'chunks of independent behaviour'.

Problem 1. Define causal semantics for SPA, and PA respectively.

3.4 Infinite-State Results

Fig. 5 gives an overview of results and open questions in the process hierarchy up to PA and PN.

PN. For *Petri nets* all equivalences are undecidable. This follows from Jančars's reduction from the halting problem of counter machines, which proves that bisimilarity as well as trace equivalence is undecidable [28]: first observe that Petri nets can simulate counter machines, but only in a weak way since they cannot check for 0; given a counter machine C one constructs two variations of the Petri net that weakly simulates C such that the difference between these two nets can only be exposed by faithfully simulating C and reaching the halting state (in one of the nets); the two nets are non-equivalent iff C halts. The proof carries over to hp-b and hhp-b. The undecidability of language equivalence was first proved by Hack [24], but Jančar's proof is stronger in that it only requires 5 unbounded places.

BPP. Building on Jančar's technique Hirshfeld managed to resolve that trace equivalence is undeciable for communication-free Petri nets, and hence BPP [25]. The result does not carry over to pomset trace equivalence on (S)BPP. In contrast, Sunesen and Nielsen prove that pomset trace equivalence is decidable for BPP [49]. The proof first shows that in the linear-time world SBPP and BPP coincide in that every BPP can effectively be translated into a SBPP such that they are pomset trace equivalent. The decidability result then follows by a reduction to the equivalence problem of recognizable tree languages. The complexity of the algorithm is left open (but should not be hard to resolve).

Problem 2. Resolve the complexity of pomset trace equivalence on (S)BPP.

While for BPP deciding classical bisimilarity is PSPACE-complete [29] we obtain polynomial-time decision procedures for both hp-b [36] and hhp-b [21]. The two bisimilarities coincide for SBPP [15] but they do *not* coincide for BPP in general. This follows by the standard example of [3]. The decision procedures rely on different techniques. However, they can both be decided due to good decomposition properties.

Hp-b has the *unique decomposition property*: every BPP can be expressed, up to hp-b, as a parallel composition of prime processes, where a process is prime when it cannot be represented as a non-trivial parallel composition, up to hp-b. Moreover, this decomposition is unique up to the permutation of the primes. Then hp-b can be decided using the general scheme of Hirshfeld, Jerrum, and Moller [26] of deciding classical bisimilarity on normed BPP [22,36]. The fastest algorithm for hp-b on BPP runs in $O(n^6)$ [20], and is based on the technique of the *distance-to-disabling functions* introduced in [29].

Hhp-b has even stronger decomposition properties: modulo trivial choices it fully reflects the structure of BPP expressions. It can then be decided similarly to the standard algorithm for solving tree isomorphism [20,21].

In [20] the decision algorithms for both hp-b and hhp-b are presented in a unified framework. Hhp-b is solved in time $O(n^3 log\ n)$, and hp-b in time $O(n^6)$ respectively. In particular, both algorithms use the fact that on BPP hp-b and hhp-b have a a fixpoint characterization in terms of local games played over BPP processes of causal depth 1.

To sum up, for SBPP and BPP we can confirm the trend that as soon as we look at system classes that have a 'tame' interplay between causality and conflict causal equivalences are better behaved than their interleaving counterparts. This concerns both the linear-time as well as the branching-time equivalences. All algorithms make use of the fact that under causal semantics BPP and SBPP have tree-like structure.

PA. While BPP and PN are well-investigated hardly anything is known for PA. Naturally, the hardness results of BPP carry over. Tackling PA in the interleaving world has turned out to be difficult. The only known positive result is that bisimilarity on normed PA is decidable in 2-NEXPTIME [27]. The proof is technically involved and 63 pages long. It is based on an exhaustive case analysis which investigates when a seqential and a parallel composition can be equivalent. This might be much easier for causal equivalences. We believe that in particular hhp-b on simple PA could have very strong decomposition properties analogously to those for BPP.

Problem 3. Does the positive trend for causal equivalences extend to SPA, and PA respectively?

4   Summary and Outlook

To sum up, we have the following trend for causal equivalences:

The computational dichotomy of causality: For finite-state systems causal equivalences are often computationally much harder than their interleaving counterparts. However, as soon as we consider classes with a restricted interplay between causality and conflict this trend may be reversed. In particular, this includes standard infinite-state classes such as BPP, and might extend to PA.

To use causal equivalences for equivalence-based security properties it is necessary to lift them into the applied pi-calculus. In particular, this means we have to introduce asynchronous message input and output. While this would destroy the causal structure enforced by the discipline of the composition operators we might be able to regain this structure when we restrict ourselves to tagged protocols. Say a composition is homogeneous when the components relate to the same protocol process, and heteregenous when the components relate to different protocol processes. Then, roughly speaking, one could say the results of [6,19] suggest:

> Wrt homogeneous compositions of *tagged* protocols, atoms only have a bounded range of decisive influence, and messages only have decisive influence up to a bounded size.

Thereby we could hope to achieve decidability of reachability- and equivalence-based properties for tagged protocols when verified in isolation. Moreover, roughly speaking, one could say the results of [2,7] indicate:

> Wrt hetergeneous compositions of *tagged* protocols, messages have only a local range of decisive influence as long as the tagging is disjoint.

One could define an appropriate calculus of tagged protocols, for which one could hope that reachability-based properties would be decidable, and causal equivalence-based properties would also be decidable by composition results based on the local effect of messages and the insights from the 'pure' causal equivalences. Altogether this should also lead to an efficient verification method as long as the component protocols are small.

It is nontrivial to put this down more formally, and it will be even less trivial to prove it (or disprove if it turns out not to be true!). However, it might be most difficult of all to carry such 'design for verification' paradigms like working with tagged protocols into the various standards. Striving to do so is essential: interpreted the other way around, this just implements the general engineering principle to only interconnect systems in a way that does not create any side effects.

Acknowledgments. Part of this work has been conducted while I was a member of Ernst-Rüdiger Olderog's group working towards my habilitation. I would like to express my sincere gratitude for his guidance and mentoring during this time. It is due to his influence that I have adopted the verification viewpoint and realized the necessity to address cyberphysical systems security.

References

1. Abadi, M., Needham, R.: Prudent engineering practice for cryptographic protocols. IEEE Trans. Softw. Eng. **22**(1), 6–15 (1996)
2. Arapinis, M., Cheval, V., Delaune, S.: Composing security protocols: from confidentiality to privacy. In: Focardi, R., Myers, A. (eds.) POST 2015. LNCS, vol. 9036, pp. 324–343. Springer, Heidelberg (2015)
3. Bednarczyk, M.: Hereditary history preserving bisimulation or what is the power of the future perfect in program logics. Technical report, Polish Academy of Sciences, Gdańsk (1991)
4. Blanchet, B., Abadi, M., Fournet, C.: Automated verification of selected equivalences for security protocols. Journal of Logic and Algebraic Programming **75**(1), 3–51 (2008)
5. Blanchet, B., Podelski, A.: Verification of Cryptographic Protocols: Tagging Enforces Termination. Theoretical Computer Science **333**(1–2), 67–90 (2005). Special issue FoSSaCS 2003
6. Chréetien, R., Cortier, V., Delaune, S.: Typing messages for free in security protocols: the case of equivalence properties. In: Baldan, P., Gorla, D. (eds.) CONCUR 2014. LNCS, vol. 8704, pp. 372–386. Springer, Heidelberg (2014)
7. Ciobâcă, Ş., Cortier, V.: Protocol composition for arbitrary primitives. In: Proceedings of the 23rd IEEE Computer Security Foundations Symposium (CSF 2010), Edinburgh, Scotland, UK, pp. 322–336. IEEE Computer Society Press, Edinburgh, July 2010
8. Cremers, C.J.F.: The scyther tool: verification, falsification, and analysis of security protocols. In: Gupta, A., Malik, S. (eds.) CAV 2008. LNCS, vol. 5123, pp. 414–418. Springer, Heidelberg (2008)
9. Cremers, C.: Key exchange in ipsec revisited: formal analysis of IKEv1 and IKEv2. In: Atluri, V., Diaz, C. (eds.) ESORICS 2011. LNCS, vol. 6879, pp. 315–334. Springer, Heidelberg (2011)
10. Darondeau, P., Degano, P.: Causal trees. In: Ausiello, G., Dezani-Ciancaglini, M., Rocca, S.R.D. (eds.) ICALP 1989. LNCS, vol. 372, pp. 234–348. Springer, Heidelberg (1989)
11. ETSI. TS 102 731 V1.1.1: ITS; security; security services and architecture, 09 2010
12. ETSI. TS 102 940 V1.1.1: ITS; security; ITS communications security architecture and security management, 06 2012
13. Fábrega, F.J.T., Herzog, J.C., Guttman, J.D.:. Strand spaces: Why is a security protocol correct? In: Symposium on Security and Privacy. IEEE Computer Society (1998)
14. Fröschle, S.: Decidability and Coincidence of Equivalences for Concurrency. PhD thesis, University of Edinburgh (2004)
15. Fröschle, S.: Composition and decomposition in true-concurrency. In: Sassone, V. (ed.) FOSSACS 2005. LNCS, vol. 3441, pp. 333–347. Springer, Heidelberg (2005)
16. Fröschle, S.: The decidability border of hereditary history preserving bisimilarity. Information Processing Letters **93**(6), 289–293 (2005)
17. Fröschle, S.: The insecurity problem: tackling unbounded data. In: Proceedings of the 20th IEEE Computer Security Foundations Symposium, pp. 370–384 (2007)
18. Fröschle, S.: Causality in Security Protocols and Security APIs: Foundations and Practical Verification, Habilitation thesis, Universität Oldenburg (2012)
19. Fröschle, S.: Leakiness is decidable for well-founded protocols. In: Focardi, R., Myers, A. (eds.) POST 2015. LNCS, vol. 9036, pp. 176–195. Springer, Heidelberg (2015)

20. Fröschle, S., Jančar, P., Lasota, S., Sawa, Z.: Non-interleaving bisimulation equivalences on basic parallel processes. Information and Computation **208**(1), 42–62 (2010)

21. Fröschle, S., Lasota, S.: Decomposition and complexity of hereditary history preserving bisimulation on BPP. In: Abadi, M., de Alfaro, L. (eds.) CONCUR 2005. LNCS, vol. 3653, pp. 263–277. Springer, Heidelberg (2005)

22. Fröschle, S., Lasota, S.: Normed processes, unique decomposition, and complexity of bisimulation equivalences. In: Proceedings of INFINITY 2006-2009, vol. 239, pp. 17–42. Elsevier (2009)

23. Fröschle, S., Sommer, N.: Reasoning with past to prove PKCS#11 keys secure. In: Degano, P., Etalle, S., Guttman, J. (eds.) FAST 2010. LNCS, vol. 6561, pp. 96–110. Springer, Heidelberg (2011)

24. Hack, M.: The equality problem for vector addition systems is undecidable. Theoret. Comput. Sci. **2**(1), 77–95 (1976)

25. Hirshfeld, Y.: Petri nets and the equivalence problem. In: Meinke, K., Börger, E., Gurevich, Y. (eds.) CSL 1993. LNCS, vol. 832, pp. 165–174. Springer, Heidelberg (1994)

26. Hirshfeld, Y., Jerrum, M., Moller, F.: A polynomial time algorithm for deciding bisimulation equivalence of normed Basic Parallel Processes. Mathematical Structures in Computer Science **6**, 251–259 (1996)

27. Hirshfeld, Y., Jerrum, M.: Bisimulation equivalence is decidable for normed process algebra (Extended abstract). In: Wiedermann, J., Van Emde Boas, P., Nielsen, M. (eds.) ICALP 1999. LNCS, vol. 1644, pp. 412–421. Springer, Heidelberg (1999)

28. Jančar, P.: Undecidability of bisimilarity for Petri nets and some related problems. Theoretical Computer Science **148**(2), 281–301 (1995). STACS 1994

29. Jančar, P.: Bisimilarity of basic parallel processes is PSPACE-complete. In: Proc. LICS 2003, pp. 218–227. IEEE Computer Society (2003)

30. Jategaonkar, L., Meyer, A.R.: Deciding true concurrency equivalences on safe, finite nets. Theoretical Computer Science **154**(1), 107–143 (1996)

31. Joyal, A., Nielsen, M., Winskel, G.: Bisimulation from open maps. Information and Computation **127**(2), 164–185 (1996)

32. Jurdziński, M., Nielsen, M., Srba, J.: Undecidability of domino games and hhp-bisimilarity. Inform. and Comput. **184**, 343–368 (2003)

33. Kamil, A., Lowe, G.: Analysing tls in the strand spaced model. Journal of Computer Security **19**(5), 975–1025 (2011)

34. Kanellakis, P.C., Smolka, S.A.: CCS expressions, finite state processes, and three problems of equivalence. Information and Computation **86**(1), 43–68 (1990)

35. Künnemann, R.: Automated backward analysis of PKCS#11 v2.20. In: Focardi, R., Myers, A. (eds.) POST 2015. LNCS, vol. 9036, pp. 219–238. Springer, Heidelberg (2015)

36. Lasota, S.: A polynomial-time algorithm for deciding true concurrency equivalences of basic parallel processes. In: Rovan, B., Vojtáš, P. (eds.) MFCS 2003. LNCS, vol. 2747, pp. 521–530. Springer, Heidelberg (2003)

37. Lowe, G.: Towards a completeness result for model checking of security protocols. Journal of Computer Security **7**(1), 89–146 (1999)

38. Madhusudan, P., Thiagarajan, P.S.: Controllers for discrete event systems via morphisms. In: Sangiorgi, D., de Simone, R. (eds.) CONCUR 1998. LNCS, vol. 1466, pp. 18–33. Springer, Heidelberg (1998)

39. Milner, R. (ed.): A calculus of communicating systems. LNCS, vol. 92. Springer, Heidelberg (1980)

40. Nielsen, M., Plotkin, G.D., Winskel, G.: Petri nets, event structures and domains, part i. Theor. Comput. Sci. **13**, 85–108 (1981)
41. Paige, R., Tarjan, R.E.: Three partition refinement algorithms. SIAM J. Comput. **16**(6), 973–989 (1987)
42. Park, D.: Concurrency and automata on infinite sequences. In: Deussen, P. (ed.) Theoretical Computer Science. LNCS, vol. 104, pp. 167–183. Springer, Heidelberg (1981)
43. Ramanujam, R., Suresh, S.P.: A decidable subclass of unbounded security protocols. In: WITS 2003, pp. 11–20 (2003)
44. Ramanujam, R., Suresh, S.P.: Tagging makes secrecy decidable with unbounded nonces as well. In: Pandya, P.K., Radhakrishnan, J. (eds.) FSTTCS 2003. LNCS, vol. 2914, pp. 363–374. Springer, Heidelberg (2003)
45. Ramanujam, R., Suresh, S.P.: Decidability of context-explicit security protocols. Journal of Computer Security **13**(1), 135–165 (2005)
46. Schmidt, B., Sasse, R., Cremers, C., Basin, D.: Automated verification of group key agreement protocols. In: Proceedings of the 2014 IEEE Symposium on Security and Privacy SP 2014, DC, USA, Washington, pp. 179–194 (2014)
47. Song, D.X.: Athena: A new efficient automatic checker for security protocol analysis. In: CSFW 1999, pp. 192–202. IEEE Computer Society (1999)
48. Srba, J.: Strong bisimilarity and regularity of basic parallel processes is PSPACE-hard. In: Alt, H., Ferreira, A. (eds.) STACS 2002. LNCS, vol. 2285, pp. 535–546. Springer, Heidelberg (2002)
49. Sunesen, K., Nielsen, N.: Behavioural equivalence for infinite systems—partially decidable!. In: Billington, J., Reisig, W. (eds.) ICATPN 1996. LNCS, vol. 1091, pp. 460–479. Springer, Heidelberg (1996)
50. van Glabbeek, R., Goltz, U.: Equivalence notions for concurrent systems and refinement of actions. In: Kreczmar, A., Mirkowska, G. (eds.) MFCS 1989. LNCS, vol. 379, pp. 237–248. Springer, Heidelberg (1989)
51. Vogler, W.: Deciding history preserving bisimilarity. In: Albert, J.L., Monien, B., Artalejo, M.R. (eds.) Automata, Languages and Programming. LNCS, vol. 510, pp. 495–505. Springer, Heidelberg (1991)

Structure Preserving Bisimilarity, Supporting an Operational Petri Net Semantics of CCSP

Rob J. van Glabbeek[1,2]([⊠])

[1] NICTA, Sydney, Australia
rvg@cs.stanford.edu
[2] Computer Science and Engineering, UNSW, Sydney, Australia

Abstract. In 1987 Ernst-Rüdiger Olderog provided an operational Petri net semantics for a subset of CCSP, the union of Milner's CCS and Hoare's CSP. It assigns to each process term in the subset a labelled, safe place/transition net. To demonstrate the correctness of the approach, Olderog established agreement (1) with the standard interleaving semantics of CCSP up to strong bisimulation equivalence, and (2) with standard denotational interpretations of CCSP operators in terms of Petri nets up to a suitable semantic equivalence that fully respects the causal structure of nets. For the latter he employed a linear-time semantic equivalence, namely having the same causal nets.

This paper strengthens (2), employing a novel branching-time version of this semantics—*structure preserving bisimilarity*—that moreover preserves inevitability. I establish that it is a congruence for the operators of CCSP.

1 Introduction

The system description languages CCS and CSP have converged to one theory of processes which—following a suggestion of M. Nielsen—was called "CCSP" in [26]. The standard semantics of this language is in terms of labelled transition systems modulo strong bisimilarity, or some coarser semantic equivalence. In the case of CCS, a labelled transition system is obtained by taking as states the closed CCS expressions, and as transitions those that are derivable from a collection of rules by induction on the structure of these expressions [24]; this is called a *(structural) operational semantics* [30]. The semantics of CSP was originally given in quite a different way [3,20], but [28] provided an operational semantics of CSP in the same style as the one of CCS, and showed its consistency with the original semantics.

Such semantics abstract from concurrency relations between actions by reducing concurrency to interleaving. An alternative semantics, explicitly modelling concurrency relations, requires models like Petri nets [33] or event structures [25,36]. In [21,36] non-interleaving semantics for variants of CCSP are

---

NICTA is funded by the Australian Government through the Department of Communications and the Australian Research Council through the ICT Centre of Excellence Program.

R. Meyer et al. (Eds.): Olderog-Festschrift, LNCS 9360, pp. 99–130, 2015.
DOI: 10.1007/978-3-319-23506-6_9

given in terms of event structures. However, infinite event structures are needed to model simple systems involving loops, whereas Petri nets, like labelled transition systems, offer finite representations for some such systems. Denotational semantics in terms of Petri nets of the essential CCSP operators are given in [16,18,35]—see [27] for more references. Yet a satisfactory denotational Petri net semantics treating recursion has to my knowledge not been proposed.

Olderog [26,27] closed this gap by giving an operational net semantics in the style of [24,30] for a subset of CCSP including recursion—to be precise: *guarded* recursion. To demonstrate the correctness of his approach, Olderog proposed two fundamental properties such a semantics should have, and established that both of them hold [27]:

- *Retrievability*. The standard interleaving semantics for process terms should be retrievable from the net semantics.
- *Concurrency*. The net semantics should represent the intended concurrency of process terms.

The second requirement was not met by an earlier operational net semantics from [5].

To formalise the first requirement, Olderog notes that a Petri net induces a labelled transition system through the firing relation between markings— the *interleaving case graph*—and requires that the interpretation of any CCSP expression as a state in a labelled transition system through the standard interleaving semantics of CCSP should be strongly bisimilar to the interpretation of this expression as a marking in the interleaving case graph induced by its net semantics.

To formalise the second requirement, he notes that the intended concurrency of process terms is clearly represented in the standard denotational semantics of CCSP operators [16,18,35], and thus requires that the result of applying a CCSP operator to its arguments according to this denotational semantics yields a similar result as doing this according to the new operational semantics. The correct representation of recursion follows from the correct representation of the other operators through the observation that a recursive call has the very same interpretation as a Petri net as its unfolding.

A crucial parameter in this formalisation is the meaning of "similar". A logical choice would be semantic equivalence according to one of the non-interleaving equivalences found in the literature, where a finer or more discriminating semantics gives a stronger result. To match the concurrency requirement, this equivalence should *respect concurrency*, in that it only identifies nets which display the same concurrency relations. In this philosophy, the semantics of a CCSP expression is not so much a Petri net, but a semantic equivalence class of Petri nets, i.e. a Petri net after abstraction from irrelevant differences between nets. For this idea to be entirely consistent, one needs to require that the chosen equivalence is a congruence for all CCSP constructs, so that the meaning of the composition of two systems, both represented as equivalence classes of nets, is independent of the choice of representative Petri nets within these classes.

Instead of selecting such an equivalence, Olderog instantiates "similar" in the above formalisation of the second requirement with *strongly bisimilar*, a new relation between nets that should not be confused with the traditional relation of strong bisimilarity between labelled transition systems. As shown in [1], strong bisimilarity fails to be an equivalence: it is reflexive and symmetric, but not transitive.

As pointed out in [27, Page37] this general shortcoming of strong bisimilarity "does not affect the purpose of this relation" in that book: there it "serves as an auxiliary notion in proving that structurally different nets are causally equivalent". Here *causal equivalence* means having the same causal nets, where *causal nets* [29,34] model concurrent computations or executions of Petri nets. So in effect Olderog does choose a semantic equivalence on Petri nets, namely having the same concurrent computations as modelled by causal nets. This equivalence fully respects concurrency.

1.1 Structure Preserving Bisimilarity

The contribution of the present paper is a strengthening of this choice of a semantic equivalence on Petri nets. I propose the novel *structure preserving bisimulation* equivalence on Petri nets, and establish that the result of applying a CCSP operator to its arguments according to the standard denotational semantics yields a structure preserving bisimilar result as doing this according to Olderog's operational semantics. The latter is an immediate consequence of the observation that structure preserving bisimilarity between two nets is implied by Olderog's strong bisimilarity.

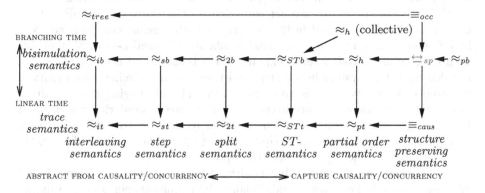

Fig. 1. A spectrum of semantic equivalences on Petri nets

Figure 1 shows a map of some equivalence relations on nets found in the literature, in relation to the new structure preserving bisimilarity, $\leftrightarrow_{sp}$. The equivalences become finer when moving up or to the right; thus coarser or less discriminating when following the arrows. The rectangle from $\approx_{it}$ to $\approx_h$ is taken from [10]. The vertical axis is the *linear time – branching time spectrum*, with *trace equivalence* at the bottom and *(strong) bisimulation equivalence*, or *bisimilarity*, at the top. A host of intermediate equivalences is discussed in [11]. The key difference is that *linear time* equivalences, like trace equivalence, only consider the set of possible executions of a process, whereas *branching time* equivalences, like bisimilarity, additionally take into account at which point the choice between two executions is made. The horizontal axis indicates to what extent concurrency information is taken into account. *Interleaving* equivalences—on the left—fully abstract from concurrency by reducing it to arbitrary interleaving; *step* equivalences additionally take into account the possibility that two concurrent actions happen at exactly the same moment; *split* equivalences recognise the beginning and end of actions, which here are regarded to be durational, thereby capturing some information about their overlap in time; *ST-* or *interval* equivalences fully capture concurrency information as far as possible by considering durational actions overlapping in time; and *partial order* equivalences capture the causal links between actions, and thereby all concurrency. By taking the product of these two axes, one obtains a two-dimensional spectrum of equivalence relations, with entries like *interleaving bisimulation* equivalence $\approx_{ib}$ and *partial order trace* equivalence $\approx_{pt}$. For the right upper corner several partial order bisimulation equivalences were proposed in the literature; according to [13] the *history preserving bisimulation* equivalence $\approx_h$, originally proposed by [32], is the coarsest one that fully captures the interplay between causality and branching time.

The causal equivalence employed by Olderog, $\equiv_{caus}$, is a linear time equivalence strictly finer than $\approx_{pt}$. Since it preserves information about the number of preplaces of a transition, it is specific to a model of concurrency based on Petri nets; i.e. there is no obvious counterpart in terms of event structures. I found only two equivalences in the literature that are finer than both $\equiv_{caus}$ and $\approx_h$, namely *occurrence net equivalence* [16]—$\equiv_{occ}$—and the *place bisimilarity* $\approx_{pb}$ of [1]. Two nets are occurrence net equivalent iff they have isomorphic unfoldings. The *unfolding*, defined in [25], associates with a given safe Petri net N a loop-free net—an *occurrence net*—that combines all causal nets of N, together with their branching structure. This unfolding is similar to the unfolding of a labelled transition system into a tree, and thus the interleaving counterpart of occurrence net equivalence is *tree equivalence* [11], identifying two transition systems iff their unfoldings are isomorphic. The place bisimilarity was inspired by Olderog's strong bisimilarity, but adapted to make it transitive, and thus an equivalence relation. My new equivalence $\leftrightarrow_{sp}$ will be shown to be strictly coarser than $\equiv_{occ}$ and $\approx_{pb}$, yet finer than both $\equiv_{caus}$ and $\approx_h$.

The equivalences discussed above (without the diagonal line in Figure 1) are all defined on safe Petri nets. Additionally, the definitions generalise to unsafe Petri nets. However, there are two possible interpretations of unsafe Petri nets, called the *collective token* and the *individual token* interpretation [12], and this leads to two versions of history preserving bisimilarity. The history preserving bisimilarity based on the individual token interpretation was first defined for Petri nets in [2], under the name *fully concurrent bisimulation* equivalence. At the level of ST-semantics the collective and individual token interpretations collapse. The unfolding of unsafe Petri nets, and thereby occurrence net equivalence, has been defined for the individual token interpretation only [7,12,23], and likewise causal equivalence can be easily generalised within the individual token interpretation. The new structure preserving bisimilarity falls in the individual token camp as well.

1.2   Criteria for Choosing This Semantic Equivalence

In selecting a new semantic equivalence for reestablishing Olderog's agreement of operational and denotational interpretations of CCSP operators, I consider the following requirements on such a semantic equivalence (with subsequent justifications):

1. it should be a branching time equivalence,
2. it should fully capture causality relations and concurrency (and the interplay between causality and branching time),
3. it should respect *inevitability* [22], meaning that if two systems are equivalent, and in one the occurrence of a certain action is inevitable, then so is it in the other,
4. it should be *real-time consistent* [16], meaning that for every association of execution times to actions, assuming that actions happen as soon as they can, the running times associated with computations in equivalent systems should be the same,
5. it should be *preserved under action refinement* [4,13], meaning that if in two equivalent Petri nets the same substitutions of nets for actions are made, the resulting nets should again be equivalent,
6. it should be finer than Olderog's causal equivalence,
7. it should not distinguish systems whose behaviours are patently the same, such as Petri nets that differ only in their unreachable parts,
8. it should be a congruence for the constructs of CCSP,
9. and it should allow to establish agreement between the operational and denotational interpretations of CCSP operators.

Requirement 1 is the driving force behind this contribution. It is motivated by the insight that branching time equivalences better capture phenomena like deadlock behaviour. Since in general a stronger result on the agreement between operational and denotational semantics is obtained when employing a finer semantics, I aim for a semantics that fully captures branching time information, and thus is at least as discriminating as interleaving bisimilarity.

Requirement 2 is an obvious choice when the goal of the project is to capture concurrency explicitly. The combination of Requirements 1 and 2 then naturally asks for an equivalence that is at least as fine as $\approx_h$. One might wonder, however, for what reason one bothers to define a semantics that captures concurrency information. In the literature, various practical reasons have been given for preferring a semantics that (partly) respects concurrency and causality over an interleaving semantics. Three of the more prominent of these reasons are formulated as requirements 3, 4 and 5 above.

Requirement 3 is manifestly useful when considering liveness properties of systems. Requirement 4 obviously has some merit when timing is an issue. Requirement 5 is useful in system design based on stepwise refinement [13].

Requirement 6 is only there so that I can truthfully state to have strengthened Olderog's agreement between the denotational and operational semantics, which was stated in terms of causal equivalence. This requirement will not be needed in my justification for introducing a new semantic equivalence—and neither will Requirement 2.

Requirement 7 is hardly in need of justification. The paper [1] lists as a desirable property of semantic equivalences—one that is not met by their own proposal $\approx_{pb}$—that they should not distinguish nets that have isomorphic unfoldings, given that unfolding a net should not be regarded as changing it behaviour. When working within the individual token interpretation of nets I will take this as a suitable formalisation of Requirement 7.

The argument for Requirement 8 has been given earlier in this introduction, and Requirement 9 underlies my main motivation for selecting a semantic equivalence in the first place.

1.3   Applying the Criteria

Table 1 tells which of these requirements are satisfied by the semantic equivalences from Section 1.1 (not considering the one collective token equivalence there). The first two rows, reporting which equivalences satisfy Requirements 1 and 2, are well-known; these results follow directly from the definitions. The third row, reporting on respect for inevitability, is a contribution of this paper, and will be discussed in Section 1.4, and delivered in Sections 11–14.

Regarding Row 4, In [16] it is established that ST-bisimilarity is real-time consistent. Moreover, the formal definition is such that if a semantic equivalence $\approx$ is real-time consistent, then so is any equivalence finer than $\approx$. Linear time equivalences are not real-time consistent, and neither is $\approx_{2b}$ [17].

In [13] it is established that $\approx_{pt}$ and $\approx_{h}$ are preserved under action refinement, but interleaving and step equivalences are not, because they do not capture enough information about concurrency. In [10] it is shown that $\approx_{STt}$ and $\approx_{STb}$ are already preserved under action refinement, whereas by [17] split semantics

Table 1. Which requirements are satisfied by the various semantic equivalences

Equivalence / Requirement	$\approx_{it}$	$\approx_{ib}$	$\approx_{tree}$	$\approx_{st}$	$\approx_{sb}$	$\approx_{2t}$	$\approx_{2b}$	$\approx_{STt}$	$\approx_{STb}$	$\approx_{pt}$	$\approx_{h}$	$\equiv_{caus}$	$\leftrightarrows_{sp}$	$\equiv_{occ}$	$\approx_{pb}$
1. Branching time	×	✓	✓	×	✓	×	✓	×	✓	×	✓	×	✓	✓	✓
2. Causality	×	×	×	×	×	×	×	×	×	✓	✓	✓	✓	✓	✓
3. Inevitability	×	×	×	×	×	×	×	×	×	×	×	×	✓	✓	✓
4. Real-time consistency	×	×	×	×	×	×	×	×	✓	×	✓	×	✓	✓	✓
5. Action refinement	×	×	×	×	×	×	×	✓	✓	✓	✓	✓?	✓?	✓?	
6. Finer than $\equiv_{caus}$	×	×	×	×	×	×	×	×	×	×	×	✓	✓	✓	✓
7. Coarser than $\equiv_{occ}$	✓	✓	✓	✓	✓	✓	✓	✓	✓	✓	✓	✓	✓	✓	×
8. Congruence	✓	✓											✓		
9. Operat. $\equiv$ denotat.	✓	✓	×	✓	✓	✓	✓	✓	✓	✓	✓	✓	✓	×	

are not. I conjecture that $\equiv_{caus}$ and $\equiv_{occ}$ are also preserved under action refinement, but I have not seen a formal proof. I also conjecture that the new $\leftrightarrow_{sp}$ is preserved under action refinement.

Rows 6 and 7 follow as soon as I have formally established the implications of Figure 1 (in Section 10). As for Row 8, I will show in Section 7 that $\leftrightarrow_{sp}$ is a congruence for the operators of CCSP. That also $\approx_{it}$ and $\approx_{ib}$ are congruences for CCSP is well known. The positive results in Row 9 follow from the fact that Olderog's strong bisimilarity implies $\leftrightarrow_{sp}$, which will be established in Section 6.

Requirements 1 and 6 together limit the search space for suitable equivalence relations to $\equiv_{occ}$, $\approx_{pb}$ and the new $\leftrightarrow_{sp}$. When dropping Requirement 6, but keeping 2, also $\approx_h$ becomes in scope. When also dropping 2, but keeping 4, I gain $\approx_{STb}$ as a candidate equivalence. However, both $\approx_h$ and $\approx_{STb}$ will fall pray to Requirement 3, so also without Requirements 2 and 6 the search space will be limited to $\equiv_{occ}$, $\approx_{pb}$ and the new $\leftrightarrow_{sp}$.

Requirement 7 rules out $\approx_{pb}$, as that equivalence makes distinctions based on unreachable parts of nets [1]. The indispensable Requirement 9 rules out $\equiv_{occ}$, since that equivalence distinguishes the operational and denotational semantics of the CCSP expression $a0 + a0$. According to the operational semantics this expression has only one transition, whereas by the denotational semantics it has two, and $\equiv_{occ}$ does not collapse identical choices. The same issue plays in interleaving semantics, where the operational and denotational transition system semantics of CCSP do not agree up to tree equivalence. This is one of the main reasons that bisimilarity is often regarded as the top of the linear time – branching time spectrum.

This constitutes the justification for the new equivalence $\leftrightarrow_{sp}$.

1.4   Inevitability

The meaning of Requirement 3 depends on which type of progress or fairness property one assumes to guarantee that actions that are due to occur will actually happen. Lots of fairness assumption are mentioned in the literature, but, as far as I can tell, they can be classified in exactly 4 groups: *progress*, *justness*, *weak fairness* and *strong fairness* [15]. These four groups form a hierarchy, in the sense that one cannot consistently assume strong fairness while objecting to weak fairness, or justness while objecting to progress.

Strong and weak fairness deal with choices that are offered infinitely often. Suppose you have a shop with only two customers A and B that may return to the shop to buy something else right after they are served. Then it is unfair to only serve customer A again and again, while B is continuously waiting to be served. In case B is not continuously ready to be served, but sometimes goes home to sleep, yet always returns to wait for his turn, it is weakly fair to always ignore customer B in favour of A, but not strongly fair.

Weak and strong fairness assumptions can be made *locally*, pertaining to *some* repeating choices of the modelled system but not to others, or *globally*, pertaining to all choices of a given type. Since the real world is largely unfair, strong and weak fairness assumptions need to be made with great caution, and they will not appear in this paper.

Justness and progress assumptions, on the other hand, come only in the global variant, and can be safely assumed much more often. A progress assumption says that if a system can do some action (that is not contingent on external input) it will do an action. In the example of the shop, if there is a customer continuously ready to be served, and the clerk stands pathetically behind the counter staring at the customer but not serving anyone, there is a failure of progress. Without assuming progress, no action is inevitable, because it is always possible that a system will remain in its initial state without ever doing anything. Hence the concept of inevitability only makes sense when assuming at least progress.

Justness [8,15] says roughly that if a parallel component can make progress (not contingent on input from outside of this component) it will do so. Suppose the shop has two counters, each manned by a clerk, and, whereas customer A is repeatedly served at counter 1, customer B is ready to be served by counter 2, but is only stared at by a pathetic clerk. This is not a failure of progress, as in any state of the system someone will be served eventually. Yet it counts as a failure of justness. In the context of Petri nets, a failure of justness can easily be formalised as an execution, during which, from some point onwards, all preplaces of a given transition remain marked, yet the transition never fires [14]. One could argue that, when taking concurrency seriously, justness should be assumed whenever one assumes progress.

Inevitability can be easily expressed in temporal logics like LTL [31] or CTL [6], and it is well known that strongly bisimilar transition systems satisfy the same temporal formulas. This suggests that interleaving bisimilarity already respects inevitability. However, this conclusion is warranted only when assuming progress but not justness, or perhaps also when assuming some form of weak or strong fairness. The system $C := \langle X | X = aX + bX \rangle$—using the CCSP syntax of Section 2—repeatedly choosing between the actions a and b, is interleaving bisimilar to the system $D := \langle Y | Y = aY \rangle \| \langle Z | Z = bZ \rangle$, which in parallel performs infinitely many as and infinitely many bs. Yet, when assuming justness but not weak fairness, the execution of the action b is inevitable in D, but not in C. This shows that when assuming justness but not weak fairness, interleaving bisimilarity does not respect inevitability. The paper [22], which doesn't use Petri nets as system model, leaves the precise formulation of a justness assumption for future work—this task is undertaken in the different context of CCS in [15]. Also, respect of inevitability as a criterion for judging semantic equivalences does not occur in [22], even though "the partial order approach" is shown to be beneficial.

In this paper, assuming justness but not strong or weak fairness, I show that neither $\approx_h$ nor $\equiv_{caus}$ respects inevitability (using infinite nets in my counterexample). Hence, respecting concurrency appears not quite enough to respect inevitability. Respect for inevitability, like real-time consistency, is a property that holds for any equivalence relation finer than one for which it is known to hold already. So also none of the ST- or interleaving equivalences respects inevitability. I show that the new equivalence $\underleftrightarrow{}_{sp}$ respects inevitability. This makes it the coarsest equivalence of Figure 1 that does so.

2   CCSP

CCSP is parametrised by the choice of an infinite set Act of actions, that I will assume to be fixed for this paper. Just like the version of CSP from Hoare [20], the version of CCSP used here is a typed language, in the sense that with every CCSP process P an explicit alphabet $\alpha(P) \subseteq Act$ is associated, which is a superset of the set of all actions the process could possibly perform. This alphabet is exploited in the definition of the parallel composition $P\|Q$: actions in the intersection of the alphabets of P and Q are required to synchronise, whereas all other actions of P and Q happen independently. Because of this, processes with different alphabets may never be identified, even if they can perform the same set of actions and are alike in all other aspects. It is for this reason that I interpret CCSP in terms of *typed* Petri nets, with an alphabet as extra component.

I also assume an infinite set V of *variable names*. A *variable* is a pair X_A with $X \in V$ and $A \subseteq Act$. The syntax of (my subset of) CCSP is given by

$$P ::= 0_A \mid aP \mid P + P \mid P\|P \mid R(P) \mid X_A \mid \langle X_A|\mathcal{S}\rangle \text{ (with } X_A \in V_{\mathcal{S}})$$

with $A \subseteq Act$, $a \in Act$, $R \subseteq Act \times Act$, $X \in V$ and $\mathcal{S}$ a *recursive specification*: a set of equations $\{Y_B = \mathcal{S}_{Y_B} \mid Y_B \in V_{\mathcal{S}}\}$ with $V_{\mathcal{S}} \subseteq V \times Act$ (the *bound variables* of $\mathcal{S}$) and $\mathcal{S}_{Y_B}$ a CCSP expression satisfying $\alpha(\mathcal{S}_{Y_B}) = B$ for all $Y_B \in V_{\mathcal{S}}$ (were $\alpha(\mathcal{S}_{Y_B})$ is defined below). The constant 0_A represents a process that is unable to perform any action. The process aP first performs the action a and then proceeds as P. The process $P + Q$ will behave as either P or Q, $\|$ is a partially synchronous parallel composition operator, R a renaming, and $\langle X_A|\mathcal{S}\rangle$ represents the X_A-component of a solution of the system of recursive equations $\mathcal{S}$. A CCSP expression P is *closed* if every occurrence of a variable X_A occurs in a subexpression $\langle Y_B|\mathcal{S}\rangle$ of P with $X_A \in V_{\mathcal{S}}$.

The constant 0 and the variables are indexed with an alphabet. The alphabet of an arbitrary CCSP expression is given by:

- $\alpha(0_A) = \alpha(X_A) = \alpha(\langle X_A|\mathcal{S}\rangle) = A$
- $\alpha(aP) = \{a\} \cup \alpha(P)$
- $\alpha(P + Q) = \alpha(P\|Q) = \alpha(P) \cup \alpha(Q)$
- $\alpha(R(P)) = \{b \mid \exists a \in \alpha(P) : (a, b) \in R\}$.

Substitutions of expressions for variables are allowed only if the alphabets match. For this reason a recursive specification $\mathcal{S}$ is declared syntactically incorrect if $\alpha(\mathcal{S}_{Y_B}) \neq B$ for some $Y_B \in V_{\mathcal{S}}$. The interleaving semantics of CCSP is given by the labelled transition relation $\rightarrow \subseteq \mathrm{T_{CCSP}} \times Act \times \mathrm{T_{CCSP}}$ on the set $\mathrm{T_{CCSP}}$ of closed CCSP terms, where the transitions $P \xrightarrow{a} Q$ (on arbitrary CCSP expressions) are derived from the rules of Table 2. Here $\langle P|\mathcal{S}\rangle$ for P an expression and $\mathcal{S}$ a recursive specification denotes the expression P in which $\langle Y_B|\mathcal{S}_{Y_B}\rangle$ has been substituted for the variable Y_B for all $Y_B \in V_{\mathcal{S}}$.

Table 2. Structural operational interleaving semantics of CCSP

$$
aP \xrightarrow{a} P
\qquad
\frac{P \xrightarrow{a} P'}{P\|Q \xrightarrow{a} P'\|Q} \ (a \notin \alpha(Q))
\qquad
\frac{P \xrightarrow{a} P'}{R(P) \xrightarrow{b} R(P')} \ ((a,b) \in R)
$$

$$
\frac{P \xrightarrow{a} P'}{P + Q \xrightarrow{a} P'}
\qquad
\frac{P \xrightarrow{a} P', \ Q \xrightarrow{a} Q'}{P\|Q \xrightarrow{a} P'\|Q'} \ (a \in \alpha(P) \cap \alpha(Q))
$$

$$
\frac{Q \xrightarrow{a} Q'}{P + Q \xrightarrow{a} Q'}
\qquad
\frac{Q \xrightarrow{a} Q'}{P\|Q \xrightarrow{a} P\|Q'} \ (a \notin \alpha(P))
\qquad
\frac{\langle \mathcal{S}_{X_A} | \mathcal{S} \rangle \xrightarrow{a} P'}{\langle X_A | \mathcal{S} \rangle \xrightarrow{a} P'}
$$

A CCSP expression is *well-typed* if for any subexpression of the form aP one has $a \in \alpha(P)$ and for any subexpression of the form $P + Q$ one has $\alpha(P) = \alpha(Q)$. Thus $a0_{\{a\}} + bX_\emptyset$ is not well-typed, although the equivalent expression $a0_{\{a,b\}} + bX_{\{a,b\}}$ is. A recursive specification $\langle X_A | \mathcal{S} \rangle$ is *guarded* if each occurrence of a variable $Y_B \in V_S$ in a term $\mathcal{S}_{Z_C}$ for some $Z_C \in V_S$ lays within a subterm of $\mathcal{S}_{Z_C}$ of the form aP. Following [27] I henceforth only consider well-typed CCSP expressions with guarded recursion.

In Olderog's subset of CCSP, each recursive specification has only one equation, and renamings must be functions instead of relations. Here I allow mutual recursion and relational renaming, where an action may be renamed into a choice of several actions—or possibly none. This generalisation does not affect any of the proofs in [27].

Example 1. The behaviour of the customer from Section 1.4 could be given by the recursive specification $\mathcal{S}_{\text{Cus}}$:

$$
\text{Cus}_{Cu} = enter \ buy \ leave \ \text{Cus}_{Cu}
$$

indicating that the customer keeps coming back to the shop to buy more things. Here $enter, buy, leave \in Act$ and $\text{Cus} \in V$. The customer's alphabet Cu is $\{enter, buy, leave\}$. Likewise, the behaviour of the store clerk could be given by the specification $\mathcal{S}_{\text{CLK}}$:

$$
\text{Clk}_{Cl} = serve \ \text{Clk}_{Cl}
$$

where $Cl = \{serve\}$. The CCSP processes representing the customer and the clerk, with their reachable states and labelled transitions between them, are displayed in Figure 2.

In order to ensure that the parallel composition synchronises the *buy*-action of the customer with the *serve*-action of the clerk, I apply renaming operators R_{Cus} and R_{CLK} with $R_{\text{Cus}}(buy) = serves$ and $R_{\text{CLK}}(serve) = serves$ and leaving all other actions unchanged, where *serves* is a joint action of the renamed customer and the renamed clerk. The total CCSP specification of a store with one clerk and one customer is

$$
R_{\text{Cus}}(\langle \text{Cus}_{Cu} | \mathcal{S}_{\text{Cus}} \rangle) \| R_{\text{CLK}}(\langle \text{Clk}_{Cl} | \mathcal{S}_{\text{CLK}} \rangle)
$$

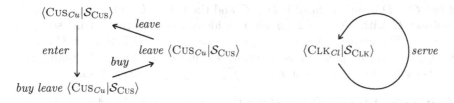

Fig. 2. Labelled transition semantics of customer and clerk

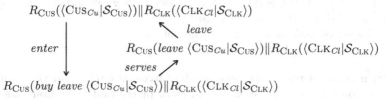

Fig. 3. Labelled transition semantics of the 1-customer 1-clerk store

and the relevant part of the labelled transition system of CCSP is displayed above.

One possible behaviour of this system is the sequence of actions *enter serves leave enter*, followed by eternal stagnation. This behaviour is ruled out by the progress assumption of Section 1.4. The only behaviour compatible with this assumption is the infinite sequence of actions $(enter\ serves\ leave)^\infty$.

To model a store with two customers (A and B) and 2 clerks (I and II), I introduce a relational renaming for each of them, defined by

$$
\begin{aligned}
R_A(enter) &= A\ enters & R_A(buy) &= \{\text{I serves } A, \text{II serves } A\} & R_A(leave) &= A\ leaves \\
R_B(enter) &= B\ enters & R_B(buy) &= \{\text{I serves } B, \text{II serves } B\} & R_B(leave) &= B\ leaves \\
& & R_I(serve) &= \{\text{I serves } A, \text{I serves } B\} \\
& & R_{II}(serve) &= \{\text{II serves } A, \text{II serves } B\}.
\end{aligned}
$$

The CCSP specification of a store with two clerks and two customers is

$$
\big(R_A(\langle \text{Cus}_{Cu}|\mathcal{S}_{\text{Cus}}\rangle) \| R_B(\langle \text{Cus}_{Cu}|\mathcal{S}_{\text{Cus}}\rangle)\big) \| \big(R_I(\langle \text{CLK}_{Cl}|\mathcal{S}_{\text{CLK}}\rangle) \| R_{II}(\langle \text{CLK}_{Cl}|\mathcal{S}_{\text{CLK}}\rangle)\big)
$$

and the part of the labelled transition system of CCSP reachable from that process has $3 \times 3 \times 1 \times 1 = 9$ states and $6 \times 4 = 24$ transitions.

3   Petri Nets

A *multiset* over a set S is a function $C : S \to \mathbb{N}$, i.e. $C \in \mathbb{N}^S$; let $|C| := \sum_{x \in X} C(x)$; $x \in S$ is an *element of* C, notation $x \in C$, iff $C(x) > 0$.

The function $\emptyset : S \to \mathbb{N}$, given by $\emptyset(x) := 0$ for all $x \in S$, is the *empty* multiset over S. For multisets C and D over S one writes $C \leq D$ iff $C(x) \leq D(x)$ for all $x \in S$; $C \cap D$ denotes the multiset over S with $(C \cap D)(x) := \min(C(x), D(x))$, $C + D$ denotes the multiset over S with $(C + D)(x) := C(x) + D(x)$; and the

multiset $C - D$ is only defined if $D \leq C$ and then $(C - D)(x) := C(x) - D(x)$.
A multiset C with $C(x) \leq 1$ for all x is identified with the (plain) set $\{x \mid C(x) = 1\}$. The construction $C := \{f(x_1, ..., x_n) \mid x_i \in D_i\}$ of a set C out of sets D_i $(i = 1, ..., n)$ generalises naturally to multisets C and D_i, taking the multiplicity $C(x)$ of an element x to be $\sum_{f(x_1, ..., x_n) = x} D_1(x_1) \cdot \ldots \cdot D_n(x_n)$.

Definition 1. A *(typed) Petri net* is a tuple $N = (S, T, F, M_0, A, \ell)$ with

- S and T disjoint sets (of *places* and *transitions*),
- $F : ((S \times T) \cup (T \times S)) \to \mathbb{N}$ (the *flow relation* including *arc weights*),
- $M_0 : S \to \mathbb{N}$ (the *initial marking*),
- A a set of *actions*, the *type* of the net, and
- $\ell : T \to A$ (the *labelling function*).

Petri nets are depicted by drawing the places as circles and the transitions as boxes, containing their label. Identities of places and transitions are displayed next to the net element. For $x, y \in S \cup T$ there are $F(x, y)$ arrows (*arcs*) from x to y. When a Petri net represents a concurrent system, a global state of this system is given as a *marking*, a multiset M of places, depicted by placing $M(s)$ dots (*tokens*) in each place s. The initial state is M_0.

The behaviour of a Petri net is defined by the possible moves between markings M and M', which take place when a transition t *fires*. In that case, t consumes $F(s, t)$ tokens from each place s. Naturally, this can happen only if M makes all these tokens available in the first place. Moreover, t produces $F(t, s)$ tokens in each place s. Definition 2 formalises this notion of behaviour.

Definition 2. Let $N = (S, T, F, M_0, A, \ell)$ be a Petri net and $x \in S \cup T$. The multisets $^\bullet x$, $x^\bullet : S \cup T \to \mathbb{N}$ are given by $^\bullet x(y) = F(y, x)$ and $x^\bullet(y) = F(x, y)$ for all $y \in S \cup T$; for $t \in T$, the elements of $^\bullet t$ and $t^\bullet$ are called *pre-* and *postplaces* of t, respectively. Transition $t \in T$ is *enabled* from the marking $M \in \mathbb{N}^S$—notation $M[t\rangle$—if $^\bullet t \leq M$. In that case firing t yields the marking $M' := M - {}^\bullet t + t^\bullet$ —notation $M[t\rangle M'$.

A *path* π of a Petri net N is an alternating sequence $M_0 t_1 M_1 t_2 M_2 t_3 \ldots$ of markings and transitions, starting from the initial marking M_0 and either being infinite or ending in a marking M_n, such that $M_k[t_k\rangle M_{k+1}$ for all k $(<n)$. A marking is *reachable* if it occurs in such a path. The Petri net N is *safe* if all reachable markings M are plain sets, meaning that $M(s) \leq 1$ for all places s. It has *bounded parallelism* [16] if there is no reachable marking M and infinite multiset of transitions U such that $\sum_{t \in U} {}^\bullet t \leq M$. In this paper I consider Petri nets with bounded parallelism only, and call them *nets*.

4   An Operational Petri Net Semantics of CCSP

This section recalls the operational Petri net semantics of CCSP, given by Olderog [26,27]. It associates a net $[\![P]\!]$ with each closed CCSP expression P.

The standard operational semantics of CCSP, presented in Section 2, yields one big labelled transition system for the entire language.[1] Each individual closed CCSP expression P appears as a state in this LTS. If desired, a *process graph*—an LTS enriched with an initial state—for P can be extracted from this system-wide LTS by appointing P as the initial state, and optionally deleting all states and transitions not reachable from P. In the same vein, an operational Petri net semantics yields one big Petri net for the entire language, but without an initial marking. I call such a Petri net *unmarked*. Each process $P \in T_{\text{CCSP}}$ corresponds with a marking $dex(P)$ of that net. If desired, a Petri net $[\![P]\!]$ for P can be extracted from this system-wide net by appointing $dex(P)$ as its initial marking, taking the type of $[\![P]\!]$ to be $\alpha(P)$, and optionally deleting all places and transitions not reachable from $dex(P)$.

The set S_{CCSP} of places in the net is the smallest set including:

		aP	*prefixing*	$\mu + \nu$	*choice*
$\mu\|_A$	*left parallel component*	$_A\|\mu$	*right component*	$R(\mu)$	*renaming*

for $P \in T_{\text{CCSP}}$, $a \in Act$, $\mu, \nu \in S_{\text{CCSP}}$, $A \subseteq Act$ and renamings R. The mapping $dex : T_{\text{CCSP}} \to \mathcal{P}(S_{\text{CCSP}})$ decomposing and expanding a process expression into a set of places is inductively defined by:

$$
\begin{aligned}
& & dex(0_A) &= \{0_A\} \\
dex(aP) &= \{aP\} & dex(R(P)) &= R(dex(P)) \\
dex(P + Q) &= dex(P) + dex(Q) & dex(\langle X_A | \mathcal{S} \rangle) &= dex(\langle S_{X_A} | \mathcal{S} \rangle) \\
dex(P\|Q) &= dex(P)\|_A \cup {}_A\|dex(Q) & \text{where } A &= \alpha(P) \cap \alpha(Q).
\end{aligned}
$$

Here $H\|_A$, $_A\|H$, $R(H)$ and $H + K$ for $H, K \subseteq S_{\text{CCSP}}$ are defined element by element; e.g. $R(H) = \{R(\mu) \mid \mu \in H\}$. The binding matters, so that $(_A\|H)\|_B \neq {}_A\|(H\|_B)$. Since I deal with guarded recursion only, dex is well-defined.

Following [27], I construct the unmarked Petri net (S, T, F, Act, ℓ) of CCSP with $S := S_{\text{CCSP}}$, specifying the triple (T, F, ℓ) as a ternary relation $\to \subseteq \mathbb{IN}^S \times Act \times \mathbb{IN}^S$. An element $H \xrightarrow{a} J$ of this relation denotes a transition $t \in T$ with $\ell(t) = a$ such that $\bullet t = H$ and $t \bullet = J$. The transitions $H \xrightarrow{\alpha} J$ are derived from the rules of Table 3.

Note that there is no rule for recursion. The transitions of a recursive process $\langle X_A | \mathcal{S} \rangle$ are taken care of indirectly by the decomposition $dex(\langle X_A | \mathcal{S} \rangle) = dex(\langle S_{X_A} | \mathcal{S} \rangle)$, which expands the decomposition of a recursive call into a decomposition of an expression in which each recursive call is guarded by an action prefix.

Example 2. The Petri net semantics of the 2-customer 2-clerk store from Section 2 is displayed in Figure 4. It is more compact than the 9-state 24-transition labelled transition system. The name of the bottommost place is $_{Ser}\|_\emptyset\| R_{II}(serve \langle \text{CLK}_{Cl} | \mathcal{S}_{\text{CLK}} \rangle)$ where Ser is the alphabet $\{\text{I } serves\, A, \text{I } serves\, B, \text{II } serves\, A, \text{II } serves\, B\}$.

---

[1] A *labelled transition system* (LTS) is given by a set S of *states* and a *transition relation* $T \subseteq S \times \mathcal{L} \times S$ for some set of labels $\mathcal{L}$. The LTS generated by CCSP has $S := T_{\text{CCSP}}$, $\mathcal{L} := Act$ and $T := \to$.

Table 3. Operational Petri net semantics of CCSP

$$\{aP\} \xrightarrow{a} dex(P)$$

$$\frac{H \xrightarrow{a} J}{R(H) \xrightarrow{b} R(J)} \ ((a,b) \in R) \qquad\qquad \frac{H \xrightarrow{a} J}{H\|_A \xrightarrow{a} J\|_A} \ (a \notin A)$$

$$\frac{H \cup K \xrightarrow{a} J}{H \cup (K + dex(Q)) \xrightarrow{a} J} \qquad\qquad \frac{H \xrightarrow{a} J \quad K \xrightarrow{a} L}{H\|_A \cup {}_A\|K \xrightarrow{a} J\|_A \cup {}_A\|L} \ (a \in A)$$

$$\frac{H \cup K \xrightarrow{a} J}{H \cup (dex(P) + K) \xrightarrow{a} J} \qquad\qquad \frac{H \xrightarrow{a} J}{{}_A\|H \xrightarrow{a} {}_A\|J} \ (a \notin A)$$

A progress assumption, as discussed in Section 1.4, disallows runs that stop after finitely many actions. So in each run some of the actions from *Ser* will occur infinitely often. When assuming strong fairness, each of those actions will occur infinitely often. When assuming only weak fairness, it is possible that II *serves* A and II *serves* B will never occur, as long as I *serves* A and I *serves* B each occur infinitely often, for in such a run the actions II *serves* A and II *serves* B are not enabled in every state (from some point onwards). However, it is not possible that I *serves* B and II *serves* B never occur, because in such a run, from some point onwards, the action I *serves* B is enabled in every state.

When assuming justness but not weak fairness, a run that bypasses any two serving actions is possible, but a run that bypasses I *serves* B, II *serves* A and II *serves* B is excluded, because in such a run, from some point onwards, the action II *serves* B is perpetually enabled, in the sense that both tokens in its preplaces never move away.

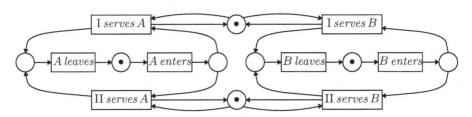

Fig. 4. Petri net semantics of the 2-customer 2-clerk store

Olderog [26, 27] shows that the Petri net $[\![P]\!]$ associated to a closed CCSP expression P is safe, and that all its reachable markings are finite; the latter implies that it has bounded parallelism. The following result, from [26, 27], shows that the standard interleaving semantics of CCSP is retrievable from the net semantics; it establishes a strong bisimulation relating any CCSP expression (seen as a state in a labelled transition system) with its interpretation as a marking in the Petri net of CCSP.

Theorem 1. There exists a relation $\mathcal{B}$ between closed CCSP expressions and markings in the unmarked Petri net of CCSP, such that

- $P \mathcal{B} \, dex(P)$ for each closed, well-typed CCSP expression with guarded recursion,
- if $P\mathcal{B}M$ and $P \xrightarrow{a} P'$ then there is a marking M' and transition t with $\ell(t) = a$, $M[t\rangle M'$ and $P\mathcal{B}M'$, and
- if $P\mathcal{B}M$ and $M[t\rangle M'$ then there is CCSP process P' with $P \xrightarrow{\ell(t)} P'$ and $P\mathcal{B}M'$.

To formalise the concurrency requirement for his net semantics Olderog defines for each n-ary CCSP operator op an n-ary operation $op_\mathcal{N}$ on safe Petri nets, inspired by proposals from [16,18,35], and requires that

$$
\begin{aligned}
(1) &\quad [\![op(P_1,\ldots,P_n)]\!] \approx op_\mathcal{N}([\![P_1]\!],\ldots,[\![P_n]\!]) \\
(2) &\quad [\![\langle X_A|\mathcal{S}\rangle]\!] \approx [\![\langle \mathcal{S}_{X_A}|\mathcal{S}\rangle]\!]
\end{aligned}
$$

for a suitable relation $\approx$. In fact, (2) turns out to hold taking for $\approx$ the identity relation. He establishes (1) taking for $\approx$ a relation he calls *strong bisimilarity*, whose definition will be recalled in Section 6. When a relation $\equiv$ includes $\approx$, and (1) holds for $\approx$, then it also holds for $\equiv$.

The operations $op_\mathcal{N}$ (i.e. $(0_A)_\mathcal{N}$ for $A \subseteq Act$, $a_\mathcal{N}$ for $a \in Act$, $R_\mathcal{N}$ for $R \subseteq Act \times Act$, $\|_\mathcal{N}$ and $+_\mathcal{N}$) are defined only up to isomorphism, but this is no problem as isomorphic nets are strongly bisimilar. The definition is recalled below—it generalises verbatim to non-safe nets, except that $+_\mathcal{N}$ is defined only for nets whose initial markings are nonempty plain sets.

Definition 3. [27] The net 0_A has type A and consists of a single place, initially marked: $(0_A)_\mathcal{N} := (\{0_A\}, \emptyset, \emptyset, \{0_A\}, A, \emptyset)$.

Given a net $N = (S, T, F, M, A, \ell)$ and $a \in Act$, take $s_0, t_a \notin S \cup T$. Then the net $a_\mathcal{N}N$ is obtained from N by the addition of the fresh place s_0 and the fresh transition t_a, labelled a, such that ${}^\bullet t_a = \{s_0\}$ and $t_a{}^\bullet = M$. The type of $a_\mathcal{N}N$ will be $A \cup \{a\}$ and the initial marking $\{s_0\}$.

Given a net $N = (S, T, F, M, A, \ell)$ and a renaming operator $R(\_)$, the net $R_\mathcal{N}(N)$ has type $R(A) := \{b \in Act \mid \exists a \in A, (a, b) \in R\}$, the same places and initial marking as N, and transitions t_b for each $t \in T$ and $b \in Act$ with $(\ell(t), b) \in R$. One has ${}^\bullet t_b := {}^\bullet t$, $t_b{}^\bullet := t^\bullet$, and the label of t_b will be b.

Given two nets $N_i = (S_i, T_i, F_i, M_i, A_i, \ell_i)$ $(i=1,2)$, their parallel composition $N_1\|_\mathcal{N}N_2 = (S, T, F, M, A, \ell)$ is obtained from the disjoint union of N_1 and N_2 by the omission of all transitions t of $T_1 \uplus T_2$ with $\ell(t) \in A_1 \cap A_2$, and the addition of fresh transitions (t_1, t_2) for all pairs $t_i \in T_i$ $(i=1,2)$ with $\ell_1(t_1) = \ell_2(t_2) \in A_1 \cap A_2$. Take ${}^\bullet(t_1, t_2) = {}^\bullet t_1 + {}^\bullet t_2$, $(t_1, t_2)^\bullet = t_1^\bullet + t_2^\bullet$, $\ell(t_1, t_2) = \ell(t_1)$, and $A := A_1 \cup A_2$.

Given nets $N_i = (S_i, T_i, F_i, M_i, A_i, \ell_i)$ with $M_i \neq \emptyset$ a plain set $(i = 1, 2)$, the net $N_1 +_\mathcal{N} N_2$ with type $A_1 \cup A_2$ is obtained from the disjoint union of N_1 and

N_2 by the addition of the set of fresh places $M_1 \times M_2$—this set will be the initial marking of $N_1 +_{\mathcal{N}} N_2$—and the addition of fresh transitions t_i^K for any $t_i \in T_i$ and $\emptyset \neq K \leq {}^\bullet t_i \cap M_i$. $\ell(t_i^K) = \ell_i(t)$, ${}^\bullet t_1^K = {}^\bullet t_1 - K + (K \times M_2)$, ${}^\bullet t_2^K = {}^\bullet t_2 - K + (M_1 \times K)$ and $(t_i^K)^\bullet = t_i{}^\bullet$.

5   Structure Preserving Bisimulation Equivalence

This section presents structure preserving bisimulation equivalence on nets.

Definition 4. Given two nets $N_i = (S_i, T_i, F_i, M_i, A_i, \ell_i)$, a *link* is a pair $(s_1, s_2) \in S_1 \times S_2$ of places. A *linking* $l \in \mathbb{N}^{S_1 \times S_1}$ is a multiset of links; it can be seen as a pair of markings with a bijection between them. Let $\pi_i(l) \in \mathbb{N}^{S_i}$ be these markings, given by $\pi_1(l)(s_1) = \sum_{s_2 \in S_2} l(s_1, s_2)$ for all $s_1 \in S_1$ and $\pi_2(l)(s_2) = \sum_{s_1 \in S_1} l(s_1, s_2)$ for all $s_2 \in S_2$. A *structure preserving bisimulation* (*sp-bisimulation*) is a set $\mathscr{B}$ of linkings, such that

- if $c \leq l \in \mathscr{B}$ and $\pi_1(c) = {}^\bullet t_1$ for $t_1 \in T_1$ then there are a transition $t_2 \in T_2$ with $\ell(t_2) = \ell(t_1)$ and $\pi_2(c) = {}^\bullet t_2$, and a linking $\bar{c}$ such that $\pi_1(\bar{c}) = t_1{}^\bullet$, $\pi_2(\bar{c}) = t_2{}^\bullet$ and $\bar{l} := l - c + \bar{c} \in \mathscr{B}$.
- if $c \leq l \in \mathscr{B}$ and $\pi_2(c) = {}^\bullet t_2$ then there are a t_1 and a $\bar{c}$ with the same properties.

N_1 and N_2 are *structure preserving bisimilar*, notation $N_1 \leftrightarrow_{sp} N_2$, if $A_1 = A_2$ and there is a linking l in a structure preserving bisimulation with $M_1 = \pi_1(l)$ and $M_2 = \pi_2(l)$.

Note that if $\mathscr{B}$ is an sp-bisimulation, then so is its downward closure $\{k \mid \exists l \in \mathscr{B}. k \leq l\}$. Moreover, if $\mathscr{B}$ is an sp-bisimulation between two nets, then the set of those linkings $l \in \mathscr{B}$ for which $\pi_1(l)$ and $\pi_2(l)$ are reachable markings is also an sp-bisimulation.

If $\mathcal{B}$ is a set of a links, let $\overline{\mathcal{B}}$ be the set of *all* linkings that are multisets over $\mathcal{B}$.

Proposition 1. Structure preserving bisimilarity is an equivalence relation.

Proof. The relation $\overline{Id}$, with Id the identity relation on places, is an sp-bisimulation, showing that $N \leftrightarrow_{sp} N$ for any net N.

Given an sp-bisimulation $\mathscr{B}$, also $\{l^{-1} \mid l \in \mathscr{B}\}$ is an sp-bisimulation, showing symmetry of $\leftrightarrow_{sp}$.

Given linkings $h \in \mathbb{N}^{S_1 \times S_3}$, $k \in \mathbb{N}^{S_1 \times S_2}$ and $l \in \mathbb{N}^{S_2 \times S_3}$, write $h \in k; l$ if there is a multiset $m \in \mathbb{N}^{S_1 \times S_2 \times S_3}$ of triples of places, with $k(s_1, s_2) = \sum_{s_3 \in S} m(s_1, s_2, s_3)$, $l(s_2, s_3) = \sum_{s_1 \in S} m(s_1, s_2, s_3)$ and $h(s_1, s_3) = \sum_{s_2 \in S} m(s_1, s_2, s_3)$. Now, for sp-bisimulations $\mathscr{B}$ and $\mathscr{B}'$, also $\mathscr{B}; \mathscr{B}' := \{h \in k; l \mid k \in \mathscr{B} \wedge l \in \mathscr{B}'\}$ is an sp-bisimulation, showing transitivity of $\leftrightarrow_{sp}$. $\square$

6   Strong Bisimilarity

As discussed in the introduction and at the end of Section 4, Olderog defined a relation of *strong bisimilarity* on safe Petri nets.

Definition 5. For $\mathcal{B} \subseteq S_1 \times S_2$ a binary relation between the places of two safe nets $N_i = (S_i, T_i, F_i, M_i, A_i, \ell_i)$, write $\widehat{\mathcal{B}}$ for the set of all linkings $l \subseteq \mathcal{B}$ such that $\pi_i(l)$ is a reachable marking of N_i for $i = 1, 2$ and $\mathcal{B} \cap \left(\pi_1(l) \times \pi_2(l) \right) = l$. Now a *strong bisimulation* as defined in [27] can be seen as a structure preserving bisimulation of the form $\widehat{\mathcal{B}}$. The nets N_1 and N_2 are *strongly bisimilar* if $A_1 = A_2$ and there is a linking l in a strong bisimulation with $M_1 = \pi_1(l)$ and $M_2 = \pi_2(l)$.

This reformulation of the definition from [27] makes immediately clear that strong bisimilarity of two safe Petri nets implies their structure preserving bisimilarity. Consequently, the concurrency requirement for the net semantics from Olderog, as formalised by Requirements (1) and (2) in Section 4, holds for structure preserving bisimilarity.

7   Compositionality

In this section I show that structure preserving bisimilarity is a congruence for the operators of CCSP, or, in other words, that these operators are compositional up to $\underline{\leftrightarrow}_{sp}$.

Theorem 2. If $N_1 \underline{\leftrightarrow}_{sp} N_2$, $a \in Act$ and $R \subseteq Act \times Act$, then $a_{\mathcal{N}} N_1 \underline{\leftrightarrow}_{sp} a_{\mathcal{N}} N_2$ and $R_{\mathcal{N}}(N_2) \underline{\leftrightarrow}_{sp} R_{\mathcal{N}}(N_2)$. If $N_1^l \underline{\leftrightarrow}_{sp} N_2^l$ and $N_1^r \underline{\leftrightarrow}_{sp} N_2^r$ then $N_1^l \|_{\mathcal{N}} N_1^r \underline{\leftrightarrow}_{sp} N_2^l \|_{\mathcal{N}} N_2^r$ and, if the initial markings of N_i^l and N_i^r are nonempty sets, $N_1^l +_{\mathcal{N}} N_1^r \underline{\leftrightarrow}_{sp} N_2^l +_{\mathcal{N}} N_2^r$.

Proof. Let $N_i = (S_i, T_i, F_i, M_i, A_i, \ell_i)$ for $i = 1, 2$, and let s_i and u_i be the fresh place and transition introduced in the definition of $a_{\mathcal{N}} N_i$. From $N_1 \underline{\leftrightarrow}_{sp} N_2$ it follows that $A_1 = A_2$ and hence $A_1 \cup \{a\} = A_2 \cup \{a\}$.

Let $\mathcal{B}$ be an sp-bisimulation containing a linking k with $M_i = \pi_i(k)$ for $i = 1, 2$. Let $\mathcal{B}_a := \mathcal{B} \cup \{h\}$, with $h = \{(s_1, s_2)\}$. Then h links the initial markings of $a_{\mathcal{N}} N_1$ and $a_{\mathcal{N}} N_2$. Hence it suffices to show that $\mathcal{B}_a$ is an sp-bisimulation. So suppose $c \le h$ and $\pi_1(c) = {}^{\bullet} t_1$ for some $t_1 \in T_1$. Then $c = h$ and $t_1 = u_1$. Take $t_2 := u_2$ and $\bar{h} := \bar{c} := k$.

To show that $R_{\mathcal{N}}(N_2) \underline{\leftrightarrow}_{sp} R_{\mathcal{N}}(N_2)$ it suffices to show that $\mathcal{B}$ also is an sp-bisimulation between $R_{\mathcal{N}}(N_2)$ and $R_{\mathcal{N}}(N_2)$, which is straightforward.

Now let $N_i^l = (S_i^l, T_i^l, F_i^l, M_i^l, A_i^l, \ell_i^l)$ and $N_i^r = (S_i^r, T_i^r, F_i^r, M_i^r, A_i^r, \ell_i^r)$ for $i = 1, 2$. Let $A := A_1^l \cap A_1^r = A_2^l \cap A_2^r$. Create the disjoint union of N_i^l and N_i^r in the definition of $N_i^l \|_{\mathcal{N}} N_i^r$ by renaming all places s and transitions t of N_i^l into $s \|_A$ and $t \|_A$, and all places s and transitions t of N_i^r into $_A \| s$ and $_A \| t$. Let $\mathcal{B}^l$ and $\mathcal{B}^r$ be sp-bisimulations containing linkings k^l and k^r, respectively, with $M_i^l = \pi_i(k^l)$ and $M_i^r = \pi_i(k^r)$, for $i = 1, 2$. Take $\mathcal{B} := \{(h^l \|_A) + (_A \| h^r) \mid h^l \in \mathcal{B}^l \wedge h^r \in \mathcal{B}^r\}$, where $h^l \|_A := \{(s_1 \|_A, s_2 \|_A) \mid (s_1, s_2) \in h^l\}$, and $_A \| h^r$ is defined likewise. Then

$\pi_i((k^l\|_A)+({}_A\|k^r)) = \pi_i(k^l)\|_A + {}_A\|\pi_i(k^r) = M_i^l\|_A + {}_A\|M_i^r$ is the initial marking of $N_i^l\|_{\mathcal{N}}N_i^r$ for $i=1,2$, so it suffices to show that $\mathscr{B}$ is an sp-bisimulation.

So suppose $c \leq (h^l\|_A)+({}_A\|h^r)\in\mathscr{B}$ with $h^l\in\mathscr{B}^l \wedge h^r\in\mathscr{B}^r$ and $\pi_1(c)={}^\bullet t_1$ for t_1 a transition of $N_1^l\|_{\mathcal{N}}N_1^r$. Then c has the form (i) $t_1^l\|_A$ for $t_1^l \in T_1^l$ with $\ell_1^l(t_1^l)\notin A$, or (ii) $(t_1^l\|_A, {}_A\|t_1^r)$ for $t_1^l \in T_1^l$ and $t_1^r \in T_1^r$ with $\ell_1^l(t_1^l) = \ell_1^r(t_1^r) \in A$, or (iii) ${}_A\|t_1^r$ for $t_1^r\in T_1^r$ with $\ell_1^r(t_1^r)\notin A$. In case (i) one has $c^r=\emptyset$ and $\pi_1(c^l)={}^\bullet t_1^l$, whereas in case (ii) $\pi_1(c^l)={}^\bullet t_1^l$ and $\pi_1(c^r)={}^\bullet t_1^r$. I only elaborate case (ii); the other two proceed likewise. Since $\mathscr{B}^l$ is an sp-bisimulation, there are a transition t_2^l with $\ell_2^l(t_2^l)=\ell_1^l(t_1^l)$ and $\pi_2(c^l)={}^\bullet t_2^l$, and a linking $\bar c^l$ such that $\pi_1(\bar c^l)=t_1^{l\,\bullet}$, $\pi_2(\bar c^l)=t_2^{l\,\bullet}$ and $\bar h^l := h^l - c^l + \bar c^l \in \mathscr{B}^l$. Likewise, since $\mathscr{B}^r$ is an sp-bisimulation, there are a transition t_2^r with $\ell_2^r(t_2^r) = \ell_1^r(t_1^r)$ and $\pi_2(c^r)={}^\bullet t_2^r$, and a linking $\bar c^r$ such that $\pi_1(\bar c^r)=t_1^{r\,\bullet}$, $\pi_2(\bar c^r)=t_2^{r\,\bullet}$ and $\bar h^r := h^r - c^r + \bar c^r\in\mathscr{B}^r$. Take $t_2 := (t_2^l\|_A, {}_A\|t_2^r)$. This transition has the same label as $t_2^l, t_2^r, t_1^l, t_1^r$ and $(t_1^l\|_A, {}_A\|t_1^r) = t_1$. Moreover, $\pi_2(c) = \pi_2(c^l)\|_A + {}_A\|\pi_2(c^r) = {}^\bullet t_2^l\|_A + {}_A\|{}^\bullet t_2^r = {}^\bullet t_2$. Take $\bar c := (\bar c^l\|_A) + ({}_A\|\bar c^r)$. Then $\pi_1(\bar c)=t_1{}^\bullet$, $\pi_2(\bar c)=t_2{}^\bullet$ and $\bar h := (h^l\|_A) + ({}_A\|h^r) - c + \bar c = (\bar h^l\|_A) + ({}_A\|\bar h^r)\in\mathscr{B}$.

Let $N_i^l = (S_i^l, T_i^l, F_i^l, M_i^l, A_i^l, \ell_i^l)$ and $N_i^r = (S_i^r, T_i^r, F_i^r, M_i^r, A_i^r, \ell_i^r)$ for $i=1,2$, with M_i^l and M_i^r nonempty plain sets, but this time I assume the nets to already be disjoint, and such that all the places and transitions added in the construction of $N_i^l +_{\mathcal{N}} N_i^r$ are fresh. Let $\mathscr{B}^l$ and $\mathscr{B}^r$ be as above. Without loss of generality I may assume that the linkings h in $\mathscr{B}^l$ and $\mathscr{B}^r$ have the property that $\pi_i(h)$ is a reachable marking for $i=1,2$, so that the restriction of $\pi_i(h)$ to M_i^l or M_i^r is a plain set. Define

$$\mathscr{B}^+ := \{h_\bullet^l + (h_+^l \otimes k^r) \mid h_\bullet^l + h_+^l \in \mathscr{B}^l \wedge h_+^l \lneqq k^l\}$$
$$\{h_\bullet^r + (k^l \otimes h_+^r) \mid h_\bullet^r + h_+^r \in \mathscr{B}^r \wedge h_+^r \lneqq k^r\} \cup \{k^l \otimes k^r\}$$

where $h^l\otimes h^r := \{((s_1^l, s_1^r), (s_2^l, s_2^r)) \mid (s_1^l, s_2^l)\in h^l \wedge (s_1^r, s_2^r)\in h^r\}$. Now $\pi_i(k^l\otimes k^r) = \pi_i(k^l)\times\pi_i(k^r) = M_i^l\times M_i^r$ is the initial marking of $N_i^l +_{\mathcal{N}} N_i^r$, so again it suffices to show that $\mathscr{B}^+$ is an sp-bisimulation.

So suppose $c \leq h_\bullet^l + (h_+^l\otimes k^r)\in\mathscr{B}^+$ with $h_\bullet^l + h_+^l\in\mathscr{B}^l$, $h_+^l \lneqq k^l$ and $\pi_1(c)={}^\bullet t_1$ for t_1 a transition of $N_1^l +_{\mathcal{N}} N_1^r$.

First consider the case that $c \leq h_\bullet^l$. Then $c \leq h_\bullet^l \leq h_\bullet^l + h_+^l\in\mathscr{B}^l$. Since $\mathscr{B}^l$ is an sp-bisimulation, there are a transition $t_2\in T_2^l$ with $\ell_2^l(t_2) = \ell_1^l(t_1)$ and $\pi_2(c) = {}^\bullet t_2$, and a linking $\bar c$ such that $\pi_1(\bar c) = t_1{}^\bullet$, $\pi_2(\bar c) = t_2{}^\bullet$ and $h_\bullet^l + h_+^l - c + \bar c\in\mathscr{B}^l$. Now $h_\bullet^l + (h_+^l\otimes k^r) - c + \bar c = (h_\bullet^l - c + \bar c) + (h_+^l\otimes k_2)\in\mathscr{B}^+$ because $(h_\bullet^l - c + \bar c) + h_+^l\in\mathscr{B}^l$.

In the remaining case $\pi_1(c)$ contains a place $(s_1^l, s_1^r)\in M_1^l\times M_1^r$, so t_1 must have either the form t_{1l}^K with $\emptyset \neq K \leq {}^\bullet t_1^l\cap M_1^l$ for some $t_1^l\in T_1^l$, or t_{1r}^K with $\emptyset \neq K \leq {}^\bullet t_1^r\cap M_1^r$ for some $t_1^r\in T_1^r$. First assume, towards a contradiction, that $t_1 = t_{1r}^K$. Then $M_1^l\times K \leq {}^\bullet t_{1r}^K = \pi_1(c) \leq \pi_1(h_\bullet^l) + \pi_1(h_+^l \otimes k^r)$. Since the places in $M_1^l\times K \subseteq M_1^l\times M_1^r$ are fresh, it follows that $M_1^l\times K \leq \pi_1(h_+^l\otimes k^r) \leq \pi_1(h_+^l)\times\pi_1(k^r) \leq \pi_1(h_+^l)\times M_1^r$, implying that $M_1^l \leq \pi_1(h_+^l)$ and $K \leq M_1^r$—here I use that $M_1^l\neq\emptyset\neq K$ and $\pi_1(h_+^l)$ and M_1^r are plain sets. However, the condition $h_+^l \lneqq k^l$ implies that $\pi_1(h_+^l) \lneqq \pi_1(k^l) = M_1^l$, yielding a contradiction. Hence t_1 is of the form t_{1l}^K.

Since $\pi_1(c) = {}^\bullet t_{1l}^K = {}^\bullet t_1^l - K + (K \times M_1^r)$, the linking c must have the form $c_\bullet + c'$ with $\pi_1(c_\bullet) = {}^\bullet t_1^l - K$ and $\pi_1(c') = K \times M_1^r$. As no place in ${}^\bullet t_1^l - K$ can be in $M_1^l \times M_1^r \supseteq \pi_1(h_+^l \otimes k^r)$, it follows that $c_\bullet \leq h_\bullet^l$. Likewise, as none of the places in $K \times M_1^r$ can be in $\pi_1(h_\bullet^l)$, it follows that $c' \leq h_+^l \otimes k^r$. Thus $K \times M_1^r = \pi_1(c') \leq \pi_1(h_+^l \otimes k^r) \leq \pi_1(h_+^l) \times \pi_1(k^r) \leq \pi_1(h_+^l) \times M_1^r$, implying $K \leq \pi_1(h_+^l)$—again using that $\pi_1(h_+^l)$ and $M_1^r \neq \emptyset$ are plain sets. The linking $h_+^l \otimes k^r$ has the property that its projection $\pi_1(h_+^l \otimes k^r)$ is a plain set. Since a subset c'' of a such linking is completely determined by its first projection $\pi_1(c'')$, it follows that $c' = c_+ \otimes k^r$ for the unique linking $c_+ \leq h_+^l$ with $\pi_1(c_+) = K$.

Now $c_\bullet + c_+ \leq h_\bullet^l + h_+^l \in \mathscr{B}^l$ and $\pi_1(c_\bullet + c_+) = ({}^\bullet t_1^l - K) + K = {}^\bullet t_1^l$. Since $\mathscr{B}^l$ is an sp-bisimulation, there are a transition $t_2^l \in T_2^l$ with $\ell_2^l(t_2^l) = \ell_1^l(t_1^l)$ and $\pi_2(c_\bullet + c_+) = {}^\bullet t_2^l$, and a linking $\bar{c}$ such that $\pi_1(\bar{c}) = t_1^{l\bullet}$, $\pi_2(\bar{c}) = t_2^{l\bullet}$ and $h_\bullet^l + h_+^l - (c_\bullet + c_+) + \bar{c} \in \mathscr{B}^l$. Let $L := \pi_2(c_+)$. Then $L \neq \emptyset$ since $K \neq \emptyset$, $L = \pi_2(c_+) \leq \pi_2(h_+^l) \leq \pi_2(k^l) = M_2^l$ and $L = \pi_2(c_+) \leq \pi_2(c_\bullet + c_+) = {}^\bullet t_2^l$. By Definition 3 $N_2^l +_{\mathcal{N}} N_2^r$ has a transition t_{2l}^L with $\ell(t_{2l}^L) = \ell_2^l(t_2^l) = \ell_1^l(t_1^l) = \ell(t_{1l}^L)$, ${}^\bullet t_{2l}^L = {}^\bullet t_2^l - L + (L \times M_2^l) = \pi_2(c_\bullet + c_+) - \pi_2(c_+) + (\pi_2(c_+) \times \pi_1(k^r)) = \pi_2(c_\bullet + (c_+ \otimes k^r)) = \pi_2(c)$ and $t_{2l}^{L\bullet} = t_2^{l\bullet} = \pi_2(\bar{c})$. Moreover, $\pi_1(\bar{c}) = t_1^{l\bullet} = t_1^{K\bullet}$. Finally, $h_\bullet^l + (h_+^l \otimes k^r) - c + \bar{c} = (h_\bullet^l - c_\bullet + \bar{c}) + ((h_+^l - c_+) \otimes k^r) \in \mathscr{B}^+$ since $(h_\bullet^l - c_\bullet + c') + (h_+^l - c_+) \in \mathscr{B}^l$ and $h_+^l - c_+ \leq h_+^l \lneq k^l$.

The case supposing $c \leq h_\bullet^r + (k^r \otimes h_+^r) \in \mathscr{B}^+$ follows by symmetry, whereas the case $c \leq k^l \otimes k^r$ proceeds by simplification of the other two cases.   □

8   Processes of Nets and Causal Equivalence

A *process* of a net N [9,19,29] is essentially a conflict-free, acyclic net together with a mapping function to N. It can be obtained by unwinding N, choosing one of the alternatives in case of conflict. It models a run, or concurrent computation, of N. The acyclic nature of the process gives rise to a notion of causality for transition firings in the original net via the mapping function. A conflict present in the original net is represented by the existence of multiple processes, each representing one possible way to decide the conflict. This notion of process differs from the one used in process algebra; there a "process" refers to the entire behaviour of a system, including all its choices.

Definition 6. A *causal net*[2] is a net $\mathcal{N} = (\mathcal{S}, \mathcal{T}, \mathcal{F}, \mathcal{M}_0, \mathcal{A}, \ell_{\mathcal{N}})$ satisfying

- $\forall s \in \mathcal{S}. |{}^\bullet s| \leq 1 \geq |s^\bullet| \wedge \mathcal{M}_0(s) = \begin{cases} 1 & \text{if } {}^\bullet s = \emptyset \\ 0 & \text{otherwise,} \end{cases}$

- $\mathcal{F}$ is acyclic, i.e., $\forall x \in \mathcal{S} \cup \mathcal{T}. (x, x) \notin \mathcal{F}^+$, where $\mathcal{F}^+$ is the transitive closure of $\{(x, y) \mid \mathcal{F}(x, y) > 0\}$,

- and $\{t \in \mathcal{T} \mid (t, u) \in \mathcal{F}^+\}$ is finite for all $u \in \mathcal{T}$.

[2] A causal net [29,34] is traditionally called an *occurrence net* [9,19,33]. Here, following [27], I will not use the terminology "occurrence net" in order to avoid confusion with the occurrence nets of [25,36]; the latter extend causal nets with forward branching places, thereby capturing all runs of the represented system, together with the branching structure between them.

A *folding* from a net $\mathcal{N} = (\mathcal{S}, \mathcal{T}, \mathcal{F}, \mathcal{M}_0, \mathcal{A}, \ell_{\mathcal{N}})$ into a net $N = (S, T, F, M_0, A, \ell)$ is a function $\rho : \mathcal{S} \cup \mathcal{T} \rightarrow S \cup T$ with $\rho(\mathcal{S}) \subseteq S$ and $\rho(\mathcal{T}) \subseteq T$, satisfying

- $\mathcal{A} = A$ and $\ell_{\mathcal{N}}(t) = \ell(\rho(t))$ for all $t \in \mathcal{T}$,
- $\rho(\mathcal{M}_0) = M_0$, i.e. $M_0(s) = |\rho^{-1}(s) \cap \mathcal{M}_0|$ for all $s \in S$, and
- $\forall t \in \mathcal{T}, s \in S.\ F(s, \rho(t)) = |\rho^{-1}(s) \cap {}^\bullet t| \wedge F(\rho(t), s) = |\rho^{-1}(s) \cap t^\bullet|.$ [3]

A pair $\mathcal{P} = (\mathcal{N}, \rho)$ of a causal net $\mathcal{N}$ and a folding of $\mathcal{N}$ into a net N is a *process* of N. $\mathcal{P}$ is called *finite* if $\mathcal{T}$ is finite.

Note that if N has bounded parallelism, than so do all of its processes.

Definition 7. [27] A net $\mathcal{N}$ is called a *causal net of* a net N if it is the first component of a process $(\mathcal{N}, \rho)$ of N. Two nets N_1 and N_2 are *causal equivalent*, notation $\equiv_{caus}$, if they have the same causal nets.

Olderog shows that his relation of strong bisimilarity is included in $\equiv_{caus}$ [27], and thereby establishes the concurrency requirement (1) from Section 4 for $\equiv_{caus}$.

For $\mathcal{N} = (\mathcal{S}, \mathcal{T}, \mathcal{F}, \mathcal{M}_0, \mathcal{A}, \ell_{\mathcal{N}})$ a causal net, let $\mathcal{N}^\circ := \{s \in \mathcal{S} \mid s^\bullet = \emptyset\}$. The following result supports the claim that finite processes model finite runs.

Proposition 2. [19, Theorems 3.5 and 3.6] M is a reachable marking of a net N iff N has a finite process $(\mathcal{N}, \rho)$ with $\rho(\mathcal{N}^\circ) = M$. Here $\rho(\mathcal{N}^\circ)(s) = |\rho^{-1}(s) \cap \mathcal{N}^\circ|$.

A process is not required to represent a completed run of the original net. It might just as well stop early. In those cases, some set of transitions can be added to the process such that another (larger) process is obtained. This corresponds to the system taking some more steps and gives rise to a natural order between processes.

Definition 8. Let $\mathcal{P} = ((\mathcal{S}, \mathcal{T}, \mathcal{F}, \mathcal{M}_0, \mathcal{A}, \ell), \rho)$ and $\mathcal{P}' = ((\mathcal{S}', \mathcal{T}', \mathcal{F}', \mathcal{M}'_0, \mathcal{A}', \ell'), \rho')$ be two processes of the same net. $\mathcal{P}'$ is a *prefix* of $\mathcal{P}$, notation $\mathcal{P}' \leq \mathcal{P}$, and $\mathcal{P}$ an *extension* of $\mathcal{P}'$, iff $\mathcal{S}' \subseteq \mathcal{S}$, $\mathcal{T}' \subseteq \mathcal{T}$, $\mathcal{M}'_0 = \mathcal{M}_0$, $\mathcal{F}' = \mathcal{F} \restriction (\mathcal{S}' \times \mathcal{T}' \cup \mathcal{T}' \times \mathcal{S}')$ and $\rho' = \rho \restriction (\mathcal{S}' \cup \mathcal{T}')$. (This implies that $\mathcal{A}' = \mathcal{A}$ and $\ell' = \ell \restriction \mathcal{T}$.)

The requirements above imply that if $\mathcal{P}' \leq \mathcal{P}$, $(x, y) \in \mathcal{F}^+$ and $y \in \mathcal{S}' \cup \mathcal{T}'$ then $x \in \mathcal{S}' \cup \mathcal{T}'$. Conversely, any subset $\mathcal{T}' \subseteq \mathcal{T}$ satisfying $(t, u) \in \mathcal{F}^+ \wedge u \in \mathcal{T}' \Rightarrow t \in \mathcal{T}'$ uniquely determines a prefix of $\mathcal{P}$. A process $(\mathcal{N}, \rho)$ of a net N is *initial* if $\mathcal{N}$ contains no transitions; then $\rho(\mathcal{N}^\circ)$ is the initial marking of N. Any process has an initial prefix.

Proposition 3. [19, Theorem 3.17] If $\mathcal{P}_i = ((\mathcal{S}_i, \mathcal{T}_i, \mathcal{F}_i, \mathcal{M}_{0i}, \mathcal{A}_i, \ell_i), \rho_i)$ $(i \in \mathbb{N})$ is a chain of processes of a net N, satisfying $\mathcal{P}_i \leq \mathcal{P}_j$ for $i \leq j$, then there exists a unique process $\mathcal{P} = ((\mathcal{S}, \mathcal{T}, \mathcal{F}, \mathcal{M}_0, \mathcal{A}, \ell), \rho)$ of N with $\mathcal{S} = \bigcup_{i \in \mathbb{N}} \mathcal{S}_i$ and $\mathcal{T} = \bigcup_{i \in \mathbb{N}} \mathcal{T}_i$—the *limit* of this chain—such that $\mathcal{P}_i \leq \mathcal{P}$ for all $i \in \mathbb{N}$. $\qquad\square$

---

[3] For $H \subseteq \mathcal{S}$, the multiset $\rho(H) \in \mathbb{N}^S$ is defined by $\rho(H)(s) = |\rho^{-1}(s) \cap H|$. Using this, these conditions can be reformulated as $\rho({}^\bullet t) = {}^\bullet \rho(t)$ and $\rho(t^\bullet) = \rho(t)^\bullet$.

In [9,19,29] processes were defined without the third requirement of Definition 6. Goltz and Reisig [19] observed that certain processes did not correspond with runs of systems, and proposed to restrict the notion of a process to those that can be obtained as the limit of a chain of finite processes [19, endofSection 3]. By [19, Theorems 3.18 and 2.14], for processes of finite nets this limitation is equivalent with imposing the third bullet point of Definition 6. My restriction to nets with bounded parallelism serves to recreate this result for processes of infinite nets.

Proposition 4. Any process of a net can be obtained as the limit of a chain of finite approximations.

Proof. Define the *depth* of a transition u in a causal net as one more than the maximum of the depth of all transitions t with tF^+u. Since the set of such transitions t is finite, the depth of a transition u is a finite integer. Now, given a process $\mathcal{P}$, the approximation $\mathcal{P}_i$ is obtained by restricting to those transitions in $\mathcal{P}$ of depth $\leq i$, together with all their pre- and postplaces, and keeping the initial marking. Clearly, these approximations form a chain, with limit $\mathcal{P}$. By induction on i one shows that $\mathcal{P}_i$ is finite. For $\mathcal{P}_0$ this is trivial, as it has no transitions. Now assume $\mathcal{P}_i$ is finite but $\mathcal{P}_{i+1}$ is not. Executing, in $\mathcal{P}_{i+1}$, all transitions of $\mathcal{P}_i$ one by one leads to a marking of $\mathcal{P}_{i+1}$ in which all remaining transitions of $\mathcal{P}_{i+1}$ are enabled. As these transitions cannot have common preplaces, this violates the assumption that $\mathcal{P}_{i+1}$ has bounded parallelism. □

9   A Process-Based Characterisation of Sp-bisimilarity

This section presents an alternative characterisation of sp-bisimilarity that will be instrumental in obtaining Theorems 4 and 5, saying that $\leftrightarrow_{sp}$ is a finer semantic equivalence than $\equiv_{caus}$ and $\approx_h$. This characterisation could have been presented as the original definition; however, the latter is instrumental in showing that $\leftrightarrow_{sp}$ is coarser than $\approx_{pb}$ and $\equiv_{occ}$, and implied by Olderog's strong bisimilarity.

Definition 9. A *process-based sp-bisimulation* between two nets N_1 and N_2 is a set $\mathcal{R}$ of triples $(\rho_1, \mathcal{N}, \rho_2)$ with $(\mathcal{N}, \rho_i)$ a finite process of N_i, for $i = 1, 2$, such that

  - $\mathcal{R}$ contains a triple $(\rho_1, \mathcal{N}, \rho_2)$ with $\mathcal{N}$ a causal net containing no transitions,
  - if $(\rho_1, \mathcal{N}, \rho_2) \in \mathcal{R}$ and $(\mathcal{N}', \rho_i')$ with $i \in \{1, 2\}$ is a fin. proc. of N_i extending $(\mathcal{N}, \rho_i)$ then N_j with $j := 3-i$ has a process $(\mathcal{N}', \rho_j') \geq (\mathcal{N}, \rho_j)$ such that $(\rho_1', \mathcal{N}', \rho_2') \in \mathcal{R}$.

Theorem 3. Two nets are sp-bisimilar iff there exists a process-based sp-bisimulation between them.

Proof. Let $\mathcal{R}$ be a process-ba sed sp-bisimulation between nets N_1 and N_2. Define $\mathcal{B} := \{\{(\rho_1(\mathfrak{s}), \rho_2(\mathfrak{s})) \mid \mathfrak{s} \in \mathcal{N}^\circ\} \mid (\rho_1, \mathcal{N}, \rho_2) \in \mathcal{R}\}$. Then $\mathcal{B}$ is an sp-bisimulation:

– Let $c \leq l \in \mathscr{B}$ and $\pi_1(c) = {}^\bullet t_1$ for $t_1 \in T_1$. Then $l = \{(\rho_1(\mathfrak{s}), \rho_2(\mathfrak{s}) \mid \mathfrak{s} \in \mathcal{N}^\circ\}$ for some $(\rho_1, \mathcal{N}, \rho_2) \in \mathscr{R}$. Extend $\mathcal{N}$ to $\mathcal{N}'$ by adding a fresh transition t and fresh places s_i for $s \in S_1$ and $i \in \mathbb{N}$ with $F_1(t_1, s) > i$; let ${}^\bullet t = \{\mathfrak{s} \in \mathcal{N}^\circ \mid \rho_1(\mathfrak{s}) \in {}^\bullet t_1\}$ and $t^\bullet = \{s_i \mid s \in S_1 \wedge i \in \mathbb{N} \wedge F_1(t_1, s) > i\}$. Furthermore, extend ρ_1 to ρ_1' by $\rho_1'(t) := t_1$ and $\rho_1'(s_i) := s$. Then ${}^\bullet \rho_1'(t) = {}^\bullet t_1 = \rho_1'({}^\bullet t)$ and $\rho_1'(t)^\bullet = t_1{}^\bullet = \rho_1'(t^\bullet)$, so $(\mathcal{N}', \rho_1')$ is a process of N_1, extending $(\mathcal{N}, \rho_1)$. Since $\mathscr{R}$ is a process-based sp-bisimulation, N_2 has a process $(\mathcal{N}', \rho_2') \geq (\mathcal{N}, \rho_2)$ such that $(\rho_1', \mathcal{N}', \rho_2') \in \mathscr{R}$. Take $t_2 := \rho_2'(t)$. Then $\ell_2(t_2) = \ell_\mathcal{N}(t) = \ell_1(t_1)$ and $c = \{(\rho_1(\mathfrak{s}), \rho_2(\mathfrak{s}) \mid \mathfrak{s} \in {}^\bullet t\}$, so $\pi_2(c) = \{\rho_2(\mathfrak{s}) \mid \mathfrak{s} \in {}^\bullet t\} = \rho_2({}^\bullet t) = \rho_2'({}^\bullet t) = {}^\bullet \rho_2'(t) = {}^\bullet t_2$. Take $c' := \{(\rho_1'(\mathfrak{s}), \rho_2'(\mathfrak{s})) \mid \mathfrak{s} \in t^\bullet\}$. Then $\pi_1(c') = t_1{}^\bullet$, $\pi_2(c') = t_2{}^\bullet$ and $l' := l - c + c' = \{(\rho_1'(\mathfrak{s}), \rho_2'(\mathfrak{s})) \mid \mathfrak{s} \in \mathcal{N}^\circ - {}^\bullet t + t^\bullet\} = \{(\rho_1'(\mathfrak{s}), \rho_2'(\mathfrak{s})) \mid \mathfrak{s} \in \mathcal{N}'^\circ\} \in \mathscr{B}$.

– The other clause follows by symmetry.

Since $\mathscr{R}$ contains a triple $(\rho_1, \mathcal{N}, \rho_2)$ with $\mathcal{N}$ a causal net containing no transitions, $\mathscr{B}$ contains a linking $l := \{(\rho_1(\mathfrak{s}), \rho_2(\mathfrak{s})) \mid \mathfrak{s} \in \mathcal{N}^\circ$ such that $\pi_i(l) = \rho_i(\mathcal{N}^\circ) = M_i$ for $i = 1, 2$, where M_i is the initial marking of N_i. Since $(\mathcal{N}, \rho_i)$ is a process of N_i, N_i must have the the same type as $\mathcal{N}$, for $i = 1, 2$. It follows that $N_1 \underline{\leftrightarrow}_{sp} N_2$.

Now let $\mathscr{B}$ be an sp-bisimulation between nets N_1 and N_2. Let $\mathscr{R} := \{(\rho_1, \mathcal{N}, \rho_2) \mid (\mathcal{N}, \rho_i)$ is a finite process of N_i $(i = 1, 2)$ and $\{(\rho_1(\mathfrak{s}), \rho_2(\mathfrak{s})) \mid \mathfrak{s} \in \mathcal{N}^\circ\} \in \mathscr{B}\}$. Then $\mathscr{R}$ is a process-based sp-bisimulation.

– $\mathscr{B}$ must contain a linking l with $\pi_i(l) = M_i$ for $i = 1, 2$, where M_i is the initial marking of N_i; let $l = \{(s_1^k, s_2^k) \mid k \in K\}$. Let $\mathcal{N}$ be a causal net with places $\mathfrak{s}^k$ for $k \in K$ and no transitions, and define ρ_i for $i = 1, 2$ by $\rho_i(\mathfrak{s}^k) = s_i^k$ for $k \in K$. Then $(\mathcal{N}, \rho_i)$ is an initial process of N_i $(i = 1, 2)$ and $(\rho_1, \mathcal{N}, \rho_2) \in \mathscr{R}$.

– Suppose $(\rho_1, \mathcal{N}, \rho_2) \in \mathscr{R}$ and $(\mathcal{N}', \rho_1')$ is a finite process of N_1 extending $(\mathcal{N}, \rho_1)$. (The case of a finite process of N_2 extending $(\mathcal{N}, \rho_1)$ will follow by symmetry.) Then $l := \{(\rho_1(\mathfrak{s}), \rho_2(\mathfrak{s})) \mid \mathfrak{s} \in \mathcal{N}^\circ\} \in \mathscr{B}$. Without loss of generality, I may assume that $\mathcal{N}'$ extends $\mathcal{N}$ by just one transition, t. The definition of a causal net ensures that ${}^\bullet t \subseteq \mathcal{N}^\circ$, and the definition of a process gives $\rho_1'({}^\bullet t) = {}^\bullet t_1$, where $t_1 := \rho_1'(t)$. Let $c := \{(\rho_1(\mathfrak{s}), \rho_2(\mathfrak{s})) \mid \mathfrak{s} \in {}^\bullet t\}$. Then $c \leq l$ and $\pi_1(c) = \rho_1({}^\bullet t) = \rho_1'({}^\bullet t) = {}^\bullet t_1$. Since $\mathscr{B}$ is an sp-bisimulation, there are a transition t_2 with $\ell(t_2) = \ell(t_1)$ and $\pi_2(c) = {}^\bullet t_2$, and a linking c' such that $\pi_1(c') = t_1{}^\bullet$, $\pi_2(c') = t_2{}^\bullet$ and $l' := l - c + c' \in \mathscr{B}$. The definition of a process gives $\rho_1'(t^\bullet) = t_1{}^\bullet$. This makes it possible to extend ρ_2 to ρ_2' so that $\rho_2'(t) := t_2$, $\rho_2'(t^\bullet) = t_2{}^\bullet$ and $c' = \{(\rho_1'(\mathfrak{s}), \rho_2'(\mathfrak{s})) \mid \mathfrak{s} \in t^\bullet\}$. Moreover, $\rho_2'({}^\bullet t) = \rho_2({}^\bullet t) = \pi_2(c) = {}^\bullet t_2$. Thus $(\mathcal{N}', \rho_2')$ is a finite process of N_2 extending $(\mathcal{N}, \rho_2)$. Furthermore, $\{(\rho_1'(\mathfrak{s}), \rho_2'(\mathfrak{s})) \mid \mathfrak{s} \in \mathcal{N}'^\circ\} = \{(\rho_1'(\mathfrak{s}), \rho_2'(\mathfrak{s})) \mid \mathfrak{s} \in \mathcal{N}^\circ - {}^\bullet t + t^\bullet\} = l - c + c' \in \mathscr{B}$. Hence $(\rho_1', \mathcal{N}', \rho_2') \in \mathscr{R}$. □

10   Relating Sp-bisimilarity to other Semantic Equivalences

In this section I place sp-bisimilarity in the spectrum of existing semantic equivalences for nets, as indicated in Figure 1.

10.1   Place Bisimilarity

The notion of a place bisimulation, defined in [1], can be reformulated as follows.

Definition 10. A *place bisimulation* is a structure preserving bisimulation of the form $\overline{B}$ (where $\overline{B}$ is defined in Section 5). Two nets $N_i = (S_i, T_i, F_i, M_i, A_i, \ell_i)$ $(i = 1, 2)$ are *strongly bisimilar*, notation $N_1 \approx_{pb} N_2$, if $A_1 = A_2$ and there is a linking l in a place bisimulation with $M_1 = \pi_1(l)$ and $M_2 = \pi_2(l)$.

It follows that $\approx_{pb}$ is finer than $\leftrightarrow_{sp}$, in the sense that place bisimilarity of two nets implies their structure preserving bisimilarity.

10.2   Occurrence Net Equivalence

Definitions of the *unfolding* for various classes of Petri nets into an *occurrence net* appear in [7,12,16,23,25,35,36]—I will not repeat them here. In all these cases, the definition directly implies that if an occurrence net $\mathcal{N}$ results from unfolding a net N then $\mathcal{N}$ is safe and there exists a folding of $\mathcal{N}$ into N (recall Definition 6) satisfying

- if $\mathcal{M}$ is a reachable marking of $\mathcal{N}$, and $t \in T$ is a transition of N with $^\bullet t \leq \rho(\mathcal{M})$ then there is a $t \in \mathcal{T}$ with $\rho(t) = t$.

Proposition 5. If such a folding from $\mathcal{N}$ to N exists, then $\mathcal{N} \leftrightarrow_{sp} N$.

Proof. The set of linkings $\mathcal{B} := \{\{(s, \rho(s)) \mid s \in \mathcal{M}\} \mid M$ a reachable marking of $\mathcal{N}\}$ is an sp-bisimulation between $\mathcal{N}$ and N. Checking this is trivial.   □

Two nets N_1 and N_2 are *occurrence net equivalent* [16] if they have isomorphic unfoldings. Since isomorphic nets are strongly bisimilar [27] and hence structure preserving bisimilar, it follows that occurrence net equivalence between nets is finer than structure preserving bisimilarity.

In [1] it is pointed out that the strong bisimilarity of Olderog "is not compatible with unfoldings": they show two nets that have isomorphic unfoldings, yet are not strongly bisimilar. However, when the net N is safe, the sp-bisimulation displayed in the proof of Proposition 5 is in fact a strong bisimulation, showing that each net is strongly bisimilar with its unfolding. This is compatible with the observation of [1] because of the non-transitivity of strong bisimilarity.

10.3   Causal Equivalence

Causal equivalence is coarser than structure preserving bisimilarity.

Theorem 4. If $N_1 \leftrightarrow_{sp} N_2$ for nets N_1 and N_2, then $N_1 \equiv_{caus} N_2$.

Proof. By Theorem 3 there exists a process-based sp-bisimulation $\mathscr{R}$ between N_1 and N_2. $\mathscr{R}$ must contain a triple $(\rho_1^0, \mathcal{N}^0, \rho_2^0)$ with $\mathcal{N}^0$ a causal net containing no transitions. So $(\mathcal{N}^0, \rho_1^0)$ and $(\mathcal{N}^0, \rho_2^0)$ are initial processes of N_1 and N_2, respectively. The net $\mathcal{N}^0$ contains isolated places only, as many as the size of the initial markings of N_1 and N_2.

Let $\mathcal{N}$ be a causal net of N_1. I have to prove that $\mathcal{N}$ is also a causal net of N_2. Without loss of generality I may assume that $\mathcal{N}^0$ is a prefix of $\mathcal{N}$, as being a causal net of a given Petri net is invariant under renaming of its places and transitions.

So N_1 has a process $\mathcal{P}_1 = (\mathcal{N}, \rho_1)$. By Proposition 4, $\mathcal{P}_1$ is the limit of a chain $\mathcal{P}_1^0 \leq \mathcal{P}_1^1 \leq \mathcal{P}_1^2 \leq \dots$ of finite processes of N_1. Moreover, for $\mathcal{P}_1^0$ one can take $(\mathcal{N}^0, \rho_1^0)$. Let $\mathcal{P}_1^i = (\mathcal{N}^i, \rho_1^i)$ for $i \in \mathbb{N}$. By induction on $i \in \mathbb{N}$, it now follows from the properties of a process-based sp-bisimulation that N_2 has processes $\mathcal{P}_2^{i+1} = (\mathcal{N}^{i+1}, \rho_2^{i+1})$, such that $(\mathcal{N}^i, \rho_2^i) \leq (\mathcal{N}^{i+1}, \rho_2^{i+1})$ and $(\rho_1^{i+1}, \mathcal{N}^{i+1}, \rho_2^{i+1}) \in \mathscr{R}$. Using Proposition 3, the limit $\mathcal{P}_2 = (\mathcal{N}, \rho_2)$ of this chain is a process of N_2, contributing the causal net $\mathcal{N}$. □

10.4 History Preserving bisimilarity

The notion of *history preserving bisimilarity* was originally proposed in [32] under the name *behavior structure bisimilarity*, studied on event structures in [13], and first defined on Petri nets, under to the individual token interpretation, in [2], under the name *fully concurrent bisimulation* equivalence.

Definition 11. [2] Let $\mathcal{N}_i = (\mathcal{S}_i, \mathcal{T}_i, \mathcal{F}_i, \mathcal{M}_{0i}, \mathcal{A}_i, \ell_i)$ $(i = 1, 2)$ be two causal nets. An *order-isomorphism* between them is a bijection $\beta : \mathcal{T}_1 \to \mathcal{T}_2$ such that $\mathcal{A}_1 = \mathcal{A}_2$, $\ell_2(\beta(t)) = \ell_1(t)$ for all $t \in \mathcal{T}_1$, and $t \, \mathcal{F}_1^+ \, u$ iff $\beta(t) \, \mathcal{F}_2^+ \, \beta(u)$ for all $t, u \in T_1$.

Definition 12. [2] A *fully concurrent bisimulation* between two nets N_1 and N_2 is a set $\mathscr{R}$ of triples $((\rho_1, \mathcal{N}_1), \beta, (\mathcal{N}_2, \rho_2))$ with $(\mathcal{N}_i, \rho_i)$ a finite process of N_i, for $i = 1, 2$, and β an order-isomorphism between $\mathcal{N}_1$ and $\mathcal{N}_2$, such that

- $\mathscr{R}$ contains a triple $((\rho_1, \mathcal{N}_1), \beta, (\mathcal{N}_2, \rho_2))$ with $\mathcal{N}_1$ containing no transitions,
- if $(\mathcal{P}_1, \beta, \mathcal{P}_2) \in \mathscr{R}$ and $\mathcal{P}_i'$ with $i \in \{1, 2\}$ is a fin. proc. of N_i extending $\mathcal{P}_i$, then N_j with $j := 3-i$ has a process $\mathcal{P}_j' \geq \mathcal{P}_j$ such that $(\mathcal{P}_1', \beta', \mathcal{P}_2') \in \mathscr{R}$ for some $\beta' \supseteq \beta$.

Write $N_1 \approx_h N_2$ or $N_1 \approx_{fcb} N_2$ iff such a bisimulation exists.

It follows immediately from the process-based characterisation of sp-bisimilarity in Section 9 that fully concurrent bisimilarity (or history preserving bisimilarity based on the individual token interpretation of nets) is coarser than sp-bisimilarity.

Theorem 5. If $N_1 \overset{\leftrightarrow}{=}_{sp} N_2$ for nets N_1 and N_2, then $N_1 \approx_{fcb} N_2$.

Proof. A process-based sp-bisimulation is simply a fully concurrent bisimulation with the extra requirement that β must be the identity relation. □

11   Inevitability for Non-reactive Systems

A run or execution of a system modelled as Petri net N can be formalised as a
path of N (defined in Section 3) or a process of N (defined in Section 8). A path
or process representing a complete run of the represented system—one that is
not just the first part of a larger run—is sometimes called a *complete* path or
process. Once a formal definition of a complete path or process is agreed upon,
an action b is *inevitable* in a net N iff each complete path (or each complete
process) of N contains a transition labelled b. In case completeness is defined
both for paths and processes, the definitions ought to be such that they give rise
to the same concept of inevitability.

The definition of which paths or processes count as being complete depends
on two factors: (1) whether actions that a net can perform by firing a transition
are fully under control of the represented system itself or (also) of the environment in which it will be running, and (2) what type of progress or fairness
assumption one postulates to guarantee that actions that are due to occur will
actually happen. In order to address (2) first, in this section I deal only with
nets in which all activity is fully under control of the represented system. In
Section 14 I will generalise the conclusions to reactive systems.

When making no progress or fairness assumptions, a system always has the
option not to progress further, and all paths and all processes are complete—
in particular initial paths and processes, containing no transitions. Consequently, no action is inevitable in any net, so each semantic equivalence respects
inevitability.

When assuming progress, but not justness or fairness, any infinite path or
process is complete, and a finite path or process is complete iff it is maximal, in
the sense that it has no proper extension. In this setting, interleaving bisimilarity,
and hence also each finer equivalence, respects inevitability. The argument is that
an interleaving bisimulation induces a relation between the paths of two related
nets N_1 and N_2, such that

- each path of N_1 is related to a path of N_2 and vice versa,
- if two paths are related, either both or neither contain a transition labelled
  b,
- if two paths are related, either both or neither of them are complete.

In the rest of this paper I will assume justness, and hence also progress,
but not (weak or strong) fairness, as explained in Section 1.4. In this setting a
process is *just* or *complete*[4] iff it is maximal, in the sense that it has no proper
extension.

Example. The net depicted on the right has a complete process performing the action a infinitely often,
but never the action b. It consumes each token that

---

[4] The term "complete" is meant to vary with the choice of a progress or fairness
   assumption; when assuming only justness, it is set to the value "just".

is initially present or stems from any firing of the transition t^a. Hence b is not inevitable. This fits with the intuition that if a transition occurrence is perpetually enabled it will eventually happen—but only when strictly adhering to the individual token interpretation of nets. Under this interpretation, each firing of t^b using a particular token is a different transition occurrence. It is possible to schedule an infinite sequence of as in such a way that none such transition occurrence is perpetually enabled from some point onwards.

When adhering to the collective token interpretation of nets, the action b would be inevitable, as in any execution scheduling as only, transition t^b is perpetually enabled. Since my structure preserving bisimulation fits within the individual token interpretation, here one either should adhere to that interpretation, or restrict attention to safe nets, where there is no difference between both interpretations.

12   History Preserving Bisimilarity does not Respect Inevitability

Consider the safe net N_1 depicted in Figure 5, and the net N_2 obtained from N_1 by exchanging for any transition t_i^b ($i>0$) the preplace s_{i-1}^1 for s^4. The net N_2 performs in parallel an infinite sequence of a-transitions (where at each step $i>0$ there is a choice between t_i^l and t_i^r) and a single b-transition (where there is a choice between t_i^b for $i>0$). In N_2 the action b is inevitable. In N_1, on the other hand, b is not inevitable, for the run of N_1 in which t_i^l is chosen over t_i^r

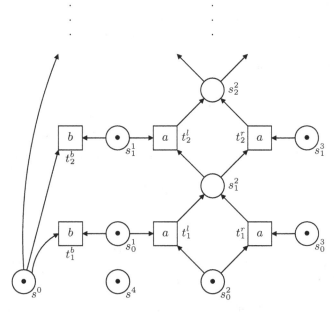

Fig. 5. A net in which the action b is not inevitable

for all $i>0$ is complete, and cannot be extended which a b-transition. Thus, each semantic equivalence that equates N_1 and N_2 fails to respect inevitability.

Theorem 6. Causal equivalence does not respect inevitability.

Proof. $N_1 \equiv_{caus} N_2$, because both nets have the same causal nets. One of these nets is depicted in Figure 6; the others are obtained by omitting the b-transition, and/or omitting all but a finite prefix of the a-transitions. □

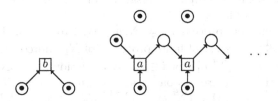

Fig. 6. A causal net of N_1 and N_2

Theorem 7. History preserving bisimilarity does not respect inevitability.

Proof. Recall that N_1 and N_2 differ only in their flow relations, and have the same set of transitions. I need to describe a fully concurrent bisimulation $\mathscr{R}$ between N_1 and N_2. $\mathscr{R}$ consists of a set of triples, each consisting of a process of N_1, a related process of N_2, and an order isomorphism between them. First of all I include all triples $(\mathcal{P}_1, \beta, \mathcal{P}_2)$ where $\mathcal{P}_1$ is an arbitrary process of N_1, $\mathcal{P}_2$ is the unique process of N_2 that induces the same set of transitions as $\mathcal{P}_1$, and β relates transition of $\mathcal{P}_1$ and $\mathcal{P}_2$ when they map to the same transition of N_i ($i=1,2$). Secondly, I include all triples $(\mathcal{P}_1, \beta, \mathcal{P}_2)$ where $\mathcal{P}_2$ is an arbitrary process of N_2 inducing both t_k^b and t_k^l for some $k>0$, and $\mathcal{P}_1$ is any process of N_1 that induces the same transitions as $\mathcal{P}_2$ except that, for some $h \geq k$ the induced transition t_h^l, if present, is replaced by t_h^r, and t_k^b is replaced by t_h^b. (β should be obvious.) It is trivial to check that the resulting relation is a fully concurrent bisimulation indeed. □

13 Structure Preserving Bisimilarity Respects Inevitability

Definition 13. A net $\mathcal{N}$ is called a *complete* causal net of a net N if it is the first component of a maximal process $(\mathcal{N}, \rho)$ of N. Two nets N_1 and N_2 are *complete causal net equivalent*, notation $\equiv_{cc}$, if they have the same complete causal nets.

Since the causal nets of a net N are completely determined by the complete causal nets of N, namely as their prefixes, $N_1 \equiv_{cc} N_2$ implies $N_1 \equiv_{caus} N_2$. It follows immediately from the definition of inevitability that $\equiv_{cc}$ respects inevitability. Thus, to prove that $\leftrightarrow_{sp}$ respects inevitability it suffices to show that $\leftrightarrow_{sp}$ is finer than $\equiv_{cc}$.

Theorem 8. If $N_1 \leftrightarrows_{sp} N_2$ for nets N_1 and N_2, then $N_1 \equiv_{cc} N_2$.

Proof. Suppose $N_1 \leftrightarrows_{sp} N_2$. By Theorem 3 there exists a process-based sp-bisimulation $\mathscr{R}$ between N_1 and N_2. $\mathscr{R}$ must contain a triple $(\rho_1^0, \mathcal{N}^0, \rho_2^0)$ with $\mathcal{N}^0$ a causal net containing no transitions. So $(\mathcal{N}^0, \rho_1^0)$ and $(\mathcal{N}^0, \rho_2^0)$ are initial processes of N_1 and N_2, respectively. The net $\mathcal{N}^0$ contains isolated places only.

Let $\mathcal{N}$ be a complete causal net of N_1. I have to prove that $\mathcal{N}$ is also a complete causal net of N_2. Without loss of generality I may assume that $\mathcal{N}^0$ is a prefix of $\mathcal{N}$, as being a complete causal net of a given Petri net is invariant under renaming of its places.

So N_1 has a complete process $\mathcal{P}_1 = (\mathcal{N}, \rho_1)$. By Proposition 4, $\mathcal{P}_1$ is the limit of a chain $\mathcal{P}_1^0 \leq \mathcal{P}_1^1 \leq \mathcal{P}_1^2 \leq \ldots$ of finite processes of N_1. Moreover, for $\mathcal{P}_1^0$ one can take $(\mathcal{N}^0, \rho_1^0)$. Let $\mathcal{P}_1^i = (\mathcal{N}^i, \rho_1^i)$ for $i \in \mathbb{N}$. By induction on $i \in \mathbb{N}$, it now follows from the properties of a process-based sp-bisimulation that N_2 has processes $\mathcal{P}_2^{i+1} = (\mathcal{N}^{i+1}, \rho_2^{i+1})$, such that $(\mathcal{N}^i, \rho_2^i) \leq (\mathcal{N}^{i+1}, \rho_2^{i+1})$ and $(\rho_1^{i+1}, \mathcal{N}^{i+1}, \rho_2^{i+1}) \in \mathscr{R}$. Using Proposition 3, the limit $\mathcal{P}_2 = (\mathcal{N}, \rho_2)$ of this chain is a process of N_2. It remains to show that $\mathcal{P}_2$ is complete.

Towards a contradiction, let $\mathcal{P}_{2u} = (\mathcal{N}_u, \rho_{2u})$ be a proper extension of $\mathcal{P}_2$, say with just one transition, u. Then ${}^\bullet u \subseteq \mathcal{N}^\circ$. By the third requirement on occurrence nets of Definition 6, their are only finitely many transitions t with $(t, u) \in \mathcal{F}_{2u}^+$. Hence one of the finite approximations $\mathcal{N}^k$ of $\mathcal{N}$ contains all these transitions. So ${}^\bullet u \subseteq (\mathcal{N}^k)^\circ$. Let, for all $i \geq k$, $\mathcal{P}_{2u}^i = (\mathcal{N}_u^i, \rho_{2u}^i)$ be the finite prefix of $\mathcal{P}_2$ that extends $\mathcal{P}_2^i$ with the single transition u. Then $\mathcal{P}_{2u}^i \leq \mathcal{P}_{2u}^{i+1}$ for all $i \geq k$, and the limit of the chain $\mathcal{P}_{2u}^k \leq \mathcal{P}_{2u}^{k+1} \leq \ldots$ is $\mathcal{P}_{2u}$. By induction on $i \in \mathbb{N}$, it now follows from the properties of a process-based sp-bisimulation that N_1 has processes $\mathcal{P}_{1u}^i = (\mathcal{N}_u^i, \rho_{1u}^i)$ for all $i \geq k$, such that $(\rho_{1u}^i, \mathcal{N}_u^i, \rho_{2u}^i) \in \mathscr{R}$, $(\mathcal{N}^k, \rho_1^k) \leq (\mathcal{N}_u^i, \rho_{1u}^i)$ and $(\mathcal{N}_u^i, \rho_{1u}^i) \leq (\mathcal{N}_u^{i+1}, \rho_{1u}^{i+1})$. Using Proposition 3, the limit $\mathcal{P}_{1u} = (\mathcal{N}_u, \rho_{1u})$ of this chain is a process of N_1. It extends $\mathcal{P}_1$ with the single transition u, contradicting the maximality of $\mathcal{P}_1$.   □

14   Inevitability for Reactive Systems

In the modelling of reactive systems, an action performed by a net is typically a synchronisation between the net itself and its environment. Such an action can take place only when the net is ready to perform it, as well as its environment. In this setting, an adequate formalisation of the concepts of justness and inevitability requires keeping track of the set of actions that from some point onwards are blocked by the environment—e.g. because the environment is not ready to partake in the synchronisation. Such actions are not required to occur eventually, even when they are perpetually enabled by the net itself. Let's speak of a Y-*environment* if Y is this set of actions. In Section 11 I restricted attention to $\emptyset$-environments, in which an action can happen as soon as it is enabled by the net in question. In [15] a path is called Y-just iff, when assuming justness, it models a complete run of the represented system in a Y-environment.

The below is a formalisation for this concept for Petri nets under the individual token interpretation.

Definition 14. A process of a net is Y-*just* or Y-*complete* it each of its proper extensions adds a transition with a label in Y.

Note that a just or complete path as defined in Section 11 is a $\emptyset$-just or $\emptyset$-complete path. In applications there often is a subset of actions that are known to be fully controlled by the system under consideration, and not by its environment. Such action are often called *non-blocking*. A typical example from process algebra [24] is the internal action τ. In such a setting, Y-environments exists only for sets of actions $Y \subseteq \mathscr{C}$, where $\mathscr{C}$ is the set of all non-non-blocking actions.

A process of a net is *complete* if it models a complete run of the represented system in some environment. This is the case iff it is Y-complete for some set $Y \subseteq \mathscr{C}$, which is the case iff it is $\mathscr{C}$-complete.

In [34], non-blocking is a property of transitions rather than actions, and non-blocking transitions are called *hot*. Transitions that are not hot are *cold*, which inspired my choice of the latter $\mathscr{C}$ above. In this setting, a process $\mathcal{P} = (\mathcal{N}, \rho)$ is complete iff the marking $\rho(\mathcal{N}^\circ)$ enables cold transitions only [34].

Definition 15. A action b is Y-*inevitable* in a net if each Y-complete process contains a transition labelled b. A semantic equivalence $\approx$ *respects* Y-*inevitability* if whenever $N_1 \approx N_2$ and b is Y-inevitable in N_1, then b is Y-inevitable in N_2. It *respects inevitability* iff it respects Y-inevitability for each $Y \subseteq \mathscr{C}$.

In Section 12 it is shown that $\equiv_{caus}$ and $\approx_h$ do not respect $\emptyset$-inevitability. From this it follows that they do not respect inevitability. In Section 13 it is shown that $\underset{sp}{\leftrightarrow}$ does respect $\emptyset$-inevitability. By means of a trivial adaptation the same proof shows that $\underset{sp}{\leftrightarrow}$ respects Y-inevitability, for arbitrary Y. All that is needed is to assume that the transition u in that proof has a label $\notin Y$. Thus $\underset{sp}{\leftrightarrow}$ respects inevitability.

15   Conclusion

This paper proposes a novel semantic equivalence for current systems represented as Petri nets: *structure preserving bisimilarity*. As a major application—the one that inspired this work—it is used to establish the agreement between the operational Petri net semantics of the process algebra CCSP as proposed by Olderog, and its denotational counterpart. An earlier semantic relation used for this purpose was Olderog's *strong bisimilarity* on safe Petri nets, but that relation failed to be transitive. I hereby conjecture that on the subclass of occurrence nets, strong bisimilarity and structure preserving bisimilarity coincide. If this it true, it follows, together with the observations of Section 6 that strong bisimilarity is included in structure preserving bisimilarity, and of Section 10.2 that each safe net is strongly bisimilar with its unfolding into an occurrence net, that on safe nets structure preserving bisimilarity is the transitive closure of strong bisimilarity.

Section 1.2 proposes nine requirements on a semantic equivalence that is used for purposes like the one above. I have shown that structure preserving bisimilarity meets eight of these requirements and conjecture that it meets the remaining one as well.

- It meets Requirement 1, that it respects branching time, as a consequence of Theorem 5, saying that it is finer than history preserving bisimilarity, which is known to be finer than interleaving bisimilarity.
- It meets Requirement 2, that it fully captures causality and concurrency (and their interplay with branching time),[5] also as a consequence of Theorem 5.
- It meets Requirement 3, that it respects inevitability (under the standard interpretation of Petri nets that assumes justness but not fairness),[5] as shown in Section 13.
- It meets Requirement 4, that it is real-time consistent, as a result of Theorem 5.
- I conjecture that it meets Requirement 5, that it is preserved under action refinement.
- It meets Requirement 6, that it is finer than causal equivalence, by Theorem 4.
- It meets Requirement 7, that it is coarser than $\equiv_{occ}$, as shown in Section 10.2.
- It meets Requirement 8, that it is a congruence for the CCSP operators, by Thm. 2.
- It meets Requirement 9, that it allows to establish agreement between the operational and denotational interpretations of CCSP operators, since it is coarser than Olderog's strong bisimilarity, as shown in Section 6.

Moreover, structure preserving bisimilarity is the first known equivalence that meets these requirements. In fact, it is the first that meets the key Requirements 3, 4, 7 and 9.

Acknowledgments. My thanks to Ursula Goltz for proofreading and valuable feedback.

References

1. Autant, C., Belmesk, Z., Schnoebelen, P.: Strong bisimilarity on nets revisited. In: Aarts, E.H.L., van Leeuwen, J., Rem, M. (eds.) Proc. PARLE 1991. LNCS, vol. 506, pp. 295–312. Springer, Heidelberg (1991)
2. Best, E., Devillers, R., Kiehn, A., Pomello, L.: Concurrent Bisimulations in Petri nets. Acta Informatica **28**, 231–264 (1991). doi:10.1007/BF01178506
3. Brookes, S.D., Hoare, C.A.R., Roscoe, A.W.: A theory of communicating sequential processes. Journal of the ACM **31**(3), 560–599 (1984). doi:10.1145/828.833
4. Castellano, L., De Michelis, G., Pomello, L.: Concurrency vs interleaving: an instructive example. Bulletin of the EATCS **31**, 12–15 (1987)

---

[5] When taking the individual token interpretation of nets, or restricting attention to safe ones

5. Degano, P., De Nicola, R., Montanari, U.: CCS is an (augmented) contact free C/E system. In: Zilli, M.V. (ed.) MMSP 1987. LNCS, vol. 280, pp. 144–165. Springer, Heidelberg (1987)

6. Emerson, E.A., Clarke, E.M.: Using Branching Time Temporal Logic to Synthesize Synchronization Skeletons. Science of Computer Programming **2**(3), 241–266 (1982). doi:10.1016/0167-6423(83)90017-5

7. Engelfriet, J.: Branching Processes of Petri Nets. Acta Informatica **28**(6), 575–591 (1991). doi:10.1007/BF01463946

8. Fehnker, A., van Glabbeek, R.J., Höfner, P., McIver, A.K., Portmann, M., Tan, W.L.: A Process Algebra for Wireless Mesh Networks used for Modelling, Verifying and Analysing AODV. Technical Report 5513, NICTA, Sydney, Australia (2013). http://arxiv.org/abs/1312.7645

9. Genrich, H., Stankiewicz-Wiechno, E.: A dictionary of some basic notions of net theory. In: Brauer, W. (ed.) Advanced Course: Net Theory and Applications. LNCS, vol. 84, pp. 519–531. Springer, Heidelberg (1980)

10. van Glabbeek, R.J.: The refinement theorem for ST-bisimulation semantics. In: Broy, M., Jones, C.B. (eds.) Proceedings IFIP TC2 Working Conference on Programming Concepts and Methods. IFIP, pp. 27–52. Springer, Heidelberg (1990)

11. van Glabbeek, R.J.: The linear time - branching time spectrum I; the semantics of concrete, sequential processes. In: Bergstra, J.A., Ponse, A., Smolka, S.A. (eds.) Handbook of Process Algebra, vol. 1, pp. 3–99. Elsevier (2001). doi:10.1016/B978-044482830-9/50019-9

12. van Glabbeek, R.J.: The individual and collective token interpretations of petri nets. In: Abadi, M., de Alfaro, L. (eds.) CONCUR 2005. LNCS, vol. 3653, pp. 323–337. Springer, Heidelberg (2005)

13. van Glabbeek, R.J., Goltz, U.: Refinement of Actions and Equivalence Notions for Concurrent Systems. Acta Informatica **37**, 229–327 (2001). doi:10.1007/s002360000041

14. van Glabbeek, R.J., Höfner, P.: CCS: It's not fair!. Acta Informatica **52**(2–3), 175–205 (2015). doi:10.1007/s00236-015-0221-6

15. van Glabbeek, R.J., Höfner, P.: Progress, Fairness and Justness in Process Algebra (2015). http://arxiv.org/abs/1501.03268

16. van Glabbeek, R.J., Vaandrager, F.W.: Petri net models for algebraic theories of concurrency. In: de Bakker, J.W., Nijman, A.J., Treleaven, P.C. (eds.) Proc. PARLE. LNCS, vol. 259, pp. 224–242. Springer, Heidelberg (1987)

17. van Glabbeek, R.J., Vaandrager, F.W.: The Difference Between Splitting in n and $n+1$. Information and Comput. **136**(2), 109–142 (1997). doi:10.1006/inco.1997.2634

18. Goltz, U., Mycroft, A.: On the relationship of CCS and Petri nets. In: Paredaens, J. (ed.) Proceedings 11th ICALP. LNCS, vol. 172, pp. 196–208. Springer, Heidelberg (1984)

19. Goltz, U., Reisig, W.: The Non-Sequential Behaviour of Petri Nets. Information and Control **57**(2–3), 125–147 (1983). doi:10.1016/S0019-9958(83)80040-0

20. Hoare, C.A.R.: Communicating Sequential Processes. Prentice Hall, Englewood Cliffs (1985)

21. Loogen, R., Goltz, U.: Modelling nondeterministic concurrent processes with event structures. Fundamenta Informaticae **14**(1), 39–74 (1991)

22. Mazurkiewicz, A.W., Ochmanski, E., Penczek, W.: Concurrent Systems and Inevitability. TCS **64**(3), 281–304 (1989). doi:10.1016/0304-3975(89)90052-2

23. Meseguer, J., Montanari, U., Sassone, V.: On the semantics of place/transition Petri nets. Mathematical Structures in Computer Science **7**(4), 359–397 (1997). doi:10.1017/S0960129597002314
24. Milner, R.: Operational and algebraic semantics of concurrent processes. In: van Leeuwen, J. (ed.) Handbook of Theoretical Computer Science, chap. 19. Elsevier Science Publishers B.V. (North-Holland), pp. 1201–1242 (1990). Alternatively see Communication and Concurrency, Prentice-Hall, Englewood Cliffs, of which an earlier version appeared as A Calculus of Communicating Systems. LNCS, vol. 92. Springer (1980). doi:10.1007/3-540-10235-3
25. Nielsen, M., Plotkin, G.D., Winskel, G.: Petri nets, event structures and domains, part I. TCS **13**(1), 85–108 (1981). doi:10.1016/0304-3975(81)90112-2
26. Olderog, E.-R.: Operational Petri net semantics for CCSP. In: Rozenberg, G. (ed.) Advances in Petri Nets 1987. LNCS, vol. 266, pp. 196–223. Springer, Heidelberg (1987)
27. Olderog, E.-R.: Nets, Terms and Formulas: Three Views of Concurrent Processes and their Relationship. Cambridge Tracts in Theor. Comp. Sc. **23**. Cambridge University Press (1991)
28. Olderog, E.-R., Hoare, C.A.R.: Specification-oriented semantics for communicating processes. Acta Informatica **23**, 9–66 (1986). doi:10.1007/BF00268075
29. Petri, C.A.: Non-sequential processes. Internal Report GMD-ISF-77.05, GMD, St. Augustin (1977)
30. Plotkin, G.D.: A Structural Approach to Operational Semantics. The Journal of Logic and Algebraic Programming **60–61**, 17–139 (2004). doi:10.1016/j.jlap.2004.05.001. Originally appeared in 1981
31. Pnueli, A.: The temporal logic of programs. In: Foundations of Computer Science (FOCS 1977), pp. 46–57. IEEE (1977). doi:10.1109/SFCS.1977.32
32. Rabinovich, A., Trakhtenbrot, B.A.: Behavior Structures and Nets. Fundamenta Informaticae **11**(4), 357–404 (1988)
33. Reisig, W. Petri nets – an introduction. In: EATCS Monographs on Theoretical Computer Science, vol. 4. Springer (1985). doi:10.1007/978-3-642-69968-9
34. Reisig, W.: Understanding Petri Nets - Modeling Techniques, Analysis Methods, Case Studies. Springer (2013). doi:10.1007/978-3-642-33278-4
35. Winskel, G.: A new definition of morphism on Petri nets. In: Fontet, M., Mehlhorn, K. (eds.) STACS 84. LNCS, vol. 166, pp. 140–150. Springer, Heidelberg (1984)
36. Winskel, G.: Event structures. In: Brauer, W., Reisig, W., Rozenberg, G. (eds.) Petri Nets: Applications and Relationships to Other Models of Concurrency. LNCS, vol. 255, pp. 325–392. Springer, Heidelberg (1986)

Logic

Translating Testing Theories
for Concurrent Systems

Jan Peleska[(⊠)]

Department of Mathematics and Computer Science,
University of Bremen, Bremen, Germany
jp@informatik.uni-bremen.de
http://informatik.uni-bremen.de/agbs

Abstract. In this article the "classical" topic of theory translation is re-visited. It is argued that the importance of this research field is currently growing fast, due to the necessity of re-using known theoretical results in the context of novel semantic frameworks. As a practical background, we consider cyber-physical systems and their development and verification in distributed collaborative environments, where multiple modelling formalisms are used for different sub-systems. For verification of the integrated system, these different views need to be integrated and consolidated as well, in order to ensure that the required emergent properties have been realised as intended. The topic is illustrated by a practical problem from the field of runtime verification. It is shown how a class of complete health monitors (i.e. checkers monitoring system behaviour) elaborated within the semantic framework of Kripke structures and LTL assertions can be re-used for runtime verification in the context of the CSP process algebra with trace/refusal specifications. We point out how crucial ideas for this theory translation have already been anticipated in Ernst-Rüdiger Olderog's early work.

Keywords: Semantics · Institutions · Cyber-physical systems · Model-based testing · Runtime verification

1 Introduction

1.1 Motivation

History is periodically re-evaluated and interpreted in the light of new events and from evolving sociological perspectives. Similarly, "classical" results from computer science are re-visited, as novel opportunities for their application arise, and growing computing power and more ingenious algorithms allow for the automated solution of more complex problems.

In this paper, we re-visit the problem of theory translation between different semantic frameworks. This problem has originally been investigated in the context of Goguen's and Burstall's *institutions*, where it was shown how assertions like *"M is a model for φ"* can be translated between different signatures, provided that certain consistency conditions for mapping semantic models M and

© Springer International Publishing Switzerland 2015
R. Meyer et al. (Eds.): Olderog-Festschrift, LNCS 9360, pp. 133–151, 2015.
DOI: 10.1007/978-3-319-23506-6_10

sentences φ are fulfilled by the signature morphisms involved [8]. The availability of such mappings allows not only for translation of single pairs of models and sentences, but also of whole theories, that is, collections of sentences already established to be valid in one semantic framework. Over the years, alternative model-theoretic approaches have been developed; we name the unifying theories of programming [10] as a prominent example.

The main motivation for us to further investigate problems of theory translation is very practical and concerns the field of cyber-physical systems (CPS) development, where distributed computational entities control physical objects. The growing complexity of CPSs and the heterogeneity of their sub-components (controllers, smart sensors, actuators, electric drives, and other electromechanical devices) suggest the application of multiple formalisms for their development. When verifying the complete system, however, methodological integration is required in order to justify that the CPS satisfies its emergent properties. Moreover, complex methods and associated tools have been developed for application in specific semantic frameworks, and it would be too expensive to re-develop these for another semantics.

1.2 Main Contributions and Overview

In this paper, theory translation techniques are illustrated by means of a problem from the field of runtime verification which is detailed in Section 2: the existence of *health monitors*[1] with specific properties, namely (1) completeness (every deviation of the monitored system from the specified behaviour is detected, while the monitor is active), (2) absence of synchronisation requirements (the health monitor can be activated at any time and will detect all errors from then on), and (3) hard realtime suitability (monitoring can be performed with bounded resources and in bounded response time). We show in the semantic framework of Kripke structures and LTL formulas, that complete, unsynchronised hard realtime health monitors exist and how they can be implemented. In Section 3, this solution is exemplified by application to nondeterministic programs as introduced by Apt, de Boer, and Olderog in [3]. These programs are interpreted in a Kripke structure semantics, so the theory elaborated before can be directly applied.

In Section 4, the existence and implementation of complete, unsynchronised hard realtime health monitors for CSP processes and trace/refusal specifications is investigated. Instead of developing a CSP-specific theory from scratch, we construct a mapping from CSP processes to nondeterministic programs, thereby introducing a model mapping associating CSP failures models with Kripke structures, and a sentence translation mapping associating LTL formulas with trace/refusal specifications. It is shown that these mappings fulfil the satisfaction condition which is required in order to consistently map theories from

---

[1] We use the term health monitor as introduced, for example, for avionic control systems in [1], for checkers monitoring the behaviour of a software component or of a complete CPS at runtime.

one semantic framework to another. As a consequence, the construction proves the existence of these health monitors for CSP processes, and the mapping from CSP to nondeterministic programs shows how CSP process simulators can be implemented as nondeterministic programs, to be monitored with the techniques introduced before.

It is noteworthy that the basic ideas of this approach have already been anticipated by Apt and Olderog around 1990 (see the first edition of [3]), when constructing a translation from distributed CSP-like sequential processes into sequential nondeterministic programs, so that properties of the former could be established by using the proof theory of the latter. The main difference between their essential ideas and the exposition presented here in this paper is that Apt and Olderog investigated this translation in a uniform framework of Kripke structures, while our construction creates mappings between the semantic domain of CSP failures models and that of Kripke structures.

Further motivation for investigating other test-related problems of theory translation is presented in the Conclusions (Section 5). References to related work are given throughout the exposition.

2   Runtime Verification and Complete, Unsynchronised Hard Realtime Health Monitors

2.1   Definition

Monitoring a software component or a complete system at runtime for the purpose of error detection is called *runtime verification* [9]. Typically, certain observations are extracted from the running system and checked against some specification of admissible behaviour. Runtime verification methods may include mechanisms for reacting on detected failures. In contrast to dynamic testing, however, it does not purposefully stimulate the SUT for provoking specific reactions to be analysed for certain test objectives. Therefore it is also called *passive testing* [2]. Methods and techniques for monitoring an SUT during its real operation have been known for a long time; they are mandatory, for example, in avionics[2] and railway control systems. In the following, we will use the shorter term *health monitor* for the error detection mechanism of an implemented runtime verification method.

The trustworthiness of health monitors can be captured by the notions of soundness and completeness, as defined, for example, in [15]:

Soundness. Whenever an observation leads to the verdict FAIL produced by the health monitor, the property under consideration has really been violated.

Completeness. The health monitor is sound, and whenever an observation could *not* have been produced by a correct property implementation, this leads to the verdict FAIL.

---

[2] Systems performing runtime verification of avionic components are called *built-in test equipment (BITE)* or *health monitors*.

When performing runtime verification on software components or very small HW/SW systems, it is usually possible to activate the checking mechanism in synchrony with the SUT. In that case the full trace (i.e. finite sequence) of observations is available to the health monitor. For large-scale systems, however, this synchrony cannot be achieved, because there does not exist a fully synchronous startup procedure, and components as well as the health monitor itself may enter the system configuration at a later point in time or may be re-started during system operation. Therefore we have suggested the concept of *unsynchronised* health monitors, whose verdict does not rely on the availability of the full observation trace, but only on a suffix thereof [13]. It is obvious that the absence of synchronisation requirements depends on the class of specifications to be checked by the health monitor against the actual system behaviour. Therefore the health monitoring problem may be re-phrased in a more formal way as

> Given a semantic framework of models $\mathcal{M}$ and specifications $\mathcal{S}$, find a health monitor design H and a subset of specifications $\mathcal{S}' \subseteq \mathcal{S}$, such that H is complete, unsynchronised, and suitable for hard realtime execution against all implementations of models from $\mathcal{M}$ and specifications from $\mathcal{S}'$.

2.2 Health Monitor Design for Kripke Structures and LTL Properties

As our first semantic framework we will now consider Kripke structures with specifications described by linear temporal logic LTL. The material presented in this section is based on [13, Section 4.5], where a complete, unsynchronised hard realtime health monitor has been constructed for a well-defined subset of LTL. This so-called $\mathbf{G}\varphi$-monitor is presented in the following paragraphs.

Recall that a *Kripke structure* is specified by a tuple $K = (S, s_0, R, L, AP)$ with state space S, initial state $s_0 \in S$, total transition relation $R \subseteq S \times S$, atomic propositions $p \in AP$, and labelling function $L : S \to 2^{AP}$. Let us suppose that K is a reference model for the expected behaviour of the SUT to be checked by means of runtime verification. The labelling function provides a means for property abstraction: instead of analysing the details of each execution state $s \in S$, it often suffices to check whether the execution fulfils the "vital properties" $L(s)$ that should hold in that state. Whole computations – that is, sequences of states $s_0.s_1.s_2 \ldots$ starting in the initial state, such that $R(s_{i-1}, s_i)$ holds for all $i > 0$ – can be abstracted to the associated sequences $L(s_0).L(s_1).L(s_2) \ldots$ of atomic proposition sets. The admissibility of these abstracted observations can be specified by LTL formulas using atomic propositions from AP as free variables. As a consequence, it is often unnecessary to explicitly specify the whole Kripke structure, including its transition relation: instead, it suffices to specify AP, s_0, an LTL formula to be respected by any valid implementation of K, and rules for extracting the validity of atomic propositions $p \in AP$ from the sequence of runtime observations.

A typical health monitor (see Fig. 1) is then structured into a lower layer abstracting concrete observations $s_0.s_1.s_2 \ldots$ (e.g. variable values) to the sets of atomic propositions they fulfil, and an upper layer checking the abstracted observation sequence $L(s_0).L(s_1).L(s_2)\ldots$ with respect to its conformity to the specification φ. The upper layer acts as a test oracle. It signals FAIL as soon as a violation of the specification formula φ has been detected and PASS if φ can be decided to be fulfilled on a finite observation trace (e.g. in the case where the SUT terminates). The output INCONCLUSIVE indicates that no violation of φ has been detected so far, but the system is still running and may still violate its specification at a later point in time.

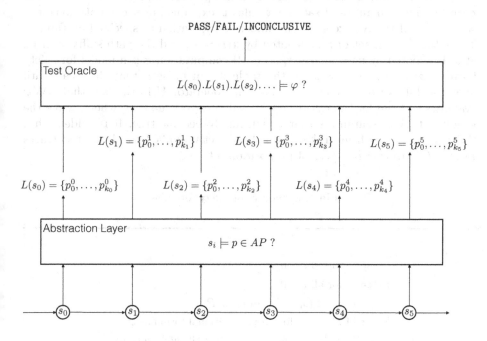

Fig. 1. Basic architecture of a health monitor.

2.3 An LTL Subclass $\mathcal{S}'$ for Unsynchronised Health Monitoring

Since runtime verification can only be performed on finite observation traces, we are only interested in specifications whose violation can be decided on a finite computation prefix, that is, we only investigate *safety formulas*. According to [20], the safety formulas of LTL can be characterised in a syntactic way as the LTL subset specified as follows.

- Every Boolean constant $\mathtt{true}(= 1)$ or $\mathtt{false}(= 0)$ is a safety formula.
- Every atomic proposition $p \in AP$ is a safety formula.

- Every propositional formula constructed from atomic propositions using $\wedge, \neg$ is a safety formula.
- If φ, ψ are safety formulas, then $\varphi \wedge \psi$, $\mathbf{X}\varphi$, and $\varphi\mathbf{W}\psi$ are safety formulas.

The above definition implies that safety LTL formulas are always *positive*: this means that none of the temporal operators $\mathbf{X}, \mathbf{W}$ (further temporal operators will be derived from these basic ones below) appear in the scope of $\neg$. Therefore negation only occurs as prefix of some propositional sub-formula. In particular, LTL formulas in negation normal form are always positive.

Temporal operator $\mathbf{X}$ is the "usual" next-operator, and $\mathbf{W}$ is the *weak until operator*. The semantics of safety formulas is explained over computations $\pi = s_0.s_1.s_2 \ldots$ of the Kripke structure under consideration as specified in Table 1. There the ith element of π is denoted by $\pi(i) = s_i$, and the path suffix starting at s_i is denoted by $\pi^i = s_i.s_{i+1}.s_{i+2}\ldots$. If computation π is a model for safety formula φ, we write $\pi \models \varphi$. In the table it can be seen that the weak until operator differs from the conventional until operator $\mathbf{U}$ in the fact that $\varphi\mathbf{W}\psi$ evaluates to `true` for the case where φ evaluates to `true` everywhere, while the validity of $\varphi\mathbf{U}\psi$ guarantees that ψ will finally become true. It is evident that the semantic rules from Table 1 can be directly applied to abstracted traces $L(\pi) = L(s_0).L(s_1).L(s_2)\ldots$ of proposition sets.

Table 1. Semantics of safety formulas.

$$\pi^i \models \texttt{true} \quad \text{for all } i \geq 0$$
$$\pi^i \not\models \texttt{false} \quad \text{for all } i \geq 0$$
$$\pi^i \models p \text{ iff } p \in L(s_i) \quad \text{for all } p \in AP$$
$$\pi^i \models \neg\varphi \text{ iff } \pi^i \not\models \varphi \quad \text{for all propositional formulas } \varphi$$
$$\pi^i \models \varphi \wedge \psi \text{ iff } \pi^i \models \varphi \text{ and } \pi^i \models \psi \quad \text{for all safety formulas } \varphi, \psi$$
$$\pi^i \models \mathbf{X}\varphi \text{ iff } \pi^{i+1} \models \varphi \quad \text{for all safety formulas } \varphi$$
$$\pi^i \models \varphi\mathbf{W}\psi \text{ iff either } \forall k \geq i : \pi^k \models \varphi$$
$$\text{or } \exists j \geq i : \pi^j \models \psi \text{ and } \forall i \leq k < j : \pi^k \models \varphi$$
$$\text{for all safety formulas } \varphi, \psi$$

More LTL operators can be defined by semantic equivalence with expressions over the safety operators specified in Table 1. These equivalences are listed in Table 2. Note that formulas involving these new operators are not necessarily safety formulas again, even if their operands are safety formulas: if φ, ψ are safety formulas, $\mathbf{G}\varphi$ is also one, but $\mathbf{F}\varphi$ and $\varphi\mathbf{U}\psi$ are not.

Table 2. LTL operators defined by semantic equivalence with safety expressions.

$$\varphi \vee \psi \equiv \neg(\neg\varphi \wedge \neg\psi) \qquad \varphi \Rightarrow \psi \equiv \neg\varphi \vee \psi \qquad \varphi \Leftrightarrow \psi \equiv (\varphi \Rightarrow \psi) \wedge (\psi \Rightarrow \varphi)$$
$$\mathbf{G}\varphi \equiv \varphi \; \mathbf{W} \; \mathtt{false} \qquad \mathbf{F}\varphi \equiv \neg\mathbf{G}\neg\varphi \qquad \varphi\mathbf{U}\psi \equiv \varphi\mathbf{W}\psi \wedge \mathbf{F}\psi$$

It is easy to see that the subclass of LTL safety formulas

$$\mathcal{S}' = \{\mathbf{G}\varphi \mid \varphi \text{ is a safety formula with temporal operators in } \{\mathbf{X}, \mathbf{W}\}\} \quad (1)$$

can be monitored starting at an arbitrary observation point $s_i.s_{i+1}.s_{i+2}\ldots$. Violations of $\mathbf{G}\varphi$ having occurred in that lost prefix $s_0.s_1.s_2\ldots s_{i-1}$ cannot be detected anymore. But starting at observation s_i, the semantics of the $\mathbf{G}$-operator implies that the monitor can start checking the validity of $\pi^i \models \varphi$, $\pi^{i+1} \models \varphi, \pi^{i+2} \models \varphi$, ..., so it will detect violations of $\mathbf{G}\varphi$ that happen at observation s_i or later on. In contrast to that, the conformity of an execution π to a proposition or a safety formula starting with $\mathbf{X}, \mathbf{W}$ can generally not be decided by a health monitor, if a prefix of π could not be observed [13, Lemma 1,p. 51]. Fortunately, the subset of safety formulas identified by Formula (1) contains the most important specification patterns, in particular, the state invariants $\mathbf{G}\varphi$ where φ is a proposition.

If π^i violates $\mathbf{G}\varphi$, then it can be seen after finitely many steps that $\pi^i \models \neg\mathbf{G}\varphi$. All negations $\neg\mathbf{G}\varphi$ can be represented as [13, Lemma 2,p. 52]

$$\neg\mathbf{G}\varphi \equiv \mathbf{F}\psi, \text{ where } \psi \text{ only contains operators in } \{\mathbf{X}, \mathbf{U}, \wedge, \vee\}$$
$$\text{or propositions containing } \{\wedge, \vee, \neg\}$$

2.4   Health Monitor H for $\mathcal{S}'$

Given an arbitrary LTL formula α, it is well known that it can be checked against a model (Kripke structure) M by transforming α into a Büchi Automaton (BA) B and checking the synchronous product of M and B with respect to emptiness, which would imply that α does *not* hold on any path of $\mathcal{M}$ [7]. This general fact is now specialised on the situation where the model M is just a computation suffix π^i, and α is a formula of the type $\mathbf{G}\varphi \in \mathcal{S}'$ as specified in (1).

Since a violation of $\mathbf{G}\varphi \in \mathcal{S}'$ according to formula (1) is uncovered if $\mathbf{F}\psi \equiv \neg\mathbf{G}\varphi$ evaluates to `true` on a finite trace segment $s_i.s_{i+1}.s_{i+2}\ldots s_k$, it suffices to check *from every state* $s_{i+j}, j \geq 0$ *on*, whether ψ is fulfilled on $s_{i+j}.s_{i+j+1}.s_{i+j+2}\ldots$. To see this, recall that the equivalence $\mathbf{F}\psi \equiv (\psi \vee \mathbf{X}\mathbf{F}\psi)$ holds and can be recursively applied to path segments $s_{i+j}.s_{i+j+1}.s_{i+j+2}\cdots$ with $j \geq 0$.

A checker for $\psi \vee \mathbf{X}\mathbf{F}\psi$ can be implemented by creating a Büchi automaton B for ψ that reads proposition sets $L(s_{i+j}) \subseteq AP, j \geq 0$ as inputs and transits into an accepting state if and only if $s_{i+j}.s_{i+j+1}.s_{i+j+2}\ldots$ fulfils ψ. The product of $s_{i+j}.s_{i+j+1}.s_{i+j+2}\ldots$ and B has path segments of state pairs

$$(s_{i+j}, \texttt{init}).(s_{i+j+1}, \mathsf{S}_{\ell_1}).(s_{i+j+2}, \mathsf{S}_{\ell_2})\ldots$$

($\texttt{init}$, S_ℓ, and $\texttt{accept}$ used below denote states of B). This product deadlocks, if B does not accept the next input $L(s_{i+j+k})$ when in state S_{ℓ_k}. If the product can execute a sequence

$$(s_{i+j}, \texttt{init}).(s_{i+j+1}, \mathsf{S}_{\ell_1}).(s_{i+j+2}, \mathsf{S}_{\ell_2})\ldots(s_{i+j+k}, \texttt{accept})$$

ending in an accepting state of B, this means that $\pi^{i+j} \models \psi$, so $\pi^i \models \mathbf{F}\psi$, and a violation of $\mathbf{G}\varphi$ has been uncovered in a finite trace segment starting at $\pi(i+j)$. Conversely, if $(s_{i+j}, \texttt{init}).(s_{i+j+1}, \mathsf{S}_{\ell_1})\ldots$ deadlocks before an accepting state of B is reached, this means that $\pi^{i+j} \not\models \psi$. But then it can still be the case that $\pi^{i+j+1} \models \psi$ or $\pi^{i+j+2} \models \psi$ and so on, so we have to check the product of *every* $\pi^{i+j}, j \geq 0$ with B.

We use the approach of [4,6] for translating LTL formulas to BA: the tools LTL2BA and LTL3BA associated with [4,6] represent the resulting automaton B in the form of a Promela Never Claim[3] which can be used as a basis for implementing a test oracle checking whether some formula ψ holds in a given state s_{i+j} of the trace segment. Instead of creating a new instance of B for every path segment $\pi^{i+j}, j \geq 0$, this can be implemented with a single representation of B, where potential actual automata states are marked, and $\texttt{FAIL}$ is raised when an accepting state of the automaton could be reached. This procedure is described by the algorithm in Table 3 which specifies the test oracle layer of the health monitor H, as indicated in the architectural model from Fig. 1.

Table 3. Test oracle algorithm of H, executing the Büchi automaton B.

---

1. **Initialisation:** set the initial state of B to 'marked', all others to 'unmarked', set output to $\texttt{INCONCLUSIVE}$
2. **Input:** the next set P of atomic propositions abstracted from the actual execution state that has been observed
3. Initialise the set M of states to be marked in this processing step with $M := \{\texttt{init}\}$
4. For all marked states S of B:
   (a) Unmark S
   (b) If S has an outgoing transition t labelled with P, add its target state to M.
   (c) If the target state of t is an accepting state, signal $\texttt{FAIL}$
5. Set all states contained in M to 'marked'
6. Continue with step 2

---

[3] See http://spinroot.com/spin/Man/never.html

By choice of formula class $\mathcal{S}'$ from (1) and the BA-based test oracle, the resulting health monitor H is complete and unsynchronised. Moreover, H's abstraction layer (see Fig. 1) can be implemented by a loop checking which propositions $p \in AP$ evaluate to **true** in the current state s_i. Since AP is finite, this can be performed with a constant amount of memory and in constant time. Finally, the test oracle operates on a single instance of the finite automaton B, and each evaluation cycle according to the algorithm in Table 3 is realised by a loop ranging over the constant number of B-states. This shows that H can be executed in hard realtime.

In the next section, an example for such a health monitor will be given.

3 Application to Nondeterministic Programs

3.1 Nondeterministic Programs in Normal Form

Let us now apply the runtime verification method introduced above to a subclass of nondeterministic programs as introduced by Apt, Olderog and de Boer in [3, pp. 349]. These programs are represented by while programs extended by Dijkstra's guarded commands

$$\textbf{if } B_1 \rightarrow S_1 \;\square\ldots\square\; B_n \rightarrow S_n \textbf{ fi}$$

representing nondeterministic choice (the B_i denote Boolean conditions, the choice fails if $\bigwedge_{i=1}^{n} \neg B_i$ holds, the S_i denote sequential program parts) and

$$\textbf{do } B_1 \rightarrow S_1 \;\square\ldots\square\; B_n \rightarrow S_n \textbf{ od}$$

representing nondeterministic repetition (the loop terminates if $\bigwedge_{i=1}^{n} \neg B_i$ holds). The formal program semantics specified in [3, Section 10.2] consists of rules $< P_1, \sigma_1 > \longrightarrow < P_2, \sigma_2 >$, where P_1, P_2 are program states and σ_1, σ_2 are valuation functions $\sigma_i : V \rightarrow D$ mapping variable symbols x to their current values $\sigma_i(x)$ in the respective program states. If the program fails due to a failure in a nondeterministic choice command, the post-state is $< E, \textbf{fail} >$, where E is the empty program and $\sigma = \textbf{fail}$ denotes the failure state of the variable valuation functions.

A subclass of interest for the investigation of health monitors consists of programs with a *normal form* structure

$$P ::= P_0; \textbf{ do } B_1 \rightarrow P_1 \;\square\; \ldots \;\square\; B_n \rightarrow P_n \textbf{ od}$$

such that the program segments P_i only contain assignments, nondeterministic choice, and terminating while-loops. This means, that there is only one nondeterministic "main loop" in the program, and no further ones inside the P_i.

3.2 Kripke Structure Semantics

While the semantics introduced in [3] was mostly applied to investigate program states on termination and possible failure conditions, we are also interested in internal "observation points", namely in program states at the beginning of each nondeterministic main loop cycle. It is assumed that we can observe each program's initial state $< P, \sigma_0 >$, its termination states $< E, \sigma >$, and the states at the beginning of each cycle, that is, at $< \textbf{do } \square_{i=1}^n B_i \rightarrow P_i \textbf{ od}, \sigma >$.

These observation points induce a Kripke structure $K = (S, s_0, R, L, AP)$ with valuation functions $s : V \cup \{pc\} \rightarrow D$ as states: each state $s \in S$, when restricted to V, corresponds to a variable valuation σ of program P. Symbol pc ("program counter") extends the domain to indicate the program state $s(pc) \in \{0, 1, 2\}$ from where the variable valuation has been obtained; $s(pc) = 0$ corresponds to initial state P, $s(pc) = 1$ corresponds to program state $\textbf{do } \square_{i=1}^n B_i \rightarrow P_i \textbf{ od}$, and $s(pc) = 2$ signifies the termination state E. We write $s = \sigma \oplus \{pc \mapsto k\}$, if s restricted to V equals program state σ, and the program counter pc has current value $k \in \{0, 1, 2\}$. If P_i is a terminating program part of P that, when activated in variable pre-state σ_i will terminate in post-state σ_j, this is denoted by $< P_i, \sigma_i > \longrightarrow^* < E, \sigma_j >$: the symbol "$\longrightarrow^*$" is the transitive closure of the transition relation for sequential while program parts containing assignments, nondeterministic choice, and terminating while-loops only. The transition relation R is then defined by

$$R = \{(\sigma_0 \oplus \{pc \mapsto 0\}, \sigma \oplus \{pc \mapsto 1\}) \mid < P_0, \sigma_0 > \longrightarrow^* < E, \sigma > \land \sigma \neq \textbf{fail}\} \cup$$
$$\{(\sigma_0 \oplus \{pc \mapsto 0\}, \textbf{fail} \oplus \{pc \mapsto 2\}) \mid < P_0, \sigma_0 > \longrightarrow^* < E, \textbf{fail} >\} \cup$$
$$\{(\sigma_1 \oplus \{pc \mapsto 1\}, \sigma_2 \oplus \{pc \mapsto 1\}) \mid \exists i \in \{1, \ldots, n\} : \sigma_1 \models B_i \land$$
$$< P_i, \sigma_1 > \longrightarrow^* < E, \sigma_2 > \land \sigma_2 \neq \textbf{fail}\} \cup$$
$$\{(\sigma_1 \oplus \{pc \mapsto 1\}, \textbf{fail} \oplus \{pc \mapsto 2\}) \mid \exists i \in \{1, \ldots, n\} : \sigma_1 \models B_i \land$$
$$< P_i, \sigma_1 > \longrightarrow^* < E, \textbf{fail} >\} \cup$$
$$\{(\sigma \oplus \{pc \mapsto 1\}, \sigma \oplus \{pc \mapsto 2\}) \mid \sigma \models \bigwedge_{i=1}^n \neg B_i\} \cup$$
$$\{(\sigma \oplus \{pc \mapsto 2\}, \sigma \oplus \{pc \mapsto 2\}) \mid \sigma = \textbf{fail} \lor \sigma \models \bigwedge_{i=1}^n \neg B_i\}$$

The transition relation R relates the initial state to every possible post-state of P_0. States where at least one of the B_i evaluates to **true** are related to all possible post-states of each of the associated P_i. States where none of the guards B_i evaluate to **true** are related to the termination state. Whenever a program fragment P_i might fail when executing from some pre-state σ_1, R relates this to the **fail**-state with pc-value 2. Since Kripke structures require total transition relations (so that all computations are infinite), we introduce stuttering for all termination states and failure states.

The elements $p \in AP$ are now atomic propositions over variable symbols from $V \cup \{pc\}$. Depending on the type of each variable, Boolean comparisons and arithmetic expressions are allowed in each p. A special atomic proposition $p = \textbf{fail}$ indicates failure. The labelling function $L : V \cup \{pc\} \rightarrow D$ is defined by the valuation functions: for all $s \in S$, atomic proposition p is in $L(s)$ if and only if $s \models p$, that is, p evaluates to **true** when replacing every symbol $x \in V \cup \{pc\}$ in p by its value $s(x)$.

3.3 Example: Health Monitor for Nondeterministic Normal Form Program

Since nondeterministic normal form programs have Kripke structures as models as explained above, the health monitors H constructed in Section 2.4 for assertions $\mathbf{G}\varphi \in \mathcal{S}'$ are directly applicable. The following example illustrates the operation of H, applied to the nondeterministic normal form program

$P ::=$ **unsigned int** $x, y, i, inputs[max]; \quad y := 0; \quad i = 0; \quad x := inputs[i];$

    do

        $i < max \wedge x > 0 \to y := 0; \quad i := i + 1; \quad x := inputs[\min\{i, max - 1\}];$

        □

        $i < max \wedge x = 0 \to y := 1; \quad i := i + 1; \quad x := inputs[\min\{i, max - 1\}];$

    od

The program operates on an array of input values; and the actual input to be processed is assigned to variable x. If the actual value of x is greater than zero, variable y (which could be a shared output variable read by another program) is set to zero, otherwise y is set to one.

    We wish to check by means of runtime verification, whether a drop of x to zero leads to a non-zero y value at the beginning of the next main loop cycle, and whether y is reset to zero after x becomes positive again. This property can be expressed by the safety formula

$$\Phi_1 \equiv \mathbf{G}\big((x > 0 \wedge \mathbf{X}y = 0)\mathbf{W}(x = 0 \wedge \mathbf{X}y \neq 0)\big)$$

which is an element of $\mathcal{S}'$ defined in (1). H observes traces

$$s_i.s_{i+1}.s_{i+2} \cdots s_k \cdots$$

of valuation functions $s_i : \{i, inputs[], x, y, pc\} \to$ **unsigned int**. The atomic propositions $AP = \{p, q, r, u\}$ define the following abstractions.

$$p \equiv x > 0, \quad q \equiv x = 0, \quad r \equiv y = 0, \quad u \equiv y > 0$$

Abstracting Φ_1 with propositions from AP results in

$$\Phi \equiv \mathbf{G}\big((p \wedge \mathbf{X}r)\mathbf{W}(q \wedge \mathbf{X}u)\big),$$

and its negation is[4]

$$\neg\Phi \equiv \mathbf{F}\big((p \vee \mathbf{X}r)\mathbf{U}((p \vee \mathbf{X}r) \wedge (q \vee \mathbf{X}u))\big)$$
$$\equiv \mathbf{F}\psi \quad \text{with}$$
$$\psi \equiv ((p \vee \mathbf{X}r) \wedge (q \vee \mathbf{X}u))$$

---

[4] Observe that $\neg(\varphi\mathbf{W}\psi) \equiv (\neg\psi\mathbf{U}\neg(\varphi \vee \psi))$ and $\mathbf{F}(\alpha\mathbf{U}\beta) \equiv \mathbf{F}\beta$, see [13, Lemma 7].

Translating ψ into a Büchi Automaton B using the LTL2BA tool results in the Promela Never Claim shown in Table 4; it is graphically depicted as a nondeterministic state machine in Fig. 2. The test oracle layer of H performs the algorithm specified above in Table 3 on B.

Table 4. Promela Never Claim for checking against formula $\psi \equiv ((p \vee \mathbf{X}r) \wedge (q \vee \mathbf{X}u))$

```
never { /* (p || X r) && (q || X u) */
accept_init:
    if
    :: (p && q) -> goto accept_all
    :: (p) -> goto accept_S1
    :: (q) -> goto accept_S2
    :: (1) -> goto accept_S3
    fi;
accept_S1:
    if
    :: (u) -> goto accept_all
    fi;
accept_S2:
    if
    :: (r) -> goto accept_all
    fi;
accept_S3:
    if
    :: (r && u) -> goto accept_all
    fi;
accept_all:
    skip
}
```

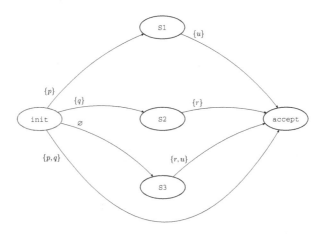

Fig. 2. Checking state machine resulting from Promela Never Claim.

The abstraction layer of H observes sequences of state valuations s_i over variables $\{i, inputs[], x, y, pc\}$ and abstracts these to sets of propositions

$P_i \subseteq AP$. The transition labels of B's state machine representation denote subsets $Q \subseteq AP$. A transition labelled by Q can be taken at observation point s_i, if its source state is active and if $Q \subseteq P_i$. Now suppose that the monitor observes the following sequence of state valuations at the beginning of each main loop, which are abstracted as shown in column P.

j	$s_{i+j}(x)$	$s_{i+j}(y)$	P	$m(\texttt{init})$	$m(\texttt{S1})$	$m(\texttt{S2})$	$m(\texttt{S3})$	$m(\texttt{accept})$
–	–	–	–	1	0	0	0	0
0	17	0	$\{p,r\}$	1	1	0	1	0
1	17	0	$\{p,r\}$	1	1	0	1	0
2	11	0	$\{p,r\}$	1	1	0	1	0
3	1	0	$\{p,r\}$	1	1	0	1	0
4	0	0	$\{q,r\}$	1	0	1	1	0
5	0	1	$\{q,u\}$	1	0	1	1	0
6	0	1	$\{q,u\}$	1	0	1	1	0
7	1	1	$\{p,u\}$	1	1	0	1	0
8	2	0	$\{p,r\}$	1	1	0	1	0

The above table also shows how the test oracle marks the states of the Büchi automaton from Fig. 2: initially, only the init-node is marked, and it will stay marked because every new observation starts an evaluation of ψ from the init-node. When the first observation is made (j = 0), $P = \{p,r\}$ enables transitions from init to S1 and from init to S3. Consequently, $m(\texttt{S1}) = m(\texttt{S3}) = 1$, and $m(\texttt{init})$ stays 1 in preparation for the next observation.

Next, observation $j = 1$ is abstracted to $P = \{p,r\}$. In S1 and S3, this observation is not accepted, so their markers are reset: this reflects the fact that $\pi^i \not\models \psi$. From init, however, $P = \{p,r\}$ enables once again transitions to S1 and S3, so these nodes are marked again. This continues until observation $j = 4$ is made. Again, the associated input $P = \{q,r\}$ is not accepted by S1 and S3, so their markers are reset. From init, P now leads to S2 and S3, and these nodes are marked accordingly. In observation $j = 5$, $P = \{q,u\}$ holds and leads again to markings of init, S2, and S3. This continues to observation $j = 7$, where P becomes $\{p,u\}$, leading again to markings of init, S1, and S3. Since y is reset to zero in observation $j = 8$, this marking is preserved, and no accepting state of the Büchi automaton can be reached.

If P contained a programming error, such as

P' ::= **unsigned int** $x, y, i, inputs[max]$; $y := 0$; $i := 0$; $x := inputs[i]$;

> **do**
>
> $\quad i < max \wedge x \geq 0 \rightarrow y := 0$; $i := i + 1$; $x := inputs[\min\{i, max - 1\}]$;
>
> $\quad \square$
>
> $\quad i < max \wedge x = 0 \rightarrow y := 1$; $i := i + 1$; $x := inputs[\min\{i, max - 1\}]$;
>
> **od**

an abstracted observation sequence $L(s_{i+j}).L(s_{i+j+1}) = \{q,r\}.\{q,r\}$ could be encountered during a program execution; and this would lead to an accepting state of the Büchi automaton, so that `FAIL` would be signalled by the test oracle.

4   Health Monitors for CSP Simulations

4.1   CSP Processes with Failures Semantics

Now suppose that we wish to verify CSP processes Q interpreted in the failures model of CSP in a semi-formal way by running simulation traces of Q and checking whether such a trace violates a given assertion

$$Q \text{ sat } S(tr, U)$$

with free variables tr denoting the trace Q has run through and U a refusal set of Q/tr. Instead of developing runtime verification support for CSP from scratch, it is more effective to make use of the existing solution elaborated above in the context of Kripke structures, the subset $\mathcal{S}'$ of LTL formulas, and nondeterministic programs.

The method to achieve this has been inspired by Apt's and Olderog's observation on distributed programs and their property-preserving transformation into nondeterministic programs [3, Chapter 11]. There it has been pointed out by the authors that certain verification objectives for CSP-like communicating sequential processes can be translated into proof obligations for nondeterministic programs. As a consequence, proof theories (as elaborated in [3, Chapter 10]) and tool support for the latter class of programs could be applied to verification tasks of the former class. Apart from the examples given in [3], this technique has been applied, for example, in [16] for the purpose of combining algebraic proof techniques with Hoare-style verification techniques using pre and post conditions.

Recall that the *failures model* of CSP introduces a denotational semantics F for each process Q as a set of failures $(tr, U) \in F$. For each failure (tr, U), tr is a trace the process may run through and U is a refusal containing events of the alphabet that may be refused by Q after having run through tr. The axioms specifying admissible semantic models F have been described and discussed in [11, 18, 19]. [5]

4.2   Model Map From Failures Models into Kripke Structures

We will now create a model map f_M from failures models into Kripke structures. To this end, we make use of the *normalisation algorithm* for CSP processes, as

---

[5] As is common in the context of testing and its formal foundations, we assume that Q is free of divergences, because black-box testing cannot distinguish internal process divergence from deadlock. As a consequence, there is no need to interpret processes Q in the more refined failures/divergence model, which is also described in [11, 18, 19].

presented in [18, Chapter 21]. There it is shown that the semantic model F of a finite state CSP process Q with alphabet A can be represented as a finite graph

$$G = (N, n_0, e : N \times A \nrightarrow N, r : N \rightarrow \mathbb{P}(\mathbb{P}(A)))$$

with nodes $n \in N$, initial node $n_0 \in N$, and labelled edges specified by a partial function e: if $(n, a) \in N \times A$ is contained in the domain of e, then there exists a uniquely defined edge from n to $e(n, a)$, which is labelled by a. Function r labels each node $n \in N$ with the set $r(n)$ of maximal refusals defined in n.[6] Every trace tr of Q is associated with exactly one finite path through G, such that edges are labelled with the events of tr. If process Q, after having performed tr, is associated with node $n \in N$, then every failure $(tr, U) \in F$ fulfils $\exists X \in r(n) : U \subseteq X$. This means that every refusal U associated with Q after tr is a subset of (at least) one of the maximal refusals n is labelled with.

Given CSP process Q with normalised transition graph $G = (N, n_0, e : N \times A \nrightarrow N, r : N \rightarrow \mathbb{P}(\mathbb{P}(A)))$, we map this graph into the nondeterministic program $\nu(Q)$ displayed in Table 5.

Table 5. Nondeterministic program $\nu(Q)$ created from CSP process Q.

$\nu(Q) ::= N\ n;\ \mathbb{P}(A)\ U;\ A^*\ tr;$

   $tr := \langle\rangle;\ n = n_0;\ \textbf{if}\ \Box_{U' \in r(n)}\textbf{true} \rightarrow U := U';\ \textbf{fi};$

do

   $\Box_{a \in A}(a \in A - U) \rightarrow$

      $tr := tr \frown \langle a \rangle;$

      $n := e(n, a);$

      $\textbf{if}\ \Box_{U' \in r(n)}\textbf{true} \rightarrow U := U';\ \textbf{fi};$

od

Program $\nu(Q)$ is in fact a simulator for CSP process Q. It uses a variable n to record the current process state Q is residing in, a variable U representing the current refusal selected at random for the current process state, and a variable tr for recording the trace the process has run through so far. The initialisation part of $\nu(Q)$ sets tr to the empty trace, n to the initial node, and selects a refusal from $r(n_0)$ – that is, a refusal associated with the initial state – at random, which is then stored in U. Program $\nu(Q)$ is constructed in such a way that (tr, U) always represents a valid failure from F, if evaluated at the beginning of a main

---

[6] Recall that the set of all refusals associated with a process state is subset closed. Therefore it is determined by the identification of its maximal refusals.

loop cycle. This is already clear for the initial arrival at the loop. Inside the loop body, an event $a \in A$ is selected at random from the subset of events that cannot be refused when in state n and refusal U applies. A deadlock situation is characterised by the condition $U = A$, whereupon the simulator terminates, since all guards of the loop evaluate to `false`. If deadlock does not occur in state n and for refusal U, one of the accepted events a is selected at random. The selection is recorded in the trace variable tr, and n is set to the uniquely defined post state $n = e(n, a)$ under event a. Then a possible maximal refusal of $e(n, a)$ is selected at random and stored again in U, and this ends the current main loop cycle.

Program $\nu(P)$ can be directly mapped into a Kripke structure with states $s : \{n, U, tr, pc\} \rightarrow D$ and atomic propositions $p \in AP$ with free variables in $\{n, U, tr, pc\}$ as described in the previous section. Thus we have constructed a model mapping f_M from CSP failures models F via normalised transition graphs and nondeterministic programs to Kripke structures.

4.3   Sentence Translation Map and Health Monitor for CSP Processes

Since $\nu(Q)$ is a nondeterministic normal form program interpreted by its Kripke structure semantics, an unsynchronised, complete, hard realtime health monitor can be constructed for $\nu(Q)$, as exemplified in Section 3.3 above. While such a health monitor exists for arbitrary formulas $\mathbf{G}\varphi \in \mathcal{S}'$ over atomic propositions with free variables from $\{n, U, tr, pc\}$, we will now restrict this class further to formulas

$$\mathbf{G}(pc = 1 \Rightarrow \varphi) \text{ such that } \varphi \text{ is a proposition with free variables in } \{tr, U\} \quad (2)$$

These formulas express assertions that are only evaluated at the beginning of each main loop cycle and can be expressed as simple propositions, without the help of temporal operators. We now define a sentence (i.e. formula) translation map f_S from safety LTL formulas of the type specified in formula (2) to CSP specifications $S(tr, U)$ by setting

$$f_S(\mathbf{G}(pc = 1 \Rightarrow \varphi)) = \varphi$$

It is straightforward to see that the whole construction process for f_M, f_S yields the following essential property:

For all formulas $\mathbf{G}(pc = 1 \Rightarrow \varphi)$ according to (2) and for all failure models F:

$$F \text{ sat } f_S(\mathbf{G}(pc = 1 \Rightarrow \varphi)) \text{ iff } f_M(F) \text{ sat } \mathbf{G}(pc = 1 \Rightarrow \varphi) \quad (3)$$

or, equivalently,

$$F \text{ sat } \varphi \text{ iff } f_M(F) \text{ sat } \mathbf{G}(pc = 1 \Rightarrow \varphi)$$

Formula (3) represents a specific case of the *satisfaction condition* required for Goguen's and Burstall's institutions [8]. The situation, however, is slightly more complicated than described there, because we map models and formulas between signatures of *different* institutions: the CSP process algebra can be considered as an institution (this has been elaborated in [14]), but this institution does not contain Kripke structures. This is typically handled by institutions with many-sorted signatures, and these can be conveniently represented by Grothendiek Institutions [5, Chapter 12]. In any case, the satisfaction condition (3) allows us to transfer theories established in the context of Kripke structures into the CSP process algebra and vice versa.

The practical consequences for our example are as follows.

– We can map any divergence-free CSP process Q via its normalised transition graph into a nondeterministic program $\nu(Q)$.
– This program can be interpreted in a Kripke structure, and therefore run-time verification can be performed in hard realtime with a complete health monitor, provided that the specification to be monitored is an LTL safety formula of the type $\mathbf{G}\varphi$ introduced above.
– The satisfaction condition (3) guarantees that original CSP process Q satisfies a specification $S(tr, U)$ over traces and refusals if and only if $\nu(Q)$ satisfies LTL safety formula $\mathbf{G}(pc = 1 \Rightarrow S(tr, U))$.
– Since $S(tr, U)$ is a proposition with free variables from $\{tr, U\}$, the unsynchronised, complete, hard realtime health monitor is applicable for checking the conformity of $\nu(Q)$'s behaviour with $\mathbf{G}(pc = 1 \Rightarrow S(tr, U))$.

For practical application, the FDR model checker[7] can be used to generate normalised transition graphs from CSP processes. Then $\nu(Q)$ is easy to program as an interpreter traversing the transition graph structure – this has been actually performed in the model-based testing tool RT-Tester for testing against CSP processes [17].

5 Conclusions

In this paper the existence of complete, unsynchronised, hard realtime health monitors for runtime verification in the semantic framework of Kripke structures and a subset of LTL formulas has been shown. The results could be directly applied to nondeterministic programs. Constructing a mapping from CSP processes interpreted in the denotational failures semantics to nondeterministic programs interpreted by Kripke models, the theory of health monitors, as well as concrete implementation guidelines could be transferred into the CSP process algebra.

The problem described in this paper is just one of a larger number of theory translation problems related to testing theories: in [12] a complete equivalence class testing theory has been developed for model-based testing against models

---

[7] http://www.cs.ox.ac.uk/projects/concurrency-tools/

interpreted as Kripke structures. The crucial step of the completeness proof consisted in creating an abstraction from Kripke structures to finite state machines (FSM). This abstraction allowed to apply an existing complete theory for FSM and "translate" it into a theory for Kripke structures. Using the theory translation approach described here we expect that it is possible to map a variety of similar valuable results from the FSM domain into more general semantic frameworks.

Acknowledgements. I would like to express my gratitude to Ernst-Rüdiger Olderog for substantial support and valuable advice during the time when I still worked in industry and became strongly attracted by formal methods in computer science.

The work presented in this paper as been elaborated within project *ITTCPS – Implementable Testing Theory for Cyber-physical Systems*[8] which has been granted by the University of Bremen in the context of the German Universities Excellence Initiative.[9]

References

1. Aeronautical Radio Inc: ARINC SPECIFICATION 653P1-2: Avionics Application Software Standard Interface, Part 1 - Required Services, December 2005
2. Andrés, C., Cavalli, A.R.: How to reduce the cost of passive testing. In: 14th International IEEE Symposium on High-Assurance Systems Engineering, HASE 2012, Omaha, NE, USA, October 25–27, 2012, pp. 209–216. IEEE Computer Society (2012). http://dx.doi.org/10.1109/HASE.2012.36
3. Apt, K.R., de Boer, F.S., Olderog, E.R.: Verification of Sequential and Concurrent Programs. Springer, Heidelberg (2010)
4. Babiak, T., Kretínský, M., Řehák, V., Strejček, J.: LTL to Büchi automata translation: fast and more deterministic. In: Flanagan, C., König, B. (eds.) TACAS 2012. LNCS, vol. 7214, pp. 95–109. Springer, Heidelberg (2012). http://link.springer.com/chapter/10.1007/978-3-642-28756-5_8
5. Diaconescu, R.: Institution-independent Model Theory. Birkhäuser Verlag AG, Basel (2008)
6. Gastin, P., Oddoux, D.: Fast LTL to Büchi automata translation. In: Berry, G., Comon, H., Finkel, A. (eds.) CAV 2001. LNCS, vol. 2102, pp. 53–65. Springer, Heidelberg (2001). http://link.springer.com/chapter/10.1007/3-540-44585-4_6
7. Gerth, R., Peled, D., Vardi, M.Y., Wolper, P.: Simple on-the-fly automatic verification of linear temporal logic. In: Dembinski, P., Sredniawa, M. (eds.) PSTV, Protocol Specification, Testing and Verification XV, Proceedings of the Fifteenth IFIP WG6.1 International Symposium on Protocol Specification, Testing and Verification, Warsaw, Poland, June 1995. IFIP Conference Proceedings, vol. 38, pp. 3–18. Chapman & Hall (1995)
8. Goguen, J.A., Burstall, R.M.: Institutions: Abstract Model Theory for Specification and Programming. J. ACM **39**(1), 95–146 (1992). http://doi.acm.org/10.1145/147508.147524

---

[8] http://www.informatik.uni-bremen.de/agbs/projects/ittcps/index.html
[9] http://en.wikipedia.org/wiki/German_Universities_Excellence_Initiative

9. Havelund, K.: Rule-based runtime verification revisited. STTT **17**(2), 143–170 (2015). http://dx.doi.org/10.1007/s10009-014-0309-2
10. Hoare, C.A.R., Jifeng, H.: Unifying Theories of Programming. Prentice-Hall (1998)
11. Hoare, T.: Communication Sequential Processes, vol. 07632. Prentice-Hall International, Englewood Cliffs (1985)
12. Huang, W.l., Peleska, J.: Complete model-based equivalence class testing. International Journal on Software Tools for Technology Transfer, 1–19 (2014). http://dx.doi.org/10.1007/s10009-014-0356-8
13. Huang, W.l., Peleska, J., Schulze, U.: Contract Support for Evolving SoS. Public Document D34.3, COMPASS (2014). http://www.compass-research.eu/deliverables.html
14. Mossakowski, T., Roggenbach, M.: Structured CSP – a process algebra as an institution. In: Fiadeiro, J.L., Schobbens, P.-Y. (eds.) WADT 2006. LNCS, vol. 4409, pp. 92–110. Springer, Heidelberg (2007). http://dx.doi.org/10.1007/978-3-540-71998-4_6
15. Netravali, A.N., Sabnani, K.K., Viswanathan, R.: Correct passive testing algorithms and complete fault coverage. In: König, H., Heiner, M., Wolisz, A. (eds.) FORTE 2003. LNCS, vol. 2767, pp. 303–318. Springer, Heidelberg (2003). http://link.springer.com/chapter/10.1007/978-3-540-39979-7_20
16. Peleska, J.: Design and verification of fault tolerant systems with csp. Distributed Computing **5**(2), 95–106 (1991). http://dx.doi.org/10.1007/BF02259751
17. Peleska, J.: Formal methods for test automation - hard real-time testing of controllers for the airbus aircraft family. In: Proc. of the Sixth Biennial World Conference on Integrated Design & Process Technology (IDPT2002), Pasadena, California, June 23–28, 2002. Society for Design and Process Science, June 2002. ISSN 1090–9389
18. Roscoe, A.W. (ed.): A Classical Mind: Essays in Honour of C.A.R. Hoare. Prentice Hall International (UK) Ltd., Hertfordshire (1994)
19. Roscoe, A.W., Hoare, C.A.R., Bird, R.: The Theory and Practice of Concurrency. Prentice Hall PTR, Upper Saddle River (1997)
20. Sistla, A.P.: Safety, liveness and fairness in temporal logic. Formal Aspects of Computing **6**(5), 495–511 (1994). http://link.springer.com/article/10.1007/BF01211865

No Need Knowing Numerous Neighbours
Towards a Realizable Interpretation of MLSL

Martin Fränzle[1], Michael R. Hansen[2], and Heinrich Ody[1]([⊠])

[1] Department of Computing Science, University of Oldenburg, Oldenburg, Germany
fraenzle@informatik.uni-oldenburg.de, heinrich.ody@uni-oldenburg.de
[2] DTU Compute, Technical University of Denmark, Lyngby, Denmark
mire@dtu.dk

Abstract. The Multi-Lane Spatial Logic MLSL introduced by Hilscher et al. in [4] is a two-dimensional spatial logic geared towards modelling and analysis of traffic situations, where the two dimensions are interpreted as the lanes of a road and the distance travelled down that road, respectively. The intended use of MLSL is for capturing (and reasoning about) guards and invariants in decision-making schemes for highly automated driving [12]. Unfortunately, the logic turns out to be undecidable [7,8,11], rendering implementability and thus the actual use of such guard conditions in real-time decision making questionable in general. We here show that under a reasonable model of technical observation of the traffic situation, the actual decidability and implementability issues take a much more pleasing form: given that an actual autonomous car can only sample state information of a finite set of environmental cars in real-time, we show that it is decidable whether truth of an arbitrary MLSL formula can be safely determined on a given sample size. For such feasible formulas, we furthermore state a procedure for determining their truth values based on such a sample.

Keywords: Highly automated driving · Real-time decision making · Spatial logic · Decidability

1 Introduction

The societal need for drastically enhancing efficiency of transportation has recently provided impetus to research on automated driving, as automation has the potential for significantly improving both the ecological footprint and the safety of road-bound traffic. While automated driving — in all its variants,

M. Fränzle—Work of the author was partially supported by Deutsche Forschungsgemeinschaft within the Transregional Collaborative Research Center SFB/TR 14 AVACS.

M.R. Hansen—Work of the author was partially supported by the Danish Research Foundation for Basic Research within the IDEA4CPS project.

H. Ody—Work of the author was partially supported by Deutsche Forschungsgemeinschaft within the Research Training Group DFG GRK 1765 SCARE.

R. Meyer et al. (Eds.): Olderog-Festschrift, LNCS 9360, pp. 152–171, 2015.
DOI: 10.1007/978-3-319-23506-6_11

ranging from low over partial to high automation — earlier was conceived as a commodity, it now is a central element of the concerted actions deemed to contribute to, e.g., the EU's "Vision Zero"[1] of eliminating fatalities in road-bound traffic. Such reliance on automation in complex and only partially observable environments, however, induces a personal and societal dependence on the reliability of automatic object detection and classification, on correctness of computerized situation interpretation, on permanent adequacy of automated decision making, and on the reliable availability and continued operation of diverse supervision and control functions. While reliability of the latter components can to quite some extent be achieved by classical means of reliability engineering, ranging from functional verification of system designs to fault-tolerance mechanisms in their implementation, the front-end tasks of object detection, object classification, and situation interpretation are prone to relatively high rates of — in general unavoidable— errors. Understanding the genesis and rigorously controlling the propagation of such errors is of utmost importance to the safety of autonomous driving, as they may lead to misconceptions and situationally inadequate decisions, which could in turn induce inappropriate control actions.

As the decision-making, being based on the situation interpretation and in turn determining which traffic maneuver to select and how to decompose it into a sequence of low-level control tasks, is the pivotal point in this propagation chain, suggestions for systematizing its design and enhancing its safety analysis by model-based principles have recently been made. The common denominator of all model-based approaches is to first establish a sufficiently detailed model of the system under design and the relevant sphere of its environment, then utilize this model for validation and verification of desired functional and non-functional properties (whereby different, viewpoint-specific models may well be used), and later on refine the design to an actual implementation. Depending on how faithful the model(s) underlying the analysis steps were — and there are many good reasons for choosing simplified, approximate models here —, the implementation step may require different forms of justification, ranging from a demonstration that it refines its design model to rigorous arguments for insensitivity of property satisfaction to the particular simplifications adopted in the analysis models. The latter problem becomes particularly prominent if the system design is not elaborated manually by domain experts, but a correct-by-construction approach is applied, where the design is (partially or fully) generated or derived from the environment model and the desired properties.

An interesting variant of this theme has been developed by Hilscher et al. in [4,5]: In a strife for separating the logical design of the decision layer from questions concerning the detailed dynamics of road objects, Hilscher et al. suggest to employ (timed) automata manipulating abstract traffic situations as their inputs and states. Such traffic situations are in turn described by a dedicated spatial logic, called Multi-Lane Spatial Logic (MLSL), able to describe quantitative spatial relations between traffic participants on the road. Using MLSL conditions

---

[1] The term "Vision Zero" originally was conceived in Sweden in 1997 as name for a long-term concerted action on road safety. It has later been adopted by the EU.

as, e.g., guards and invariants in the automata [4,8], traditional means of program verification, like reasoning over pre- and postconditions of actions, can be generalized to methods for proving maintenance of desired functional properties by the decision-making algorithms (e.g., maintaining spatial separation between cars, i.e., collision freedom). The desired separation of concerns between logical design and detailed dynamics however comes at the price of deferring questions concerning the observability and controllability of the dynamics of road objects to later stages of design; the analysis at the logical level essentially adopts a simplified model in order to realize a useful divide-and-conquer approach to design.

Demonstrating appropriate controllability can be tackled by well-established methods from control: it amounts to devising adequate low-level controllers and proving that these, whenever activated in a situation satisfying the precondition, will establish the postcondition (within the given time frame, if applicable). Such proofs can be conducted with respect to the low-level dynamics, which mostly is kinematic and hence well-understood. The problem of observability, i.e., of reliably determining whether the guard (and thus precondition) of an action actually is met, however is a fundamentally different one. It involves reasoning about situational awareness. The central question here is whether the guard can reliably be evaluated at all, i.e., whether an evaluation mechanism can be implemented which reliably determines the truth value of the guard, as necessary for actually using the decision-making strategy encoded in the control automaton. It is that later problem we will address in this article, thereby formalizing some (necessary) criteria for suitability as an implementable guard and answering decidability questions concerning these criteria.

Following the lines of Hilscher et al. [4] and extending their framework towards implementation, we exemplify our ideas on MLSL. The logic MLSL is a multi-dimensional interval logic with a discrete dimension of lanes and a continuous dimension of travel distance. It is tailored towards reasoning about traffic manoeuvres, a very specific use case, and the design of MLSL is primarily inspired by Duration Calculus [2], Propositional Interval Temporal Logic [10] and Shape Calculus [13]. The first two logics are usually used to describe temporal properties, while Shape Calculus is considered as a spatio-temporal logic. Other interval logics having some similarities with MLSL are CDT, a modal logic for chopping intervals [14] and Halpern-Shoham-logic [3], a logic based on Allen's interval relations [1]. While Duration Calculus and Shape Calculus allow quantitative reasoning, the other logics only permit qualitative reasoning.

Being inspired by interval logics, MLSL has similar strengths and weaknesses as such logics, that is, MLSL is expressive and satisfiability problems for MLSL are typically undecidable [7,8,11]. The satisfiability (and model-checking) problem for a fragment of MLSL is shown decidable in [11] when a fixed maximal bound on the number of cars in a traffic situation is imposed.

A primary application of MLSL exemplifying the above ideas of logical design of decision strategies is in the definition of lane change controllers for motorways [4,8]. These controllers are defined as extended timed automata which, additionally to clock constraints, may have MLSL formulas as transition guards.

A central implementability constraint then is that MLSL reasons about a countable infinite set of car identifiers, yet technical surveillance of the traffic situation can in real-time only harvest information about a finite set (also called a sample) of neighbouring cars. Evaluation of guard or invariant conditions employed in the lane change controller can consequently only resort to such a finite sample. Due to random effects, including arbitration protocols as well as disturbances and imperfect observation, different samples may be drawn from the same traffic situation (or it may at least be practically impossible to unambiguously describe the sample to be drawn). This immediately provokees two questions:

1. Is the evaluation of an MLSL formula used as a guard independent from the particular sample drawn, which may vary within reasonable bounds?
2. Will the evaluation of the guard on the finite sample provide reliable information on its validity over all cars, including the hidden cars, i.e., cars that are not currently observed by the equipment?

Extending MLSL with a so-called scope formula $\{c_1, c_2, \ldots, c_n\} : \phi$ which restricts quantification over cars in ϕ to a finite subset $c_1, c_2, \ldots, c_n$, we arrive at a logic in which the above two questions can be formalized and studied. The aim of this work is to get decidability results for the above questions under reasonable assumptions, thus providing a mechanism for checking suitability of an MLSL condition as a guard.

In Sect. 2 we therefore introduce our extension of MLSL called Multi-Lane Spatial Logic with Scope (MLSLS), and in Sect. 3 we formalize the above questions. In Sect. 4, it is shown that the satisfiability problem for so-called well-scoped MLSLS formulas is decidable. The technique for showing this decidability result is strongly based on [11], where the satisfiability problem for MLSL is reduced to the satisfiability problem for quantified linear mixed integer-real arithmetic (QLIRA). A difference is that while the decidability result of [11] is based on a bound on the number of cars considered, i.e., on a constraint on traffic scenes which may or may not apply in a particular situation, the decidability for well-scoped MLSLS formulas is based on a syntactical restriction on formulas. Here it suffices that every existentially quantified formula $\exists c.\psi$ occurs within the context of a scoped formula $\{c_1, c_2, \ldots, c_n\} : \phi$, thereby restricting the range of c to a finite set given by $\{c_1, c_2, \ldots, c_n\}$. In Sect. 5 we show how this decision procedure can be used in connections with reasonable model and formula assumptions to decide the questions above. Sect. 6 contains a brief summary.

2   Multi-lane Spatial Logic with Scope

In this section we introduce Multi-Lane Spatial Logic (MLSL) due to Hilscher et al. together with a novel extension called Multi-Lane Spatial Logic with Scope (MLSLS). In this extension it is possible to confine the *scope* for the cars considered in a given traffic situation. MLSLS is a conservative extension of MLSL. The definition of MLSLS is based on the definition of MLSL in [4,8]. It is simpler in the sense that we only consider spatial properties of static traffic configurations in this paper, and more complex because we introduce a scope component.

2.1   The Model

Only motorway traffic is considered here and a motorway is modelled as a two-dimensional world; the vertical discrete dimension represents the different *lanes* and the horizontal dense dimension represents the *extension* of the lanes. A *traffic snapshot* contains for every car information about the current lane of the car, which we call *reservation* and the position along the lane. Usually, a car only has a reservation for one lane, but when it is changing lanes it has reservations on two adjacent lanes. Additionally, when a car would like to change to another lane it has a *claim* for that lane.

We assume a countably infinite set of *car identifiers* $\mathbb{I}$ and an arbitrary but fixed set of lanes $\mathbb{L} = \{0, \ldots, k\}$, for some $k \in \mathbb{N}_{\geq 1}$ to be given. Let $\mathcal{P}(\mathbb{L})$ denote the powerset of $\mathbb{L}$.

Definition 1 (Traffic snapshot [4,8]). *A traffic snapshot $\mathcal{TS}$ is a structure $\mathcal{TS} = (\mathrm{res}, \mathrm{clm}, \mathrm{pos})$, where*

- res : $\mathbb{I} \to \mathcal{P}(\mathbb{L})$ *maps cars to their reserved lanes,*
- clm : $\mathbb{I} \to \mathcal{P}(\mathbb{L})$ *maps cars to their claimed lanes and*
- pos : $\mathbb{I} \to \mathbb{R}$ *maps cars to the position of their rear along the lanes.*

Furthermore, we require the following sanity conditions *to hold for all $C \in \mathbb{I}$:*

1. *Car C cannot both reserve and claim the same lane:* $\mathrm{res}(C) \cap \mathrm{clm}(C) = \emptyset$
2. *Car C can reserve at most two lanes:* $1 \leq |\mathrm{res}(C)| \leq 2$
3. *Reserved lanes must be next to each other:*

$$|\mathrm{res}(c)| = 2 \ \text{implies} \ \exists n \in \mathbb{L}.\mathrm{res}(C) = \{n, n+1\}$$

4. *Car C can claim at most one lane:* $0 \leq |\mathrm{clm}(C)| \leq 1$
5. *Car C can reserve or claim at most two lanes:* $1 \leq |\mathrm{res}(C)| + |\mathrm{clm}(C)| \leq 2$
6. *A claimed lane must be next to a reserved lane for car C:*

$$\mathrm{clm}(C) \neq \emptyset \ \text{implies} \ \exists n \in \mathbb{L}.\mathrm{res}(C) \cup \mathrm{clm}(C) = \{n, n+1\}$$

7. *Only finitely many cars participate or initiate in lane changing manoeuvres:*

$$|\mathrm{res}(C)| = 2 \ \text{or} \ |\mathrm{clm}(C)| = 1 \ \text{holds only for finitely many} \ C \in \mathbb{I}$$

We denote the set of all traffic snapshots by $\mathbb{TS}$.

To address the safety of given traffic situation, a notion of *safety envelope* of a car is introduced in [4] to capture the necessary space for a safe stop of the car. No car should interfere with the safety envelope of another car during an overtake manoeuvre, for example. The safety envelope of C in the traffic snapshot $\mathcal{TS}$ is

$$\mathrm{se}(C, \mathcal{TS}) = [\mathrm{pos}(C), \mathrm{pos}(C) + \mathrm{spacing}_C],$$

where spacing$_C$ is derived from the current speed (both absolute and relative) and an accepted temporal spacing Δt between cars, which may transiently be considerably shorter than the legal minimum spacing of 1.8 seconds required by traffic laws. Determining Δt is beyond the scope of this article, but has extensively been studied by traffic psycholgy, cf. e.g. [6].

In MLSL, properties from the perspective of a specific car called *ego* are considered. The notion *view* captures this perspective, where a view has information about the lanes, their extension and the identity of ego. Intuitively, a view is a window through which ego perceives a traffic snapshot.

Definition 2 (View). *A* view *is a structure* $V = (L, X, E)$, *where*

- $L = [l, n] \subseteq \mathbb{L}$ *is an interval of lanes that are visible in the view,*
- $X = [r, t] \subseteq \mathbb{R}$ *is the extension of the lanes that is visible in the view and*
- $E \in \mathbb{I}$ *is the identifier of the car under consideration, that is,* ego.

A subview *of V is obtained by restricting the lanes and extension we observe. Let L', X' be subintervals of L and X, then we define*

$$V^{L'} = (L', X, E) \qquad \text{and} \qquad V_{X'} = (L, X', E).$$

If $l > n$ or $r = t$ we say that the view is empty.

Let *CVar* be a set of variables ranging over car identifiers. In the logic we use a special constant ego to refer to the owner of the current view. A valuation maps variables and the special symbol ego to car identifiers, i.e., a valuation is a function $\nu : CVar \cup \{ego\} \rightarrow \mathbb{I}$. Further, we define valuation updates with the override notation $\oplus$ from Z [16] as $\nu \oplus \{c \mapsto C\}(c') = C$ if $c = c'$ and $\nu(c')$ otherwise.

A view narrows down the spatial part of the motorway to a possibly restricted set of lanes with a possibly restricted extent. We introduce the notion *scope* to the model to be able to narrow down the considered cars in a given situation. This leads to the following definition of a *model with scope*.

Definition 3 (Model with Scope). *Let $CS \subseteq \mathbb{I}$ be a set of cars, $\mathcal{TS}$ be a traffic snapshot, V be a view and ν be a valuation. Then we call $\mathcal{M} = (CS, \mathcal{TS}, V, \nu)$ a model of MLSLS with scope CS.*

Notice that a model $\mathcal{M}$ with scope $\mathbb{I}$ is a model of MLSL in the sense of [4, 8].

2.2   The Logic: MLSLS

The logic MLSL is a multi-modal, first-order logic with modalities inspired by interval logic, i.e., a *vertical chop* modality for partitioning a view into an upper and a lower subview, and a *horizontal chop* modality for partitioning a view into a left and a right subview. MLSLS extends MLSL with formulas of the form:

$$cs : \phi, \text{ for } cs \subseteq CVar,$$

where the set of cars considered when determining the truth value of ϕ is narrowed down to cars denoted by variables in cs.

Definition 4 (Syntax). *The set of* MLSLS *formulas* $\phi \in \Phi$ *is given as*

$$\phi ::= \gamma = \gamma' \mid \texttt{free} \mid \text{re}(\gamma) \mid \text{cl}(\gamma) \mid \ell = k \mid \neg\phi \mid \phi \wedge \phi \mid \exists c.\phi \mid \phi \frown \phi \mid \genfrac{}{}{0pt}{}{\phi}{\phi} \mid cs{:}\phi,$$

where $c \in CVar$, k *is a rational number, i.e.,* $k \in \mathbb{Q}$, $\gamma, \gamma' \in CVar \cup \{\text{ego}\}$, $cs \subseteq CVar$ *and* ℓ *is a special symbol denoting the length of the lanes' extension.*

Contrary to [4,8] we use rational constants as they are machine representable.

The formula $\texttt{free}$ is true for one-lane views containing no cars, $\text{re}(c)$ and $\text{cl}(c)$ are true for one-lane views that are fully covered by the safety envelope of a reservation or claim, respectively, by c. Furthermore, $\phi \frown \psi$ denotes horizontal partitioning of a view and $\genfrac{}{}{0pt}{}{\phi}{\psi}$ vertical chop of a view.

Let freeVar(ϕ) denote the set of free variables occurring in an MLSLS formula ϕ. The definition of this function is standard for the first-order fragment, so we just give the parts for ego, chopped and scoped formulas:

$$\text{freeVar}(\text{ego}) = \{\text{ego}\}$$

$$\text{freeVar}(\phi \frown \psi) = \text{freeVar}\begin{pmatrix}\phi \\ \psi\end{pmatrix} = \text{freeVar}(\phi) \cup \text{freeVar}(\psi)$$

$$\text{freeVar}(xs{:}\phi) = xs \cup \text{freeVar}(\phi)$$

Definition 5 (Semantics). *Let* $c \in CVar$, $k \in \mathbb{Q}$ *and* $\gamma, \gamma' \in CVar \cup \{\text{ego}\}$. *Given a scope* $CS \subseteq \mathbb{I}$, *a traffic snapshot* $\mathcal{TS}$, *a view* $V = ([l, n], [r, t], E)$ *and a valuation* ν *with* $\nu(\text{ego}) = E$ *we define the* satisfaction *of a formula by a model* $\mathcal{M} = (CS, \mathcal{TS}, V, \nu)$ *as follows:*

$$\mathcal{M} \models \gamma = \gamma' \quad \Leftrightarrow \quad \nu(\gamma) = \nu(\gamma')$$

$$\mathcal{M} \models \texttt{free} \quad \Leftrightarrow \quad (l \notin \text{res}(C) \cup \text{clm}(C) \text{ or } \text{se}(C, \mathcal{TS}) \cap (r, t) = \emptyset)$$
$$\qquad\qquad\qquad \textit{for every } C \in CS, \textit{ and } l = n \textit{ and } r < t$$

$$\mathcal{M} \models \text{re}(\gamma) \quad \Leftrightarrow \quad l \in \text{res}(\nu(\gamma)) \text{ and } [r, t] \subseteq \text{se}(\nu(\gamma), \mathcal{TS}) \text{ and } l = n \text{ and } r < t$$

$$\mathcal{M} \models \text{cl}(\gamma) \quad \Leftrightarrow \quad l \in \text{clm}(\nu(\gamma)) \text{ and } [r, t] \subseteq \text{se}(\nu(\gamma), \mathcal{TS}) \text{ and } l = n \text{ and } r < t$$

$$\mathcal{M} \models \ell = k \quad \Leftrightarrow \quad t - r = k$$

$$\mathcal{M} \models cs{:}\phi \quad \Leftrightarrow \quad (\{\nu(c) \mid c \in cs\}, \mathcal{TS}, V, \nu) \models \phi$$

$$\mathcal{M} \models \neg\phi \quad \Leftrightarrow \quad \mathcal{M} \not\models \phi$$

$$\mathcal{M} \models \phi_0 \wedge \phi_1 \quad \Leftrightarrow \quad \mathcal{M} \models \phi_0 \text{ and } \mathcal{M} \models \phi_1$$

$$\mathcal{M} \models \exists c.\phi \quad \Leftrightarrow \quad (CS, \mathcal{TS}, V, \nu \oplus \{c \mapsto C\}) \models \phi, \text{ for some } C \text{ in } CS$$

$$\mathcal{M} \models \phi_0 \frown \phi_1 \quad \Leftrightarrow \quad (CS, \mathcal{TS}, V_{[r,s]}, \nu) \models \phi_0 \text{ and } (CS, \mathcal{TS}, V_{[s,t]}, \nu) \models \phi_1,$$
$$\qquad\qquad\qquad \textit{for some } s, \textit{ where } r \leq s \leq t$$

$$\mathcal{M} \models \genfrac{}{}{0pt}{}{\phi_1}{\phi_0} \quad \Leftrightarrow \quad l \leq n \text{ implies}$$
$$\qquad\qquad\qquad (CS, \mathcal{TS}, V^{[l,m]}, \nu) \models \phi_0 \text{ and } (CS, \mathcal{TS}, V^{[m+1,n]}, \nu) \models \phi_1$$
$$\qquad\qquad\qquad \textit{for some } m, \textit{ where } l - 1 \leq m \leq n, \textit{ and}$$
$$\qquad\qquad\qquad l > n \text{ implies } (CS, \mathcal{TS}, V, \nu) \models \phi_0 \text{ and } (CS, \mathcal{TS}, V, \nu) \models \phi_1$$

In the semantics of the vertical chop operator $\begin{smallmatrix} \phi_1 \\ \phi_0 \end{smallmatrix}$, we deviate from the classical semantics and distinguish two cases. If the current view contains at least one lane we split the view into a lower and an upper subview and evaluate ϕ_0 on the lower subview and ϕ_1 on the upper subview. Otherwise, when the view is empty, we do not chop the view and instead evaluate both formulas on the same view. The intuition here is that all subviews of an empty view are empty and we can not distinguish different empty views with MLSLS. This special handling is necessary, because if we chop along a lane into a lower and an upper subview the lanes of the two subviews should be disjoint. However, for horizontal chops the endpoint of left subview and the startpoint of the right subview are shared.

The scope component CS of a model (CS, TS, V, ν) is used in the semantics for the formulas $\mathtt{free}$ and $\exists c. \phi$. The formula $\mathtt{free}$ holds if no car from the scope CS occupies a part of the lane under consideration, and $\exists c. \phi$ holds if ϕ holds for some car C in the scope CS.

Definition 6 (Satisfiability and Validity).

- *An MLSLS formula ϕ is satisfiable iff $(CS, TS, V, \nu) \models \phi$ holds for some scope $CS \subseteq \mathbb{I}$, traffic snapshot TS, view V, and valuation ν.*
- *An MLSLS formula ϕ is valid iff $(\mathbb{I}, TS, V, \nu) \models \phi$ holds for every traffic snapshot TS, view V, and valuation ν.*

If we disregard formulas of the form $cs : \phi$ and use $\mathbb{I}$ as scope component in models, then the above semantics coincides with that for MLSL.

We make use of the standard first order abbreviations such as $\mathrm{true}, \mathrm{false}, \vee, \forall$. In addition we define

$$\ell \geq k \equiv \ell = k \frown \mathrm{true} \ ,$$

denoting the fact that the extension is longer than or equal to k. It is now easy to define $\ell < k, \ell \leq k$, and $\ell > k$.

To derive a similar constraint for the lane dimension: $\mathcal{L} = 1$, i.e., the number of lanes in the current view is one, we use the formula $\mathtt{free}$:

$$\mathcal{L} = 1 \equiv \{\} : \mathtt{free}$$

together with the empty scope. Inspecting the semantics we see that this formula is true for a model with view $([l, n], [r, t], E)$ iff $l = n$ and $r < t$, i.e., it is required that a lane has a positive extent. Further relations on the number of lanes can be derived using vertical chop, for example:

$$\mathcal{L} = 2 \equiv \begin{pmatrix} \mathcal{L} = 1 \\ \mathcal{L} = 1 \end{pmatrix}$$

The abbreviation $\langle \phi \rangle$ expresses that there is a subview satisfying ϕ:

$$\langle \phi \rangle \equiv \mathrm{true} \frown \begin{pmatrix} \mathrm{true} \\ \phi \\ \mathrm{true} \end{pmatrix} \frown \mathrm{true}$$

The following example of a specification formula expresses that any two cars on the same lane should always keep clear by at least 4 distance units (generally taken to be meters):

$$\neg \langle \exists c, d. \mathrm{re}(c) \frown \ell < 4 \frown \mathrm{re}(d) \rangle$$

3   Technical Observability and Stable Models

While MLSL reasons about a countable infinite set $\mathbb{I}$ of (unique) car identifiers, technical surveillance of the traffic situation by the ego car can in real-time and thus in situ only harvest information about a finite set $\mathbb{S} \subset \mathbb{I}$ of neighbouring cars. Evaluation of guard or invariant conditions employed in decision making can consequently only resort to state information pertaining to the perceived set $\mathbb{S}$ of cars. Let us assume that the particular sample $\mathbb{S} \subset \mathbb{I}$ drawn satisfies some reasonable constraints that the on-board sensing and data-harvesting subsystems of the ego car can guarantee. Then this leads to the question

1. whether evaluating such a guard or invariant condition instrumental to decision making is independent from the particular sample $\mathbb{S} \subset \mathbb{I}$ drawn and
2. whether evaluating that condition on a sample $\mathbb{S} \subset \mathbb{I}$ provides reliable information on its validity over $\mathbb{I}$ itself, including the hidden states of cars present in $\mathbb{I} \setminus \mathbb{S}$, yet not detected and observed.

Note that the constraints do not determine a single sample, but rather exclude samples that we do not expect to observe. In the rest of this section we first formalize these properties and then give an example.

We assume that $\mathbb{S}$ has a fixed maximal size $|\mathbb{S}| \leq N \in \mathbb{N}$ imposed by the real-time constraints on harvesting environmental information via measurements by the ego car and via car2x communication. We furthermore assume that we know which sample sets $\mathbb{S}$ may arise in a given situation, which in turn is represented by an omniscient traffic snapshot $\mathcal{TS}$. That is, we assume a relation $consistent \subset \mathbb{TS} \times \mathcal{P}(\mathbb{I})$, where $consistent(\mathcal{TS}, \mathbb{S})$ captures the relation between overall traffic situations $\mathcal{TS}$ and samples $\mathbb{S} \subset \mathbb{I}$ that may arise due to technical surveillance within that particular situation. We use MLSLS to express the consistency relation.

Definition 7 (Consistency Constraint). *Let $\bar{c} \equiv \mathrm{ego}, c_2, \ldots, c_N$ be a vector of car variables. A* consistency constraint *is an MLSLS formula $consistent(\bar{c})$ which has $c_2, \ldots, c_N$ as free variables. In any satisfying model of $consistent(\bar{c})$, the assignments to $\bar{c}$ constitute a* consistent sample *for the traffic snapshot.*

Note that a sample may contain less than N identifiers. In this case some of the variables from $\bar{c}$ are mapped to the same car. This consistency formula can be considered a requirements specification for the in-car equipment used for sensing cars in the neighbourhood and for data harvesting.

Now we can formalize properties 1. and 2. from above in MLSLS. A formula is *stable* iff on all models, the evaluation of its truth value does not depend on the particular consistent sample drawn. Further, a formula is *strongly stable* iff it always evaluates to the same value on any consistent sample as over the omniscient traffic snapshot.

Definition 8 (Stability under sampling). *Let* $\bar{c} \equiv$ *ego*$, c_2, \ldots, c_N$ *and* $\bar{c}' \equiv$ *ego*$, c'_2, \ldots, c'_N$ *be two vectors of car variables, where* $c_2, \ldots, c_N$ *and* $c'_2, \ldots, c'_N$ *are mutually distinct and let consistent*$(\bar{c})$ *be a consistency constraint, as defined above. Then an MLSLS formula* ϕ *is called* stable *iff*

$$(consistent(\bar{c}) \wedge consistent(\bar{c}')) \implies (\{\bar{c}\} : \phi \iff \{\bar{c}'\} : \phi) \text{ is valid,} \qquad (1)$$

and it is called strongly stable *iff*

$$consistent(\bar{c}) \implies (\{\bar{c}\} : \phi \iff \phi) \text{ is valid.} \qquad (2)$$

3.1 An Example of Stability Under Sampling

In the following, we give an example of a consistency constraint and an MLSL formula and argue that that particular formula is stable under sampling. As abbreviations we introduce

$$\text{notObs}(\bar{c}, c) \equiv \bigwedge\nolimits_{c' \in \bar{c}} c \neq c' \wedge (\text{re}(c) \vee \text{cl}(c)),$$

$$\text{someObs}(\bar{c}) \equiv \bigvee\nolimits_{c' \in \bar{c}} (\text{re}(c') \vee \text{cl}(c')),$$

where notObs$(\bar{c}, c)$ holds if the extension $[r, t]$ of the considered lane is covered by a reservation or claim of a car c that does not belong to the sample, and someObs$(\bar{c})$ holds if the extension $[r, t]$ of the considered lane is covered by a reservation or claim of some sampled car (possibly ego itself).

The following onservability constraint, later used for defining consistency of a sample, is organized into three groups: The first concerns cars that are definitely not observed, the second and third groups concern observed cars ahead of the ego car, where the second addresses the same lane as the ego lane, while the last concerns the lanes next to the ego lane:

1. Cars beyond a certain distance, e.g. 500 m or two lanes, are never observed, be it due to physical limits of sensors or to filtering mechanisms in car2x communication aiming at confining communication bandwidth.

$$\text{Con}_1(\bar{c}) \equiv \neg \left(\begin{array}{c} (\langle \text{re}(ego) \rangle \frown \ell \geq 500 \frown \langle \text{someObs}(\bar{c}) \rangle) \\ \vee \left(\begin{array}{c} \langle \text{re}(ego) \rangle \\ \mathcal{L} \geq 2 \\ \langle \text{someObs}(\bar{c}) \rangle \end{array} \right) \vee \left(\begin{array}{c} \langle \text{someObs}(\bar{c}) \rangle \\ \mathcal{L} \geq 2 \\ \langle \text{re}(ego) \rangle \end{array} \right) \end{array} \right)$$

2. Within a distance of 250 m from ego, it is not the case that a car on the own lane is not sampled while another car further away from ego is sampled.[2]

$$\text{Con}_2(\bar{c}) \equiv \neg(\langle \text{re}(ego) \frown \ell \leq 250 \frown \exists c.\text{notObs}(\bar{c}, c) \rangle \frown \langle \text{someObs}(\bar{c}) \rangle)$$

---

[2] The intuition is that within a range of 250m, all cars are observed unless occluded by another car closer to ego.

3. All cars on neighboring lanes within 100 m of ego belong to the sample.

$$\mathrm{Con}_3(\bar{c}) \equiv \neg \left(\begin{array}{c} \left\langle \left(\begin{array}{c} \mathcal{L} = 1 \\ \mathrm{re(ego)} \end{array} \right) \frown \ell \leq 100 \frown \left(\begin{array}{c} \exists c.\mathrm{notObs}(\bar{c}, c) \\ \mathcal{L} = 1 \end{array} \right) \right\rangle \\ \vee \\ \left\langle \left(\begin{array}{c} \mathrm{re(ego)} \\ \mathcal{L} = 1 \end{array} \right) \frown \ell \leq 100 \frown \left(\begin{array}{c} \mathcal{L} = 1 \\ \exists c.\mathrm{notObs}(\bar{c}, c) \end{array} \right) \right\rangle \end{array} \right)$$

Constraints concerning the region behind ego could be organized in a similar fashion. We omit them here for the sake of brevity. Given the above constraints on possible observations, we define the overall consistency constraint as

$$\mathrm{consistent}(\bar{c}) \equiv \bigwedge_{i=1}^{3} \mathrm{Con}_i(\bar{c}) \tag{3}$$

Consider next a guard on a transition of a control automaton that should be taken when an overtake manoeuvre is initiated. This guard should ensure that an overtake manoeuvre is meaningful and safe. It is meaningful when there is a car in front of ego (on the same lane) within 35 m, and it is safe when the lane to the left of ego is free for at least 100 m.[3] This leads to the following specification:

$$\mathrm{guard} \equiv \left\langle \begin{array}{c} (\mathbf{free} \wedge \ell \geq 100) \frown \mathrm{true} \\ \mathrm{re(ego)} \frown (\mathbf{free} \wedge \ell < 35) \frown \exists c.\mathrm{re}(c) \frown \mathrm{true} \end{array} \right\rangle$$

In order to draw a driving decision for or against overtaking, this guard will have to be evaluated on a given sample for a given traffic snapshot.

In order to check whether the guard is stable when evaluated over consistent samples up to size $N = 20$, we let $\bar{c} \equiv \mathrm{ego}, c_2, \ldots, c_{20}$ and $\bar{c}' \equiv \mathrm{ego}, c_2', \ldots, c_{20}'$ be two vectors of car variables. We instantiate the stability formula from Definition 8 as

$$(\mathrm{consistent}(\bar{c}) \wedge \mathrm{consistent}(\bar{c}')) \implies (\{\bar{c}\}:\mathrm{guard} \iff \{\bar{c}'\}:\mathrm{guard})$$

and check whether it is satisfied by all models of the form $(\mathbb{I}, \mathcal{TS}, V, \nu)$.

To see that the MLSL formula guard is stable w.r.t. the consistency constraint from (3), observe that when $(\mathbb{I}, \mathcal{TS}, V, \nu) \models \mathrm{consistent}(\bar{c}) \wedge \mathrm{consistent}(\bar{c}')$ holds, then ν contains an assignments to $\bar{c}, \bar{c}'$, which induces two consistent samples $\mathbb{S}$ and $\mathbb{S}'$. Furthermore, we know that Con_3 ensures that $\mathbb{S}$ and $\mathbb{S}'$ both contain all cars on adjacent lanes within 100 m in front of ego. Similarly, we know from Con_2 that a car in front of us will be detected if in the 35 m range. As guard reasons about space at most 100 m from ego on the left lane, and at most 35 m from ego on the own lane, we can deduce that $(\mathbb{I}, \mathcal{TS}, V, \nu) \models \{\bar{c}\}:\mathrm{guard} \iff \{\bar{c}'\}:\mathrm{guard}$ holds for each consistent pair of samples $\mathbb{S}$ and $\mathbb{S}'$. The MLSL formula guard consequently is stable. A similar argument shows that it is strongly stable.

---

[3] Safety under these conditions obviously applies to the low-speed regime only. This example specification therefore is not a complete one, yet should be considered being part of a larger overall scenario.

4   Satisfiability of MLSLS

In this section, we give a decision procedure for deciding a subset of MLSLS. To do so we transform formulas to constraints belonging to quantified linear integer-real arithmetic (QLIRA), for which the satisfiability problem is decidable [9,15]. In the considered fragment, scoped formulas are used to enforce that there is a fixed bound on the number of cars that need consideration. In particular, it is required that the formulas $\texttt{free}$ and $\exists c.\phi$ occur only inside a scoped formula. Such formulas are called *well-scoped formulas*.

Definition 9 (Well-scoped MLSLS formulas). *The set of* well-scoped MLSLS *formulas* $\phi \in \Phi_W$ *is generated by the following grammar:*

$$\phi ::= A \mid \neg\phi \mid \phi \wedge \phi \mid \phi \frown \phi \mid \begin{smallmatrix} \phi \\ \phi \end{smallmatrix} \mid cs\!:\!\phi',$$

$$A ::= \ell = k \mid \gamma = \gamma' \mid \mathrm{re}(\gamma) \mid \mathrm{cl}(\gamma)$$

$$\phi' ::= \texttt{free} \mid \exists c.\phi' \mid A \mid \neg\phi' \mid \phi' \wedge \phi' \mid \phi' \frown \phi' \mid \begin{smallmatrix} \phi' \\ \phi' \end{smallmatrix} \mid cs\!:\!\phi',$$

where $c \in CVar$, $cs \subseteq CVar$ *is finite,* $k \in \mathbb{Q}$ *and* $\gamma, \gamma' \in CVar \cup \{\text{ego}\}$.

In QLIRA we use variables ranging over the real numbers, as well as the operations of linear arithmetic and rounding to the next smaller integer.

Definition 10 (Formulas of QLIRA). *The set of* QLIRA *formulas* $\psi \in \Psi$ *is generated by the following grammar:*

$$\psi ::= \exists x \in \mathbb{R}.\psi \mid term \leq term \mid \neg\psi \mid \psi \wedge \psi,$$
$$term ::= k \mid x \mid \lfloor x \rfloor \mid term + term,$$

where $x \in RVar$ *(a set of variables ranging over real numbers) and* $k \in \mathbb{Q}$.

We use the remaining propositional connectives and $=, <, \geq$ and $>$ as abbreviations. Furthermore, $\exists i \in \mathbb{N}.\psi$ is an abbreviation for $\exists i \in \mathbb{R}.i = \lfloor i \rfloor \wedge 0 \leq i \wedge \psi$. When the constraint $i = \lfloor i \rfloor \wedge 0 \leq i$ is associated with a variable i, then i ranges over natural numbers and we say that $i \in NVar$, that is, $NVar$ is a set of variables ranging over natural numbers.

For terms $term_j, j \in [0,3]$, and terms $term_0, \ldots, term_k$, $k \in \mathbb{N}_{\geq 1}$, we define

$$[term_0, term_1] \subseteq [term_2, term_3] \equiv term_2 \leq term_0 \wedge term_1 \leq term_3,$$

$$term_0 \in \{term_1, \ldots, term_k\} \equiv \bigvee_{term_j \in \{term_1, \ldots, term_k\}} term_j = term_0,$$

$$[term_0, term_1] \cap (term_2, term_3) = \emptyset \equiv term_1 \leq term_2 \vee term_3 \leq term_0.$$

Note that both intervals in the subset definition are closed, but for the intersection definition the interval $(term_2, term_3)$ is open.

4.1   A QLIRA Representation of a Traffic Snapshot $\mathcal{TS}$

It is now described how the satisfiability problem for well-scoped formulas ϕ is reduced to satisfiability of QLIRA formulas. Variables of QLIRA are introduced so that the various components of a model $(CS, \mathcal{TS}, V, \nu)$, for $V = ([l, n], [r, t], E)$, can be represented in QLIRA, and so that the translation function "mimics" the definition of the semantics relation $\models$ in Definition 5. A key issue is the QLIRA representation of a traffic snapshot.

Let ϕ be a well-scoped formula with free variables $cs = \{c_0, c_1, \ldots, c_{n-1}\} = \text{freeVar}(\phi)$. Then, due to the structure of well-scoped formulas, the number of free variables n is a bound on the number of cars necessary to consider when checking for the satisfiability of ϕ. Hence, only a finite traffic snapshot needs to be represented when checking for satisfiability. To do so, we introduce n natural number variables of QLIRA $C^0, C^1, \ldots, C^{n-1}$ representing n cars. Furthermore, let $f_{\text{init}} : cs \to NVar$ be defined by

$$f_{\text{init}}(c_i) = C^i \tag{4}$$

The spatial information for each of these cars, say $C^i \in NVar$, is represented by five QLIRA variables: C^i_{pos}, C^i_{res}, $C^i_{\text{res}'}$, C^i_{clm}, and C^i_{spacing}, for the position, lane reserved, alternative lane reserved, lane claimed, and size of the safety envelope. Hence, prior to the translation of ϕ a table is created containing n entries $(C^i, (C^i_{\text{pos}}, C^i_{\text{res}}, C^i_{\text{res}'}, C^i_{\text{clm}}, C^i_{\text{spacing}}))$ with QLIRA variables for n cars.

Variables C^i_{pos} and C^i_{spacing} range over the reals and variables C^i_{res} range over natural numbers. Variables $C^i_{\text{res}'}, C^i_{\text{clm}}$ range over natural numbers denoting a lane or may take a special value, say $\text{nil} = -2$ denoting no reservation or no claim. Technically this is enforced by associating a constraint of the form $(x = \lfloor x \rfloor \wedge 0 \leq x) \vee x = -2$ with each such variable.

To meaningfully represent a traffic snapshot, these variables must satisfy properties such as $C^i_{\text{spacing}} > 0$ and if two distinct variables C^i and C^j denote the same car, then the characterizing variables for C^i and C^j must agree. Such properties can be formulated in QLIRA, for example:

$$C^i = C^j \implies \left(\begin{array}{c} C^i_{\text{pos}} = C^j_{\text{pos}} \wedge C^i_{\text{spacing}} = C^j_{\text{spacing}} \\ \wedge\ C^i_{\text{res}} = C^j_{\text{res}} \wedge C^i_{\text{res}'} = C^j_{\text{res}'} \wedge C^i_{\text{clm}} = C^j_{\text{clm}} \\ \wedge\ (C^i_{\text{res}'} = \text{nil} \vee C^i_{\text{clm}} = \text{nil}) \end{array} \right)$$

Notice that also the sanity constraints of Definition 1 can be expressed in QLIRA using the variables introduced. We will not give further details here but just assume the existence of a QLIRA formula, named "sanity", capturing the sanity constraints for the QLIRA representation (like the formula above) as well as the sanity constraints for traffic snapshots.

4.2   Translating Well-Scoped MLSLS Formulas to QLIRA

The translation function from well-scoped formulas to QLIRA should "mimic" the definition of the semantic relation $\models$ in Definition 5. Inspecting this semantics, it is observed that the traffic snapshot $\mathcal{TS}$ and the ego part E of a view

$V = ([l,n],[r,t],E)$ remain constants throughout the recursive definition of $(CS,\mathcal{TS},V,\nu) \models \phi$.

Hence, the translation function must keep track of the

- scope part CS, i.e. a subset of $\{C^0, C^1, \ldots, C^{n-1}\}$,
- the lane part $[l,n]$ of V, i.e. two natural number variables of QLIRA,
- the extent part $[r,t]$ of V, i.e. two real number variables of QLIRA, and
- the valuation part ν, i.e. a function with type: $f : cs \to \{C^0, C^1, \ldots, C^{n-1}\}$.

The part that may change during translation is modelled by the type T:

$$T = \mathcal{P}(NVar) \times NVar \times NVar \times RVar \times RVar \times ((CVar \cup \{ego\}) \to NVar)$$

Definition 11 (Transformation). *The transformation is given by a function*

$$tr : T \times \varPhi_W \to \varPsi.$$

Let $\varUpsilon = (CS, i, i', x, x', f) \in T$, $k \in \mathbb{Q}$, $\gamma, \gamma' \in CVar \cup \{ego\}$ and $c \in CVar$. Then the transformation is given as:

$$
\begin{aligned}
tr(\varUpsilon, \mathrm{re}(\gamma)) :={}& x' > x \wedge [x, x'] \subseteq [C_{\mathrm{pos}}, C_{\mathrm{pos}} + C_{\mathrm{spacing}}] \wedge \\
& i = i' \wedge (i = C_{\mathrm{res}} \vee i = C_{\mathrm{res}'}), \text{ where } C = f(\gamma) \\
tr(\varUpsilon, \mathrm{cl}(\gamma)) :={}& x' > x \wedge [x, x'] \subseteq [C_{\mathrm{pos}}, C_{\mathrm{pos}} + C_{\mathrm{spacing}}] \wedge \\
& i = i' \wedge i = C_{\mathrm{clm}}, \text{ where } C = f(\gamma) \\
tr(\varUpsilon, \mathbf{free}) :={}& i = i' \wedge x' > x \wedge \bigwedge_{C \in CS} \left(\begin{array}{l} i \notin \{C_{\mathrm{res}}, C_{\mathrm{res}'}, C_{\mathrm{clm}}\} \\ \vee \\ {[x,x']} \cap [C_{\mathrm{pos}}, C_{\mathrm{pos}} + C_{\mathrm{spacing}}] = \emptyset \end{array} \right) \\
tr(\varUpsilon, \ell = k) :={}& x' - x = k \\
tr(\varUpsilon, \gamma = \gamma') :={}& f(\gamma) = f(\gamma') \\
tr(\varUpsilon, cs : \phi) :={}& tr((\{f(c) \mid c \in cs\}, i, i', x, x', f), \phi) \\
tr(\varUpsilon, \phi_0 \wedge \phi_1) :={}& tr(\varUpsilon, \phi_0) \wedge tr(\varUpsilon, \phi_1) \\
tr(\varUpsilon, \neg\phi) :={}& \neg tr(\varUpsilon, \phi) \\
tr(\varUpsilon, \exists c.\phi) :={}& \bigvee_{C \in CS} tr((CS, i, i', x, x', f \oplus \{c \mapsto C\}), \phi) \\
tr(\varUpsilon, \phi_0 \frown \phi_1) :={}& \exists x'' \in \mathbb{R}. x \le x'' \le x' \wedge \\
& \quad tr((CS, i, i', x, x'', f), \phi_0) \wedge tr((CS, i, i', x'', x', f), \phi_1) \\
& \text{where } x'' \text{ is a fresh QLIRA variable}
\end{aligned}
$$

$$
tr\left(\varUpsilon, \frac{\phi_1}{\phi_0}\right) := \left(\begin{array}{l} \left(i \le i' \implies \exists i'' \in \mathbb{N}. \left(\begin{array}{l} i - 1 \le i'' \le i' \\ \wedge\, tr((CS, i, i'', x, x', f), \phi_0) \\ \wedge\, tr((CS, i'' + 1, i', x, x', f), \phi_1) \end{array} \right) \right) \\ \wedge \\ \left(i > i' \implies \left(\begin{array}{l} tr((CS, i, i', x, x', f), \phi_0) \\ \wedge\, tr((CS, i, i', x, x', f), \phi_1) \end{array} \right) \right) \end{array} \right)
$$

where i'' is a fresh QLIRA variable

Notice that the translation function is a direct reflection of the definition of semantic relation $\models$.

We define

$$F(\phi) \equiv \text{sanity} \wedge tr((\emptyset, i, i', x, x', f_{\text{init}}), \phi),$$

where the formula "sanity" is the QLIRA formula mentioned above that expresses the sanity constraints on traffic snapshots and sanity constraints on the encoding and f_{init} is given by (4). Since the QLIRA constraint $F(\phi)$ follows the structure of Definition 5, it is equisatisfiable to the well-scoped MLSLS formula ϕ, which is stated in the following theorem.

Corollary 1. *Given a well-scoped MLSLS formula ϕ we can effectively create QLIRA constraints $F(\phi)$ such that*

$$\phi \text{ is satisfiable} \quad \textit{iff} \quad F(\phi) \text{ is satisfiable}.$$

A direct consequence is decidability of well-scoped MLSLS:

Theorem 1 (Decidability of well-scoped MLSLS). *Satisfiability and validity of well-scoped MLSLS are decidable.*

Proof. Well-scoped MLSLS is closed under negation. Hence, both satisfiability and validity can be reduced to satisfiability problems. Given a well-scoped MLSLS formula ϕ, Corollary 1 permits generating a QLIRA constraint $F(\phi)$ which is equisatisfiable to ϕ. Satisfiabilty of $F(\phi)$, and thus equivalently satisfiability of ϕ, is decidable due to decidability of QLIRA. $\square$

5   Deciding Stability

We now turn to the problem of deciding whether an MLSLS formula ϕ is stable, as defined by condition (1) in Definition 7, i.e., of deciding whether the following formula is valid:

$$(consistent(\bar{c}) \wedge consistent(\bar{c}')) \implies (\{\bar{c}\}:\phi \iff \{\bar{c}'\}:\phi)$$

Observing that the satisfiablity problem of well-scoped MLSLS is decidable, we address the above problem by adequately scoping formula (1) by successive introduction of scope operators. These introduced scopes must be in connection with subformulas of the form: $\exists c.\xi$ (cf. Definition 9), as such existential quantifiers may otherwise range over an unbounded number of cars. When scopes are introduced for such formulas, special care must be taken concerning possible occurrences of the formula free in ξ. The reason is the universal quantification over all cars in the semantic definition of free (cf. Definition 5), which causes problems when the particular occurrence is outside of any scope operator.

We note two properties of the scope operator which permit its introduction:

Lemma 1 (Scope introduction). *Let ϕ be an MLSLS formula containing a positive (or negative, respectively) occurrence of some subformula $\psi \equiv \exists c.\xi$. Furthermore assume that ψ occurs outside any scope operator in ϕ and that the formula* free *does not occur in ξ outside a scope operator. Let ϕ' be the formula that is obtained by replacing ψ in ϕ by $\{d_1, \ldots, d_n\} : \psi$, where $d_1, \ldots, d_n$ are fresh variable names. Then validity of ϕ' is a sufficient (necessary, resp.) condition for validity of ϕ.*

Proof. The freshly introduced scope operator confines the range of the existential quantifier to a subrange of the car identifiers, which strengthens or weakens, respectively, the overall formula depending on polarity of the quantifier occurrence. □

Due to decidability of well-scoped formulas (Theorem 1), we obtain a safe, yet incomplete method for checking stability under sampling: in our exemplary consistency formulas Con_1 to Con_3, we do only encounter universal statements over objects outside the sample,[4] which we consider to be the general form.[5] Furthermore, observe that the formula free does not occur in the example consistency formula at all and that it occurs only properly scoped within our exemplary stability formula.

For such consistency formulas, we can proceed as follows:

1. Build formula (1). Within this formula, existential quantifiers outside scope operators do only occur positively.
2. Scope these quantifiers by arbitrarily large scopes according to Lemma 1.
3. Decide validity of the resulting well-scoped formula using the procedure of Theorem 1.
4. If the formula is valid then report "stable" and stop. This is sound as validity of the scoped formula is a sufficient condition for validity of the original stability condition (1) according to Lemma 1.
5. Else go back to Step 2 and repeat with larger scopes.

Whenever this procedure terminates, it constitutes a constructive proof of stability of the formula under investigation.

In most cases, we can however do better than such blind search for sufficiently large scopes, as the range of technical perception tends to be bounded. This means that there is a fixed range around the ego car outside which we do not expect cars to show up in the sample. In such cases, a lossless scoping of the consistency predicate by introduction of (generally very large) scope operators is possible, which in

---

[4] The only quantifiers not ranging over samples are existential quantifiers in negative context.

[5] To this end please note that there is no need to express that a sampled object actually exists in the outside world, as this has been built into the semantics. Existential statements about objects outside the sample therefore seem to be of no practical value.

turn will be exploited for deciding whether the actual (and generally much smaller) sample still is large enough for stably determining truth values.

To get there, we first note some semantic properties of scope operators characterizing situations where introduction of scope operators does not affect satisfaction.

Lemma 2. *Let $\mathcal{TS}$ be a traffic snapshot, $V = ([l, n], [r, t], E)$ be a view and ν be a valuation, and denote by*

$$I_V = \{C \mid C \in \mathbb{I} \wedge [r, t] \cap \text{se}(C, \mathcal{TS}) \neq \emptyset \wedge [l, n] \cap (\text{res}(C) \cup \text{clm}(C)) \neq \emptyset\}$$

the set of cars visible within the current view. Let ϕ be an MLSLS formula. Then

$$(\mathbb{I}, \mathcal{TS}, V, \nu) \models \phi \iff \exists CS = I_V \uplus S \subset \mathbb{I}.|S| = m \wedge (CS, \mathcal{TS}, V, \nu) \models \phi,$$

where m is the number of quantifiers in ϕ and $\uplus$ denotes disjoint union. That is ϕ holds over all cars iff there is a satisfying sample containing all cars in the view plus as many extra cars as there are quantifiers.

Proof. By induction on the structure of ϕ. For the induction start observe that the only atomic formula which is influenced by scoping is **free**. It is easy to check that the theorem holds when ϕ itself is the formula **free**, because the semantics of **free** depends only on the cars in the view. The only interesting case left for the induction step is quantification, as the semantics of the remaining constructs is not influenced by scoping. Therefore assume in the remainder that $\phi = \exists c.\psi$.

To show the implication from left to right, assume that $(\mathbb{I}, \mathcal{TS}, V, \nu) \models \exists c.\psi$. Then there exists $C \in \mathbb{I}$ such that $(\mathbb{I}, \mathcal{TS}, V, \nu \oplus \{c \to C\}) \models \psi$. By induction hypothesis, as ψ has one quantifier less, there is $CS' = I_V \uplus S' \subset \mathbb{I}$ with $|S'| = m - 1$ such that $(CS', \mathcal{TS}, V, \nu \oplus \{c \to C\}) \models \psi$. If $C \in CS'$ then $C \in CS' \cup \{D\}$ for an arbitrary $D \in \mathbb{I} \setminus CS'$ and consequently $(CS' \cup \{D\}, \mathcal{TS}, V, \nu) \models \exists c.\psi$. If $C \notin CS'$ then especially $C \notin I_V$. Thus, from $(\mathbb{I}, \mathcal{TS}, V, \nu) \models \exists c.\psi$ we can conclude that binding c to C does not affect any of the spatial subformulas of ψ. For the equations between car identifiers observe that C is distinct from all the identifiers in CS'. Consequently $(CS' \cup \{C\}, \mathcal{TS}, V, \nu) \models \exists c.\psi$ holds again.

Let us now assume that $(\mathbb{I}, \mathcal{TS}, V, \nu) \not\models \exists c.\psi$. Then for all $C \in \mathbb{I}$ we have $(\mathbb{I}, \mathcal{TS}, V, \nu \oplus \{c \to C\}) \not\models \psi$. By induction hypothesis, as ψ has one quantifier less, $(CS', \mathcal{TS}, V, \nu \oplus \{c \to C\}) \not\models \psi$ holds for all $CS' = I_V \uplus S' \subset \mathbb{I}$ with $|S'| = m - 1$. As this does in particular apply for all such CS' with $C \notin CS'$, we can conclude $(CS, \mathcal{TS}, V, \nu) \not\models \exists c.\psi$ for all $CS = I_V \uplus S$ with $|S| = m$.

Consequently the bi-implication holds. □

To achieve decidability of stability, we introduce the assumption that an area of bounded size contains at most a bounded number of cars, which is trivially true due to the given geometric extent of cars.[6]

---

[6] The critical reader may object that the geometry of a car may change rather arbitrarily upon a severe crash. This argument, while true, however is irrelevant here, as the decision-making aims at maintaining safety such that its verification conditions invariantly apply to states in the pre-crash phase. Once a crash is encountered, the safety system has failed anyhow and there is no need to analyse any further.

Assumption 1. *Assume that a part of a motorway of length s with k lanes contains at most $n = \text{bound}(s, k)$ different cars, for some monotonic function* bound.

A direct consequence of this assumption is the following.

Corollary 2. *Let $\mathcal{TS}$ be a traffic snapshot, V be a view, ν be a valuation and n be the maximal number of cars fitting into V. Let ϕ be an MLSLS formula. Then $(\mathbb{I}, \mathcal{TS}, V, \nu) \models \phi$ iff there is a valuation ν' extending ν such that $(\mathbb{I}, \mathcal{TS}, V, \nu') \models \{c_1, \ldots, c_n, d_1, \ldots, d_m\} : \phi$, where m is the number of quantifiers in ϕ.*

We say that a quantifier $\exists c.\phi$ is unscoped if it ranges over the full range of car identifiers rather than just a finite sample. The crucial point of the previous corollary is to evaluate unscoped quantifiers only on a fixed area. We fix this area to ego and then we can introduce scopes to a large subset of MLSLS without losing completeness for this subset.

Lemma 3 (Exact scope introduction). *Let ϕ be an MLSLS formula where each unscoped quantifier occurs within a context of the form $(\eta \wedge \ell \sim_1 k_1 \wedge \mathcal{L} \sim_2 k_2) \odot \psi$, which in turn occurs positively in the overall formula. Here, $\eta \in \{\text{re(ego)}, \text{cl(ego)}, \langle \text{re(ego)} \rangle, \langle \text{cl(ego)} \rangle\}$, each k_i is a constant and $\sim_i \in \{<, \leq, =\}$, for $i \in \{1, 2\}$, and $\odot \in \{\implies, \wedge\}$. Let K_1 and K_2 be the largest such constants occurring in the formula. Then*

$$\phi \text{ is equisatisfiable to } \{c_1, \ldots, c_n, d_1, \ldots, d_m\} : \phi$$

where n is the maximum number of cars fitting a view V of length $2K_1$ and width $2K_2 - 1$ according to Assumption 1 and m is the number of quantifiers in ϕ.

Proof. Note that all subviews satisfying the "guarding" conditions ($\eta \wedge \ell \sim_1 k_1 \wedge \mathcal{L} \sim_2 k_2$) have to contain the ego car due to η and are thus within a range of $[-K_1, K_1]$ around the ego car position and within the range of $[-K_2 + 1, K_2 - 1]$ lanes around the ego car lane. The statement then follows from the previous. □

Note that this lemma permits to convert certain partially scoped formulas into equisatisfiable well-scoped formulas. Due to decidability of well-scoped formulas, this gives rise to the following decidability result.

Theorem 2 (Decidability of stability). *Let ϕ be an MLSLS formula and let* consistent$(\bar{c})$ *be a consistency predicate for which each unscoped quantifier occurs within a context of the form: $(\eta \wedge \ell \sim_1 k_1 \wedge \mathcal{L} \sim_2 k_2) \odot \psi$, which in turn occurs negatively in the overall formulas. Here, $\eta \in \{\text{re(ego)}, \text{cl(ego)}, \langle \text{re(ego)} \rangle, \langle \text{cl(ego)} \rangle\}$, k_i is constant and $\sim_i \in \{<, \leq, =\}$, for $i \in \{1, 2\}$, and $\odot \in \{\implies, \wedge\}$. Then it is decidable whether ϕ is stable under sampling with the consistency predicate* consistent$(\bar{c})$.

Proof. Stability under sampling is logically characterized by validity of formula (1). Under the preconditions of the theorem, the negation of formula (1) can be rewritten to an equisatisfiable MLSLS formula in scoped form according to Lemma 3. As satisfiablity of scoped MLSLS formulas is decidable due to Theorem 1, the claim follows. □

We show that the exemplary consistency conditions from Sect. 3 fall into the fragment of MLSLS such that stability under sampling can be decided according to Theorem 2. We first define $\phi \equiv \mathrm{re(ego)} \frown \ell \leq 250 \frown \exists c.\mathrm{notObs}(\bar{c}, c)$ and then we can introduce upper bounds into $\mathrm{Con}_2(\bar{c}) \equiv \neg(\langle \langle \phi \rangle \frown \langle \mathrm{someObs}(\bar{c}) \rangle)$ from Sect. 3 such that

$$\mathrm{Con}_2(\bar{c}) \iff \neg((\langle \langle \phi \rangle \wedge \ell \leq 251 \wedge \mathcal{L} = 1) \frown \langle \mathrm{someObs}(\bar{c}) \rangle) \ ,$$

which holds because views satisfying reservations can be arbitrary small, as long as the view has length greater zero, and the formula after the last chop is a somewhere formula. Note that instead of 251 we can use any number greater 250. Therefore, consistency constraint $\mathrm{Con}_2(\bar{c})$ can be properly scoped according to Lemma 3, and so can consistency constraint $\mathrm{Con}_3(\bar{c})$ using a similar technique.

In many cases, it even is possible to decide strong stability

$$consistent(\bar{c}) \implies (\{\bar{c}\} : \phi \iff \phi) \ .$$

As an example consider this formula instantiated with $\phi \equiv \mathrm{guard}$, as given in Sect. 3.1. This formula contains an unscoped occurrence of the formula

$$\mathrm{guard} \equiv \left\langle \begin{array}{c} (\mathbf{free} \wedge \ell \geq 100) \frown \mathrm{true} \\ \mathrm{re(ego)} \frown (\mathbf{free} \wedge \ell < 35) \frown \exists c.\mathrm{re}(c) \frown \mathrm{true} \end{array} \right\rangle \ .$$

According to to Lemma 1, $\{\bar{c}\} : \mathrm{guard}$ is a sufficient condition for guard s.t. we can reduce the strong stability condition (2) to

$$consistent(\bar{c}) \implies (\{\bar{c}\} : \phi \impliedby \phi) \ .$$

Here, the unscoped ϕ occurs in negative context only and the formula can thus be decided according to Theorem 2 after adequately rewriting ϕ by the means demonstrated above on the consistency formulas.

6   Conclusion

Multi-Lane Spatial Logic (MLSL) has been suggested as a means of increasing the level of abstraction in the design of decision-making algorithms for automated driving. This abstractness, however, comes at the price of raising concerns about implementability of its concepts. A crucial such concern is whether safe evaluation of guard conditions formulated in MLSL is technically feasible, given that technical observation of the environment in a car can only represent part of the environmental objects, which is in stark contrast to the omniscient perspective taken by MLSL's standard semantics. To address this problem, we have defined a conservative extension of MLSL called MLSL with scope (MLSLS), which permits to formulate observability constraints in the logic itself. As the relevant subset of well-scoped MLSL formulas is decidable, as demonstrated by a reduction to QLIRA developed herein, questions of suitability of an MLSL constraint as a technically realizable guard can be answered mechanically.

Acknowledgments. The authors are grateful to their dear colleague Ernst-Rüdiger Olderog for many years of friendship, support, and scientific inspiration. Without him as a continuous source of intriguing ideas, not only the work reported herein, but many results that we personally build our scientific reputation upon would not have come about.

We would also like to thank the editors of this volume, R. Meyer, A. Platzer, and H. Wehrheim, for the opportunity to prepare this article and publish it in a volume dedicated to Ernst-Rüdiger on the coccasion of his 60th birthday.

References

1. Allen, J.F.: Maintaining knowledge about temporal intervals. Communications of the ACM **26**(11), 832–843 (1983)
2. Chaochen, Z., Hoare, C.A.R., Ravn, A.P.: A calculus of durations. Information Processing Letters **40**(5), 269–276 (1991)
3. Halpern, J.Y., Shoham, Y.: A propositional modal logic of time intervals. Journal of the ACM (JACM) **38**(4), 935–962 (1991)
4. Hilscher, M., Linker, S., Olderog, E.-R., Ravn, A.P.: An Abstract Model for Proving Safety of Multi-lane Traffic Manoeuvres. In: Qin, S., Qiu, Z. (eds.) ICFEM 2011. LNCS, vol. 6991, pp. 404–419. Springer, Heidelberg (2011)
5. Hilscher, M., Linker, S., Olderog, E.-R.: Proving Safety of Traffic Manoeuvres on Country Roads. In: Liu, Z., Woodcock, J., Zhu, H. (eds.) Theories of Programming and Formal Methods. LNCS, vol. 8051, pp. 196–212. Springer, Heidelberg (2013)
6. Klebelsberg, D.: Verkehrspsychologie. Springer (2013)
7. Linker, S.: Proofs for traffic safety : combining diagrams and logic. Ph.D. thesis, Carl von Ossietzky University of Oldenburg (2015)
8. Linker, S., Hilscher, M.: Proof Theory of a Multi-Lane Spatial Logic. In: Liu, Z., Woodcock, J., Zhu, H. (eds.) ICTAC 2013. LNCS, vol. 8049, pp. 231–248. Springer, Heidelberg (2013)
9. Monniaux, D.: A Quantifier Elimination Algorithm for Linear Real Arithmetic. In: Cervesato, I., Veith, H., Voronkov, A. (eds.) LPAR 2008. LNCS (LNAI), vol. 5330, pp. 243–257. Springer, Heidelberg (2008)
10. Moszkowski, B.: A temporal logic for multi-level reasoning about hardware. IEEE Computer **18**(2), 10–19 (1985)
11. Ody, H.: Analysing decision problems of multi-lane spatial logic (2015) (manuscript). http://theoretica.informatik.uni-oldenburg.de/ sefie/files/decidability.pdf
12. Olderog, E.-R., Ravn, A.P., Wisniewski, R.: Linking Discrete and Continuous Models (2014) (manuscript)
13. Schäfer, A.: A Calculus for Shapes in Time and Space. In: Liu, Z., Araki, K. (eds.) ICTAC 2004. LNCS, vol. 3407, pp. 463–477. Springer, Heidelberg (2005)
14. Venema, Y.: A modal logic for chopping intervals. Journal of Logic and Computation **1**(4), 453–476 (1991)
15. Weispfenning, V.: Mixed real-integer linear quantifier elimination. In: ISSAC, pp. 129–136. ACM (1999)
16. Woodcock, J., Davies, J.: Using Z – Specification, Refinement, and Proof. Prentice Hall (1996)

Automated Reasoning Building Blocks

Christoph Weidenbach[(✉)]

Max Planck Institute for Informatics, Saarbrücken, Germany
weidenbach@mpi-inf.mpg.de

Abstract. There are automated reasoning building blocks shared between the prime calculi for propositional and first-order logic with equality, conflict driven clause learning (CDCL) and superposition, respectively. In this paper I identify these building blocks by a projection of superposition to propositional logic. Underlying both calculi is a partial model assumption guiding ordered resolution inferences that are not redundant.

1 Introduction

Superposition [3,4,13,19] and CDCL (Conflict Driven Clause Learning) [6,10, 11,16] are the prime calculi for first-order and propositional logic, respectively. While the superposition calculus was the result of an evolution of first-order logic calculi theory research, the progress on CDCL was driven by system development and experimental evaluation. Whereas superposition semi-decides unsatisfiability of first-order clause sets, CDCL decides satisfiability of propositional clause sets. Satisfiability of a first-order clause set is undecidable. Because of the latter, syntactic restrictions on inferences and effective syntactic redundancy criteria are state-of-the-art for today's superposition implementations. Still, the technique for proving superposition completeness relies on an explicit model assumption, a candidate model, that is unfortunately not effective in the first-order context. The candidate model is build with respect to a fixed, total, well-founded ordering on atoms, literals, and clauses. If sufficiently many superposition inferences are performed, the candidate model turns into a model for the overall clause set. For the success of CDCL an explicit, partial model assumption, again a candidate model, is one key ingredient. The candidate model is built via decisions, i.e., guessing the truth value of a literal, and propagations, i.e., forcing the truth value of a literal in order to satisfy a certain clause with respect to the candidate model so far. During a CDCL run the partial model assumption can then either be extended to an overall model or there exists a clause that is false in the model assumption guiding a resolution inference.

It seems that the two calculi don't have much in common. However, in this paper I show that CDCL can be seen as a variant of a projection of superposition to propositional logic. The main contributions of this paper are: (i) a generalized superposition model operator preserving important superposition properties (Definition 17, Lemma 18, Theorem 19), (ii) clauses learned by the CDCL calculus are not redundant, (iii) resolution inferences performed by the

© Springer International Publishing Switzerland 2015
R. Meyer et al. (Eds.): Olderog-Festschrift, LNCS 9360, pp. 172–188, 2015.
DOI: 10.1007/978-3-319-23506-6_12

CDCL calculus are actually superposition inferences, (iv) the superposition and CDCL model assumptions are identical (all Theorem 20). Ordering restrictions, model guided inferences and a compatible redundancy notion are important building blocks for automated reasoning systems. The results are discussed in more detail in Section 6. The paper starts with fixing the basic notions and notation, Section 2. The following Section 3 projects the standard superposition calculus to propositional logic. Next CDCL is introduced in form of an abstract rewrite system, Section 4, where as a first contribution I show that CDCL does not learn the same clause twice, Lemma 13. The two sections 3 and 4 are then the basis for a detailed comparison of CDCL and superposition, Section 5.

The paper focuses on the relationship between a candidate model, resolution inferences, a redundancy notion, and ordering restrictions for superposition and CDCL. In order to turn both calculi into sate-of-the-art systems a lot more is needed including sophisticated algorithms and data-structures enabling efficient implementations. These aspects are far beyond the scope of this paper.

2   Preliminaries

The background for CDCL as well as superposition is propositional clause logic. A propositional clause language is built over a set Σ of propositional variables, $\neg$ denotes negation, $\vee$ disjunction, $\wedge$ conjunction, $\perp$ false and $\top$ true. For propositional variables I write P, Q. A propositional variable is an *atom*. An atom $P \in \Sigma$ or its negation $\neg P$ is a *literal*. For literals I write L, K. The function atom maps a literal to its respective atom and the function comp a literal to its complementary literal, i.e., $\text{comp}(P) = \neg P$ and $\text{comp}(\neg P) = P$ for all $P \in \Sigma$. A *clause* is a finite disjunction of literals where I identify the disjunction $L_1 \vee \ldots \vee L_n$ and the multiset $\{L_1, \ldots, L_n\}$. The empty clause $\emptyset$ corresponds to $\perp$. For clauses I write C, D. A clause set is interpreted as the conjunction of all its clauses. I identify clause sets and clause sequences depending on whether the ordering plays a role or not. For clause sets (sequences) I write N, M, U.

An (partial) interpretation $\mathcal{I}$ is a set (or a sequence) of literals such that no complementary literals occur in $\mathcal{I}$. The relation "entails" $\models$ and the notions of a model, (un)satisfiability, validity are defined as usual. So $\mathcal{I} \models L$ if $L \in \mathcal{I}$ and the relation is not defined if neither L nor its complement occurs in $\mathcal{I}$. The relation $\models$ is extended accordingly to clauses, clause sets, e.g., $N \models C$ holds if for every interpretation $\mathcal{I}$ with $\mathcal{I} \models N$ it holds $\mathcal{I} \models C$. An interpretation $\mathcal{I}$ can be partial in the sense that not all atoms from Σ or atoms contained in a clause set N are defined by $\mathcal{I}$ or that $\mathcal{I}$ is only a partial model for some clause set N, i.e., it satisfies only a subset of N. A Herbrand interpretation $\mathcal{H}$ is a set (or a sequence) of atoms. A Herbrand interpretation is always total because $\mathcal{H} \models P$ if $P \in \mathcal{H}$ and $\mathcal{H} \models \neg P$ if $P \notin \mathcal{H}$. For example, $\mathcal{H} = \{P\}$ then $\mathcal{H} \models \neg Q$ and thus corresponds to the interpretation $\mathcal{I} = \{P, \neg Q\}$ for $\Sigma = \{P, Q\}$.

The calculi in the paper are defined in the form of rewrite systems with respect to a rewrite relation $\Rightarrow$. A calculus is *complete* with respect to satisfiability, if for any clause set N that is satisfiable, it finds a model. It is *strongly complete* if for any clause set N and interpretation $\mathcal{I} \models N$ the calculus finds $\mathcal{I}$. It is *sound* if whenever the calculus finds a model for N it actually is a model.

3 Propositional Superposition

Superposition was originally developed for first-order logic with equality [3,4,13,19]. It can be seen as an refinement of the traditional resolution calculus, where, in particular, resolution inferences are restricted to maximal literals with respect to an ordering. Here I introduce its projection to propositional logic. Superposition tests unsatisfiability of a finite clause set. For propositional clauses it is guaranteed to terminate either by deriving $\perp$, or by generating a so called saturated clause set where no more inferences are needed to be performed, see Definition 6, below. Compared to the resolution calculus [15] superposition adds (i) ordering restrictions on inferences, (ii) an abstract redundancy notion, (iii) the notion of a (partial) model, based on the ordering for inference guidance, and (iv) a saturation concept.

Definition 1 (Clause Ordering). Let $\prec$ be a total strict ordering on Σ. Then $\prec$ can be lifted to a total ordering on literals $\prec_L$ by $\prec \subseteq \prec_L$ and $P \prec_L \neg P$ and $\neg P \prec_L Q$, $\neg P \prec_L \neg Q$ for all $P \prec Q$. The ordering $\prec_L$ can be lifted to a total ordering on clauses $\prec_C$ by considering the multiset extension of $\prec_L$ for clauses.

For example, if $P \prec Q$, then $P \prec_L \neg P \prec_L Q \prec_L \neg Q$ and $P \vee Q \prec_C P \vee Q \vee Q \prec_C \neg Q$ because $\{P, Q\} \prec_L^{mul} \{P, Q, Q\} \prec_L^{mul} \{\neg Q\}$.

Eventually, I overload $\prec$ with $\prec_L$ and $\prec_C$. So if $\prec$ is applied to literals it denotes $\prec_L$, if it is applied to clauses, it denotes $\prec_C$. Recall that $\prec$ is a total ordering on literals and clauses as well. Eventually we will restrict inferences to maximal literals with respect to $\prec$. A literal $L \in C$ is *maximal* in C if there is no larger literal with respect to $\prec$ in C. It is *strictly maximal* if it is maximal and there are no duplicate occurrences of L in C. For a clause set N, I define $N^{\prec C} = \{D \in N \mid D \prec C\}$.

Proposition 2 (Properties of the Clause Ordering).
1. The orderings on literals and clauses are total and well-founded.
2. Let C and D be clauses with $P = \mathrm{atom}(\max(C))$, $Q = \mathrm{atom}(\max(D))$, where $\max(C)$ denotes the maximal literal in C with respect to $\prec_L$:
  1. If $Q \prec_L P$ then $D \prec_C C$.
  2. If $P = Q$, $\neg P = \max(C)$, $P = \max(D)$, then $D \prec_C C$.

Definition 3 (Abstract Redundancy). A clause C is *redundant* with respect to a clause set N if $N^{\prec C} \models C$.

Tautologies are redundant, because they are entailed by any clause set. Subsumed clauses are redundant for strict subset relationship. If $C \subset D$, then D is redundant in the presence of C. Duplicate clauses are anyway eliminated quietly because the calculus operates on sets of clauses.

Note that for finite propositional N, and any $C \in N$ redundancy $N^{\prec C} \models C$ can be decided but is as hard as testing unsatisfiability for a clause set N. So the goal is to invent useful redundancy notions that can be efficiently decided.

Definition 4 (Partial Model Construction). Given a clause set N and an ordering $\prec$ the (partial) Herbrand model $N_\mathcal{I}$ for N is constructed inductively as follows:

$$N_C := \bigcup_{D \prec C}^{D \in N} \delta_D$$

$$\delta_D := \begin{cases} \{P\} & \text{if } D = D' \vee P, P \text{ strictly maximal, and } N_D \models \neg D' \\ \emptyset & \text{otherwise} \end{cases}$$

$$N_\mathcal{I} := \bigcup_{C \in N} \delta_C$$

Clauses C with $\delta_C \neq \emptyset$ are called *productive*. Recall that atoms not contained in $N_\mathcal{I}$ are false. The operator only extends the model by an atom P if it is forced to by a clause $C \vee P$ where C is false and P maximal. Therefore, it constructs minimal models with respect to the subset relationship.

Please properly distinguish: N is a set of clauses interpreted as the conjunction of all clauses. $N^{\prec C}$ is of set of clauses from N strictly smaller than C with respect to $\prec$. $N_\mathcal{I}$, N_C are Herbrand interpretations. $N_\mathcal{I}$ is the overall (partial) Herbrand model for N, whereas N_C is generated from all clauses from N strictly smaller than C.

Proposition 5. *Some properties of the partial model construction.*

1. For every D with $(C \vee \neg P) \prec D$ we have $\delta_D \neq \{P\}$.
2. If $\delta_C = \{P\}$ then $N_C \cup \delta_C \models C$.
3. If $N_C \models D$ and $D \prec C$ then for all C' with $C \prec C'$ we have $N_{C'} \models D$ and in particular $N_\mathcal{I} \models D$.
4. If $N_C \models C$ then $N_\mathcal{I} \models C$.
5. If $N_\mathcal{I} \models N$ then there is no $\mathcal{H} \subset \mathcal{I}$ such that $\mathcal{H} \models N$.

The superposition calculus operates on a set of clauses. It is defined below as a set of non-deterministic, don't care rewrite rules on clause sets. The left hand side of a rule is matched against the clause set and the clause set is replaced by the right hand side of the rule. The symbol $\uplus$ denotes disjoint union.

The result of this rewrite rule presentation is a separation between the ordering in which a clause set is changed by superposition from the actual changes. Then properties of the calculus can be shown by induction and independent case analysis on the rules. Furthermore, properties relying on a particular rule application strategy can be distinguished from properties relying on the manipulations only. For example, see the different assumptions of Proposition 12 and Lemma 13 in the next section.

Superposition Left $(N \uplus \{C_1 \vee P, C_2 \vee \neg P\}) \Rightarrow_{\mathrm{SUP}} (N \cup \{C_1 \vee P, C_2 \vee \neg P\} \cup \{C_1 \vee C_2\})$

where (i) P is strictly maximal in $C_1 \vee P$ (ii) $\neg P$ is maximal in $C_2 \vee \neg P$

Factoring $(N \uplus \{C \vee P \vee P\}) \Rightarrow_{\text{SUP}} (N \cup \{C \vee P \vee P\} \cup \{C \vee P\})$

where P is maximal in $C \vee P \vee P$

Note that the superposition factoring rule differs from the resolution factoring rule in that it only applies to positive literals. In contrast to CDCL (Section 4), duplicate literal occurrences are not silently merged but either condensed (see below) or factorized. In fact, in propositional logic Condensation can effectively replace Factoring. However, as soon as the logic gets (a little) more expressive, Factoring and Condensation need to be distinguished (see also Section 6).

Definition 6 (Saturation). A set N of clauses is called *saturated up to redundancy*, if any clause generated by Superposition Left or Factoring from non-redundant clauses in N is redundant with respect to N or contained in N.

Examples for specific rules that eliminate redundant clauses or replace clauses by clauses making them redundant are:

Subsumption $(N \uplus \{C_1, C_2\}) \Rightarrow_{\text{SUP}} (N \cup \{C_1\})$

provided $C_1 \subset C_2$

Tautology
Deletion $(N \uplus \{C \vee P \vee \neg P\}) \Rightarrow_{\text{SUP}} (N)$

Condensation $(N \uplus \{C_1 \vee L \vee L\}) \Rightarrow_{\text{SUP}} (N \cup \{C_1 \vee L\})$

Subsumption
Resolution $(N \uplus \{C_1 \vee L, C_2 \vee \neg L\}) \Rightarrow_{\text{SUP}} (N \cup \{C_1 \vee L, C_2\})$

where $C_1 \subseteq C_2$

A clause C where Condensation is not applicable is called *condensed*. Note that there are no ordering restrictions on the literals involved in any of the rules. All clauses removed by Subsumption, Tautology Deletion, Condensation and Subsumption Resolution are redundant with respect to the kept or added clauses. In an implementation the redundancy elimination rules are priorized over Superposition Left and Factoring.

Theorem 7. *If N is saturated up to redundancy and $\perp \notin N$ then N is satisfiable and $N_{\mathcal{I}} \models N$.*

Proof. The proof is by contradiction. So I assume: (i) for any clause D derived by Superposition Left or Factoring from N that D is redundant, i.e., $N^{\prec D} \models D$, (ii) $\perp \notin N$ and (iii) $N_{\mathcal{I}} \not\models N$. Then there is a minimal, with respect to $\prec$, clause $C \vee L \in N$ such that $N_{\mathcal{I}} \not\models C \vee L$ and L is maximal. This clause must exist because $\perp \notin N$.

The clause $C \vee L$ is not redundant. For otherwise, $N_{C \vee L} \models C \vee L$ and hence $N_{\mathcal{I}} \models C \vee L$, Proposition 5.4, a contradiction.

I distinguish the case where L is a positive or L is a negative literal. Firstly, assume L is positive, i.e., $L = P$ for some propositional variable P. Now if P is strictly maximal in $C \vee P$ then actually $\delta_{C \vee P} = \{P\}$ and hence $N_{\mathcal{I}} \models C \vee P$, a contradiction. So P is not strictly maximal. But then actually $C \vee P$ has the form $C_1' \vee P \vee P$ and Factoring derives $C_1' \vee P$ where $(C_1' \vee P) \prec (C_1' \vee P \vee P)$. Now $C_1' \vee P$ is not redundant, strictly smaller than $C \vee L$, we have $C_1' \vee P \in N$ and $N_{\mathcal{I}} \not\models C_1' \vee P$, a contradiction against the choice that $C \vee L$ is minimal.

Secondly, assume L is negative, i.e., $L = \neg P$ for some propositional variable P. Then, since $N_{\mathcal{I}} \not\models C \vee \neg P$ it holds $P \in N_{\mathcal{I}}$. So there is a clause $D \vee P \in N$ where $\delta_{D \vee P} = \{P\}$ and P is strictly maximal in $D \vee P$ and $(D \vee P) \prec (C \vee \neg P)$. So Superposition Left derives $C \vee D$ where $(C \vee D) \prec (C \vee \neg P)$ by Proposition 2. The derived clause $C \vee D$ cannot be redundant, because for otherwise either $N_{C \vee D} \models D$ or $N_{C \vee D} \models C$. So $C \vee D \in N$ and $N_{\mathcal{I}} \not\models C \vee D$, a contradiction against the choice that $C \vee L$ is the minimal false clause. □

Note that the reverse of the above theorem does not hold. If $N_{\mathcal{I}} \models N$ the set N is not necessarily saturated. For example, consider the clause set $N = \{P, \neg P \vee \neg Q\}$ with ordering $Q \prec P$. Then $N_{\mathcal{I}} \models N$ but for saturation a Superposition Left inference between the two clauses is required. Maintaining an explicit model assumption can be beneficial for deciding satisfiability of N, because it enables to detect satisfiability before the clause set is saturated.

For a superposition implementation, the proof actually implies that at any point in the derivation only either a Superposition Left inference between a minimal false clause and a productive clause or a Factoring inference on a minimal false clause need to be considered. This principle is shared by the CDCL calculus, see the section below.

4   CDCL – Conflict Driven Clause Learning

The CDCL calculus tests satisfiability of a finite set N of propositional clauses. I assume that $\bot \notin N$ and that the clauses in N do not contain duplicate literal occurrences.

The CDCL calculus explicitely builds a candidate model for a clause set. If such a sequence of literals $L_1, \ldots, L_n$ satisfies the clause set N, it is done. If not, there is a false clause $C \in N$ with respect to $L_1, \ldots, L_n$. Now instead of just backtracking through the literals $L_1, \ldots, L_n$, CDCL generates in addition a new clause that actually guarantees that the subsequence of $L_1, \ldots, L_n$ that caused C to be false will not be generated anymore. This causes CDCL to be exponentially more powerful in proof length than its predecessor DPLL [7] or classical Tableau [17].

A CDCL problem state is a five-tuple $(M; N; U; k; C)$ where M a sequence of annotated literals representing a partial model, called a *trail*, N and U are sets of clauses, $k \in \mathbb{N}$, and C is a non-empty clause or $\top$ or $\bot$, called the *mode* of the state. In particular, the following states can be distinguished:

$(\epsilon; N; \emptyset; 0; \top)$ is the start state for some clause set N

$(M; N; U; k; \top)$ is a final state, if $M \models N$ and all literals from N are
   defined in M

$(M; N; U; k; \bot)$ is a final state, where N has no model

$(M; N; U; k; \top)$ is an intermediate model search state if $M \not\models N$ or
   not all literals from N are defined in M

$(M; N; U; k; D)$ is a backtracking state if $D \notin \{\top, \bot\}$

Literals in $L \in M$ are either annotated with a number, a level, i.e., they have
the form L^k meaning that L is the $k - th$ guessed decision literal, or they are
annotated with a clause that forced the literal to become true. A literal L is of
level k with respect to a problem state $(M; N; U; j; C)$ if L or comp(L) occurs
in M and the first decision literal left from L (comp(L)) in M is annotated with
k. If there is no such decision literal then $k = 0$. A clause D is of *level* k with
respect to a problem state $(M; N; U; j; C)$ if k is the maximal level of a literal
in D. Recall that the mode C is a non-empty clause or $\top$ or $\bot$. The rules are

Propagate $(M; N; U; k; \top) \Rightarrow_{\text{CDCL}} (ML^{C \vee L}; N; U; k; \top)$

provided $C \vee L \in (N \cup U)$, $M \models \neg C$, and L is undefined in M

Decide $(M; N; U; k; \top) \Rightarrow_{\text{CDCL}} (ML^{k+1}; N; U; k + 1; \top)$

provided L is undefined in M

Conflict $(M; N; U; k; \top) \Rightarrow_{\text{CDCL}} (M; N; U; k; D)$

provided $D \in (N \cup U)$ and $M \models \neg D$

Skip $(ML^{C \vee L}; N; U; k; D) \Rightarrow_{\text{CDCL}} (M; N; U; k; D)$

provided $D \notin \{\top, \bot\}$ and comp(L) does not occur in D

Resolve $(ML^{C \vee L}; N; U; k; D \vee \text{comp}(L)) \Rightarrow_{\text{CDCL}} (M; N; U; k; D \vee C)$

provided D is of level k

Backtrack $(M_1 K^{i+1} M_2; N; U; k; D \vee L) \Rightarrow_{\text{CDCL}} (M_1 L^{D \vee L}; N; U \cup \{D \vee L\}; i; \top)$

provided L is of level k and D is of level i.

Restart $(M; N; U; k; \top) \Rightarrow_{\text{CDCL}} (\epsilon; N; U; 0; \top)$

provided $M \not\models N$

Forget $(M; N; U \uplus \{C\}; k; \top) \Rightarrow_{\text{CDCL}} (M; N; U; k; \top)$

provided $M \not\models N$

Compared to expositions of this calculus in the literature, e.g. [12], the above
rule set is more concrete. It does not need a Fail rule anymore and 1UIP backtracking [6] is build in. The clause $D \vee L$ immediately propagates after Backtracking. Recall that $\bot$ denotes the empty clause, hence failure of searching
for a model. The level of the empty clause $\bot$ is 0. The clause $D \vee L$ added in
rule Backtrack to U is called a *learned* clause. When applying Resolve I silently
assume that duplicate literal occurrences are merged, i.e., the clause $D \vee C$ is
always condensed (see Section 3). Compared to superposition, condensation is

always applied eagerly without mentioning. The CDCL algorithm stops with a model M if neither Propagate nor Decide nor Conflict are applicable to a state $(M; N; U; k; \top)$, hence $M \models N$ and all literals of N are defined in M. The only possibility to generate a state $(M; N; U; k; \bot)$ is by the rule Resolve. So in case of detecting unsatisfiability the CDCL algorithm actually generates a resolution proof as a certificate. I will discuss this aspect in more detail in Section 5. In the special case of a unit clause L, the rule Propagate actually annotates the literal L with itself. So the propagated literals on the trail are annotated with the respective propagating clause and the decision literals with the respective level.

Obviously, the CDCL rule set does not terminate in general for a number of reasons. For example, starting with $(\epsilon; N; \emptyset; 0; \top)$ any combination of the rules Propagate, Decide and eventually Restart yields the start state again. Even after a successful application of Backtrack, exhaustive application of Forget followed by Restart again may produce the start state. So why these rules Forget and Restart? Actually, any modern SAT solver makes use of the two rules. The rule Forget is needed to get rid of "redundant" clauses. For otherwise, the number of clauses in $N \cup U$ may get too large to be processed anymore in an efficient way. The rule Restart makes sense with respect to a suitable heuristic for selecting the decision literals. If applied properly, it helps the calculus to focus on a part of N where it currently can make progress [6].

The original SAT literature [6,10,11,16] does not contain a redundancy notion for CDCL. A huge part of the results were found out via system design, such as the early Chaff or RelSAT, and experimental evaluation. I will develop a theoretical foundation in Section 5.

The following examples show that if the CDCL rules are applied in an arbitrary order, then unwanted phenomena can happen. The rules produce stuck states and clauses are learned that are already contained in the set $N \cup U$. In order to overcome all these situations, a strategy prioritizing certain rule applications is eventually added.

Example 8 (CDCL Proof). Consider the clause set $N = \{P \vee Q, \neg P \vee Q, \neg Q \vee P, \neg P \vee \neg Q\}$. For the following CDCL derivation the rules Conflict and Propagate are preferred over the other rules.

$$(\epsilon; N; \emptyset; 0; \top)$$
$$\Rightarrow_{\text{CDCL}}^{\text{Decide}} \quad (Q^1; N; \emptyset; 1; \top)$$
$$\Rightarrow_{\text{CDCL}}^{\text{Propagate}} (Q^1 P^{\neg Q \vee P}; N; \emptyset; 1; \top)$$
$$\Rightarrow_{\text{CDCL}}^{\text{Conflict}} (Q^1 P^{\neg Q \vee P}; N; \emptyset; 1; \neg P \vee \neg Q)$$
$$\Rightarrow_{\text{CDCL}}^{\text{Resolve}} (Q^1; N; \emptyset; 1; \neg Q)$$
$$\Rightarrow_{\text{CDCL}}^{\text{Backtrack}} (\neg Q^{\neg Q}; N; \{\neg Q\}; 0; \top)$$
$$\Rightarrow_{\text{CDCL}}^{\text{Propagate}} (\neg Q^{\neg Q} P^{P \vee Q}; N; \{\neg Q\}; 0; \top)$$
$$\Rightarrow_{\text{CDCL}}^{\text{Conflict}} (\neg Q^{\neg Q} P^{P \vee Q}; N; \{\neg Q\}; 0; \neg P \vee Q)$$
$$\Rightarrow_{\text{CDCL}}^{\text{Resolve}} (\neg Q^{\neg Q}; N; \{\neg Q\}; 0; Q)$$
$$\Rightarrow_{\text{CDCL}}^{\text{Resolve}} (\epsilon; N; \{\neg Q\}; 0; \bot)$$

For the clause set $N \setminus \{\neg P \vee Q\}$ the fourth last state $(\neg Q^{\neg Q} P^{P \vee Q}; N; \{\neg Q\}; 0; \top)$ is terminal, representing the model $\neg Q\, P$.

Example 9 (CDCL Stuck). The CDCL calculus can even get stuck, i.e., a sequence of rule applications leads to a state where no rule is applicable anymore, but the state does neither indicate satisfiability, nor unsatisfiability. Consider a clause set $N = \{Q \vee P, \neg P \vee \neg R, \ldots\}$ and the derivation

$$(\epsilon; N; \emptyset; 0; \top)$$
$$\Rightarrow_{\mathrm{CDCL}}^{\mathrm{Decide}} \quad (P^1; N; \emptyset; 1; \top)$$
$$\Rightarrow_{\mathrm{CDCL}}^{\mathrm{Decide}} \quad (P^1 R^2; N; \emptyset; 2; \top)$$
$$\Rightarrow_{\mathrm{CDCL}}^{\mathrm{Decide}} \quad (P^1 R^2 Q^3; N; \emptyset; 3; \top)$$
$$\Rightarrow_{\mathrm{CDCL}}^{\mathrm{Conflict}} \quad (P^1 R^2 Q^3; N; \emptyset; 3; \neg P \vee \neg R).$$

Obviously, neither Skip nor Resolve are applicable to the final state. Backtracking is not applicable as well because $\neg P \vee \neg R$ is of level 2 and the actual level of the final state is 3.

Example 10 (CDCL Redundancy). The CDCL calculus can also produce redundant clauses, in particular learn a clause that is already contained in $N \cup U$. Consider again a clause set $N = \{Q \vee P, \neg P \vee \neg R, \ldots\}$ and the derivation

$$(\epsilon; N; \emptyset; 0; \top)$$
$$\Rightarrow_{\mathrm{CDCL}}^{\mathrm{Decide}} \quad (P^1; N; \emptyset; 1; \top)$$
$$\Rightarrow_{\mathrm{CDCL}}^{\mathrm{Decide}} \quad (P^1 R^2; N; \emptyset; 2; \top)$$
$$\Rightarrow_{\mathrm{CDCL}}^{\mathrm{Conflict}} \quad (P^1 R^2; N; \emptyset; 2; \neg P \vee \neg R).$$
$$\Rightarrow_{\mathrm{CDCL}}^{\mathrm{Backtrack}} \quad (P^1 \neg R^{\neg P \vee \neg R}; N; \{\neg P \vee \neg R\}; 1; \top)$$

where the clause $\neg P \vee \neg R$ is learned although it is already contained in N.

In an implementation the rule Conflict is preferred over the rule Propagate and both over all other rules. Exactly this strategy has been used in Example 8 and is called *reasonable* below. A further ingredient of a state-of-the-art implementation is a dynamic heuristic suggesting which literal is actually used by the rule Decide. This heuristic typically depends on the literals resolved by the rule Resolve or that are contained in an eventually learned clause. All these literals "get a bonus", e.g., see [6].

Definition 11 (Reasonable CDCL Strategy). A CDCL strategy is *reasonable* if the rule Conflict is always preferred over the rule Propagate which is always preferred over all other rules.

Proposition 12 (CDCL Basic Properties). *Consider a CDCL state $(M; N; U; k; C)$ derived from a start state $(\epsilon, N, \emptyset, 0, \top)$ by any strategy but without using the rules Restart and Forget. Then the following properties hold:*

1. *M is consistent.*

2. All C is entailed by N.
3. If $C \notin \{\top, \bot\}$ then $M \models \neg C$.
4. If $C = \top$ and M contains only propagated literals then for each interpretation $\mathcal{I}$ with $\mathcal{I} \models N$ it holds that $\mathcal{I} \models M$, i.e., $M \subseteq \mathcal{I}$.
5. If $C = \top$, M contains only propagated literals and $M \models \neg D$ for some $D \in (N \cup U)$ then N is unsatisfiable.
6. If $C = \bot$ then CDCL terminates and N is unsatisfiable.
7. k is the maximal level of a literal in M.
8. Each infinite derivation

$$(\epsilon; N; \emptyset; 0; \top) \Rightarrow_{\mathrm{CDCL}} (M_1; N; U_1; k_1; D_1) \Rightarrow_{\mathrm{CDCL}} \dots$$

contains an infinite number of Backtrack applications.

Lemma 13 (CDCL Redundancy). *Consider a CDCL derivation by a reasonable strategy. Then CDCL never learns a clause contained in $N \cup U$.*

Proof. By contradiction. Assume CDCL learns the same clause twice, i.e., it reaches a state $(M; N; U; k; D \vee L)$ where Backtracking is applicable and $D \vee L \in (N \cup U)$. More precisely, the state has the form $(M_1 K^{i+1} M_2' K_1^k K_2 \dots K_n; N; U; k; D \vee L)$ where the K_i, $i > 1$ are propagated literals that do not occur complemented in D, as for otherwise D cannot be of level i. Furthermore, one of the K_i is the complement of L. But now, because $D \vee L$ is false in $M_1 K^{i+1} M_2' K_1^k K_2 \dots K_n$ and $D \vee L \in (N \cup U)$ instead of deciding K^k the literal L should have been propagated by a reasonable strategy. A contradiction. $\square$

Lemma 14 (CDCL Soundness). *In a reasonable CDCL derivation, CDCL can only terminate in two different final states: $(M; N; U; k; \top)$ where $M \models N$ and $(M; N; U; k; \bot)$ where N is unsatisfiable.*

Proof. If CDCL terminates with $(M; N; U; k; \top)$ then all literals of N are defined in M and Conflict is not applicable, i.e., for all clauses $C \in N$ it holds $M \models C$, so $M \models N$. In addition if CDCL terminates with $(M; N; U; k; \bot)$ then by Proposition 12.2 the clause set N is unsatisfiable.

What remains is to show that with a reasonable strategy CDCL cannot get stuck, see Example 9. I prove that no stuck state can be reached by contradiction. Assume that CDCL terminates in a state $(M_1 K^{i+1} M_2' K_1^k K_2 \dots K_n; N; U; k; D \vee L)$, where the K_i, $i > 1$, are propagated literals. If $\mathrm{comp}(K_n) \neq L$ and $n > 1$ then Skip is applicable. If $\mathrm{comp}(K_n) = L$ then either Resolve or Backtrack is applicable. Since neither Skip, Resolve, or Backtrack are applicable, it holds $n = 1$ and the complement of K_1^k does not occur in $D \vee L$. But then $M_1 K^{i+1} M_2' \models \neg (D \vee L)$ so the decision on K_1^k contradicts a reasonable strategy. $\square$

Proposition 15 (CDCL Strong Completeness). *The CDCL rule set is strongly complete: for any interpretation M restricted to the variables occurring in N with $M \models N$, there is a reasonable sequence of rule applications generating $(M'; N; U; k; \top)$ as a final state, where M and M' only differ in the order of literals.*

Proof. By induction on the length of M. Assume we have already reached a state $(M'; N; U; k; \top)$ where $M' \subset M$. If Propagate is applicable to $(M'; N; U; k; \top)$ extending it to $(M'L^{C \vee L}; N; U; k; \top)$ then $L \in M$. For otherwise, I pick a literal $L \in M$ that is not defined in M' and apply Decide yielding $(M'L^{k+1}; N; U; k + 1; \top)$. The rule Conflict is not applicable, because $M \models N$ and $M' \subset M$. $\square$

Proposition 16 (CDCL Termination). *Assume the algorithm CDCL with all rules except Restart and Forget is applied using a reasonable strategy. Then it terminates in a state $(M; N; U; k; D)$ with $D \in \{\top, \bot\}$.*

Proof. By Lemma 14 if CDCL terminates using a reasonable strategy then $D \in \{\top, \bot\}$. I show termination by contradiction. By Proposition 12.8 an infinite run includes infinitely many Backtrack applications. By Lemma 13 each learned clause does not occur in $N \cup U$. But there are only finitely many different condensed clauses with respect to the finite signature contained in N. A contradiction. $\square$

5   Superposition and CDCL

At the time of this writing it is often believed that the superposition (resolution) calculus is not a good choice on SAT problems in practice. Most of the successful SAT solvers implemented in 2015 are based on CDCL. In this section I will develop some relationships between superposition and CDCL. Actually, CDCL can be considered as a superposition calculus with respect to a generalized model operator.

The goal of the original superposition model operator, Definition 4, is to create minimal models with respect to positive literals, see Proposition 5.5. However, if the goal of generating minimal models is dropped, then there is more freedom to construct models while preserving the general properties of the superposition calculus, in particular, the notion of redundancy. The gained freedom can be used to be more flexible when generating a partial model with respect to a given set of clauses. For example, consider the two clauses in the clause set
$$N = \{P \vee Q, \neg P \vee R\}$$
with precedence $R \prec Q \prec P$. The superposition model operator generates $N_{\mathcal{I}} = \{P\}$ which is not a model for N. However, this model can be extended to a model for N by adding R to it. The superposition model operator does not include R because it is not maximal in the second clause. Starting with a decision on P, the CDCL calculus derives the model $P R$ via propagation. In the sequel, I show that a generalized superposition model operator can in fact generate this model as well.

In addition to an ordering $\prec$ I assume a decision heuristic H that selects whether a literal should be productive, i.e., included in the model, or not.

Definition 17 (Heuristic-Based Partial Model Construction). Given a clause set N, an ordering $\prec$ and a variable heuristic $H : \Sigma \to \{0, 1\}$, the (partial) model N_Σ^H for N and signature Σ, with $P, Q \in \Sigma$ is inductively constructed as follows:

$$N_P^H := \bigcup_{Q \prec P} \delta_Q^H$$

$$\delta_P^H := \begin{cases} \{P\} & \text{if there is a clause } (D \vee P) \in N, \text{ such that } N_P^H \models \neg D \\ & \text{and either } P \text{ is strictly maximal} \qquad \text{or} \\ & H(P) = 1 \text{ and there is no clause } (D' \vee \neg P) \in N, D' \prec P \\ & \text{such that } N_P^H \models \neg D' \\ \emptyset & \text{otherwise} \end{cases}$$

$$N_\Sigma^H := \bigcup_{P \in \Sigma} \delta_P^H$$

Please note that $N_\mathcal{I}$ is defined inductively over the clause ordering $\prec$ whereas N_Σ^H is defined inductively over the atom ordering $\prec$. For each atom P, the heuristic model construction N_Σ^H considers all clauses with maximal P, and clauses with maximal $\neg P$ at once.

Lemma 18 (N_Σ^H generalizes $N_\mathcal{I}$). *If $H(P) = 0$ for all $P \in \Sigma$ then $N_\mathcal{I} = N_\Sigma^H$ for any N.*

Proof. The proof is by contradiction. Assume $N_\mathcal{I} \neq N_\Sigma^H$, i.e., there is a minimal $P \in \Sigma$ such that P occurs only in one set out of $N_\mathcal{I}$ and N_Σ^H.

Case 1: $P \in N_\mathcal{I}$ but $P \notin N_\Sigma^H$.
Then there is a productive clause $D = D' \vee P \in N$ such that P is strictly maximal in this clause and $N_D \models \neg D'$. Since P is strictly maximal in D the clause D' only contains literals strictly smaller than P. Since both interpretations agree on all literals smaller than P from $N_D \models \neg D'$ it follows $N_P^H \models \neg D'$ and therefore $\delta_P^H = \{P\}$ contradicting $P \notin N_\Sigma^H$.

Case 2: $P \notin N_\mathcal{I}$ but $P \in N_\Sigma^H$.
Then there is a minimal productive clause $D = D' \vee P \in N$ such that P is strictly maximal in this clause and $N_P^H \models \neg D'$ because $H(P) = 0$. The atom P is strictly maximal in D, so the clause D' only contains literals strictly smaller than P. Since both interpretations agree on all literals smaller than P from $N_P^H \models \neg D'$ it follows $N_D \models \neg D'$ and therefore $\delta_D = \{P\}$ contradicting $P \notin N_\mathcal{I}$. $\square$

So the new model operator N_Σ^H is a generalization of $N_\mathcal{I}$. Next, I will show that with the help of N_Σ^H a close relationship between the model assumptions generated by the CDCL calculus and the superposition model operator can be established. This result can then further be used to apply the abstract superposition redundancy criteria to CDCL. But before going into the relationship I first show that the generalized superposition partial model operator N_Σ^H supports the standard superposition completeness result, analogous to Theorem 7. Recall that the same notion of redundancy, Definition 3, is used.

Theorem 19. *If N is saturated up to redundancy with eager Condensation and $\perp \notin N$ then N is satisfiable and $N_\Sigma^H \models N$.*

Proof. The proof is by contradiction. So I assume: (i) any clause C derived by Superposition Left or Factoring from N is redundant, i.e., $N^{\prec C} \models C$, (ii) $\bot \notin N$ and (iii) $N_{\Sigma}^{H} \not\models N$. Then there is a minimal, with respect to $\prec$, clause $C_1 \vee L \in N$ such that $N_{\Sigma}^{H} \not\models C_1 \vee L$ and L is a maximal literal in $C_1 \vee L$. This clause must exist because $\bot \notin N$.

The clause $C_1 \vee L$ is not redundant. For otherwise, $N_{\text{atom}(L)}^{H} \cup \delta_P^H \models C_1 \vee L$ and hence $N_{\Sigma}^{H} \models C_1 \vee L$, a contradiction.

I distinguish the case whether L is a positive or a negative literal. Firstly, assume L is positive, i.e., $L = P$ for some propositional variable P. Now if P is strictly maximal in $C_1 \vee P$ then actually $\delta_P^H = \{P\}$ and hence $N_{\Sigma}^{H} \models C_1 \vee P$, a contradiction. So P is not strictly maximal. But then actually $C_1 \vee P$ has the form $C_1' \vee P \vee P$ and Factoring derives $C_1' \vee P$ where $(C_1' \vee P) \prec (C_1' \vee P \vee P)$. The clause $C_1' \vee P$ is not redundant, strictly smaller than $C_1 \vee L$, we have $C_1' \vee P \in N$ and $N_{\Sigma}^{H} \not\models C_1' \vee P$, a contradiction against the choice that $C_1 \vee L$ is minimal.

Secondly, assume L is negative, i.e., $L = \neg P$ for some propositional variable P. Then, since $N_{\Sigma}^{H} \not\models C_1 \vee \neg P$ we know $P \in N_{\Sigma}^{H}$, i.e., $\delta_P^H = \{P\}$. There are two cases to distinguish. Firstly, there is a clause $C_2 \vee P \in N$ where P is strictly maximal, $N_P^H \models \neg C_2$, and by definition $(C_2 \vee P) \prec (C_1 \vee \neg P)$. Since $C_1 \prec \neg P$ and $C_1 \vee \neg P$ is not a tautology, it holds $C_1 \prec P$. So a Superposition Left inference derives $C_1 \vee C_2$ where $(C_1 \vee C_2) \prec (C_1 \vee \neg P)$. The derived clause $C_1 \vee C_2$ cannot be redundant, because for otherwise either $N_P^H \models C_2$ or $N_P^H \models C_1$. So $C_1 \vee C_2 \in N$ and $N_{\Sigma}^{H} \not\models C_1 \vee C_2$, a contradiction against the choice that $C_1 \vee L$ is minimal. Secondly, there is no clause $C_2 \vee P \in N$ where P is strictly maximal but $H(P) = 1$. But a further condition for this case is that there is no clause $(C_1 \vee \neg P) \in N$, $\neg P$ strictly maximal in $C_1 \vee \neg P$, because the clause is condensed, such that $N_P^H \not\models C_1$ contradicting the above choice of $C_1 \vee \neg P$.   □

Recalling Section 3, Superposition is based on an ordering $\prec$. It relies on a model assumption $N_{\mathcal{I}}$, Definition 4, or its generalization N_{Σ}^{H}, Definition 17. Given a set N of clauses, either $N_{\mathcal{I}}$ (N_{Σ}^{H}) is a model for N, N contains the empty clause, or there is a superposition inference on the minimal false clause with respect to $\prec$, see the proof of Theorem 7 or Theorem 19, respectively.

CDCL is based on a variable decision heuristic. It computes a model assumption via decision variables and propagation. Either this assumption is a model of N, N contains the empty clause, or there is a backjump clause that is learned.

For a CDCL state (M, N, U, k, D) generated by an application of the rule Conflict, where $M = L_1, \dots, L_n$, the subsequent Resolve step actually corresponds to a superposition step between a minimal false clause and its productive counterpart, with respect to the atom ordering $\text{atom}(L_1) \prec \text{atom}(L_2) \prec \dots \prec \text{atom}(L_n)$. The decision heuristic H is defined by $H(\text{atom}(L_m)) = 1$ if there is a positive decision literal L_m^i occurring in M and $H(\text{atom}(L_m)) = 0$ otherwise. Then the learned CDCL clause is in fact generated by a superposition inference with respect to the model operator N_{Σ}^{H}. The below theorem presents this relationship between Superposition and CDCL in full detail.

Theorem 20. *Let $(M, N, U, k, C \vee K)$ be a CDCL state generated by rule Conflict and a reasonable strategy where $M = L_1, \ldots, L_n$. Let $H(\text{atom}(L_m)) = 1$ for any positive decision literal L_m^i occurring in M and $H(\text{atom}(L_m)) = 0$ otherwise. Furthermore, I assume that if CDCL could propagate both P and $\neg P$ in some state, then P is propagated. The superposition precedence is $\text{atom}(L_1) \prec \text{atom}(L_2) \prec \ldots \prec \text{atom}(L_n)$. Let K be maximal in $C \vee K$ and $C \vee K$ be the minimal false clause with respect to $\prec$. Then*

1. *L_n is a propagated literal and $K = \text{comp}(L_n)$.*
2. *The clause generated by $C \vee K$ and the clause propagating L_n is the result of a Superposition Left inference between the clauses and it is not redundant.*
3. *$N^H_{\{L_1, \ldots, L_n\}} = \{P \mid P \in M\}$*

Proof. 1. Assume $K \neq \text{comp}(L_n)$. Since $C \vee K$ was derived by rule Conflict it is false with respect to M. Since K is maximal in $C \vee K$ it is the complement of some L_i from M with $i < n$ contradicting a reasonable strategy. So $K = \text{comp}(L_n)$. Assume L_n is a decision literal. But then at trail $L_1, \ldots, L_{n-1}$ the clause $C \vee K$ propagates K with respect to $L_1 \ldots L_{n-1}$, so with a reasonable strategy, the literal L_n cannot be a decision literal but its complement was propagated by the clause $C \vee K$.

2. Let $D \vee L_n$ be the clause propagating L_n. Both C and D only contain literals with variables from $\text{atom}(L_1), \ldots, \text{atom}(L_{n-1})$. Since in CDCL duplicate literals are (silently) removed, the literal L_n is strictly maximal in $D \vee L_n$ and $\text{comp}(L_n)$ is strictly maximal in $C \vee K$. So resolving on L_n is a superposition inference with respect to the atom ordering $\text{atom}(L_1) \prec \text{atom}(L_2) \ldots \prec \text{atom}(L_n)$. Now assume $C \vee D$ is redundant, i.e., there are clauses $D_1, \ldots, D_n$ from $N \cup U$ with $D_i \prec C \vee D$ and $D_1, \ldots, D_n \models C \vee D$. Since $C \vee K$ is the minimal false clause, the resolvent $C \vee D$ is false in $L_1 \ldots L_{n-1}$ and there is at least one D_i that is also false in $L_1 \ldots L_{n-1}$. A contradiction against the assumption that $L_1 \ldots L_{n-1}$ does not falsify any clause in $N \cup U$, i.e., rule Conflict was not applied eagerly contradicting a reasonable strategy.

3. Firstly, note that if CDCL can propagate both P and $\neg P$ then either way the next applicable reasonable rule is Conflict, so propagating P in favor of $\neg P$ is not a restriction on the propagation order. I prove the equality of the atom sets by induction on n.

"$\supseteq$" For $n = 1$ and $L_1 = [\neg]P$ propagated in M, there are two cases: (i) $L_1 = P$, so $P \in N$ and $\delta_P^H = \{P\}$; (ii) $L_1 = \neg P$, so $\neg P \in N$ and $P \notin N$, therefore $\delta_P^H = \emptyset$. If $L_1 = [\neg]P$ is a decision literal then $\neg P \notin N$ and $P \notin N$. Again there are two cases: (i) $L_1 = P$, so $H(P) = 1$ and hence $\delta_P^H = \{P\}$; (ii) $L_1 = \neg P$, so $H(P) = 0$ and hence $\delta_P^H = \emptyset$.

For the step $(n - 1) \to n$, I do the same case analysis as for the base case $n = 1$. If $L_n = [\neg]P$ is propagated in M, there are two cases: (i) $L_n = P$, so $D \vee P \in N$ and $L_1 \ldots L_{n-1} \models \neg D$. By induction hypothesis $N^H_{\{L_1, \ldots, L_{n-1}\}} \models \neg D$, $L_i \prec P$, so $\delta_P^H = \{P\}$; (ii) $L_n = \neg P$, so $D \vee \neg P \in N$, $H(P) = 0$ and there is no clause $D' \vee P \in N$ propagating P, hence $\delta_P^H = \emptyset$. If $L_n = [\neg]P$ is a decision literal in M then due to the reasonable strategy, there is no clause propagating L_n on the basis of the trail $L_1 \ldots L_{n-1}$. Again two cases: (i) $L_n = P$, so $H(P) = 1$ and there is no clause $C_1 \vee \neg P$ such that $L_1, \ldots, L_{n-1} \models \neg C_1$, hence $\delta_P^H = \{P\}$; (ii) if $L_n = \neg P$, $H(P) = 0$, and there is no clause $C_1 \vee P$ such that $L_1 \ldots L_{n-1} \models \neg C_1$, so $\delta_P^H = \emptyset$.

"$\subseteq$" By construction. $\square$

Theorem 20 is actually a nice explanation for the efficiency of the CDCL procedure: a learned clause is never redundant. Recall that redundancy here means that the learned clause C is not entailed by smaller clauses in $N \cup U$. Furthermore, the ordering underlying Theorem 20 is based on the trail, i.e., it changes during a CDCL run. For superposition it is well known that changing the ordering is not compatible with the notion of redundancy, i.e., superposition is incomplete when the ordering may be changed infinitely often and the superposition redundancy notion is applied.

Example 21. Consider the Superposition Left inference between the clauses $P \vee Q$ and $R \vee \neg Q$ with ordering $P \prec R \prec Q$ generating $P \vee R$. Changing the ordering to $Q \prec P \prec R$ the inference $P \vee R$ becomes redundant. So flipping infinitely often between $P \prec R \prec Q$ and $Q \prec P \prec R$ is already sufficient to prevent any saturation progress.

6   Conclusion

Although Example 21 shows that changing the ordering is not compatible with redundancy and superposition completeness, Theorem 20 proves that any CDCL learned clause is not redundant in the superposition sense. This relationship shows the power of reasoning with respect to a (partial) model assumption. The model assumption actually prevents the generation of redundant clauses. Nevertheless, also in the CDCL framework completeness would be lost if redundant clauses are eagerly removed, in general. Either the ordering is not changed and the superposition redundancy notion can be eagerly applied or only a weaker notion of redundancy is possible while keeping completeness for both calculi.

The crucial point is that for the superposition calculus the ordering is the basis for redundancy, termination and completeness. If the completeness proof can be decoupled from the ordering, then the ordering might be changed infinitely often and other notions of redundancy become available. However, these new notions of redundancy need to be compatible with the completeness and termination proof. For example, if the redundancy ordering requires in addition a length restriction, i.e., C is redundant if $D_1, \ldots, D_n \models C$ for $D_i \prec C$ and $|D_i| < |C|$ then CDCL and superposition remain complete with respect to this weaker notion of redundancy, even if the atom ordering or the heuristics H are dynamically changed. The ordering $<$ is already well-founded on clauses independently from the choice of $\prec$. Eager application of Forget on redundant clauses preserves completeness for CDCL. However, the above length restriction is very strong and as said the redundancy ordering is not the only parameter that can be adjusted. An interesting future research question is to investigate further the relationship between decoupled orderings, redundancy, termination, and completeness.

A "slightly" more expressive logic than propositional logic is the first-order logic Bernays-Schönfinkel (BS) fragment. Satisfiability of a BS clause set remains decidable, because the problem can be actually reduced, with a worst-case exponential blow up, to propositional satisfiability. The overall worst-case complexity moves from NP-completeness (propositional) to NEXPTIME-completeness (BS). A natural question is what happens if the principles developed in this paper are applied to calculi for the Bernays-Schönfinkel fragment. Actually, the calculi of all systems developed in particular for the BS fragment apply inferences with respect to a (partial) model assumption [5,8,14], partly in the spirit of this paper. Recently, we have shown [1] that the principles of this paper can be lifted to the BS fragment, but not in a straightforward way. An explicit Factoring inference is needed. The model assumption gets far more complicated, in order to receive compact representations of models and to guarantee non-redundant learned clauses. Actually, I see this as an important direction of future research. For all model assumption representations we know today, there are BS clause sets where the model representation becomes exponential in the size of the clause set. An important difference to SAT.

Finally, if aiming at full first-order logic available results are at a very early stage. A key operation with respect to a model representation is deciding whether a clause is false. It is open how the known representations providing decidability of this operation for first-order models can be integrated in an overall automated reasoning approach in the style of this paper, although some steps have been done into this direction [2,9,18]. The situation becomes even more difficult if first-order logic with equality is considered. Here no "expressive" representations with the above property is known.

Acknowledgments. I thank the anonymous reviewers for their detailed and constructive comments and Marco Voigt for his careful proof reading.

References

1. Alagi, G., Weidenbach, C.: NRCL - a model building approach to the Bernays-Schönfinkel fragment. In: Lutz, C., Ranise, S. (eds.) FroCoS 2015. LNAI, vol. 9322. Springer (2015)
2. Aravantinos, V., Echenim, M., Peltier, N.: A resolution calculus for first-order schemata. Fundamenta Informaticae **125**(2), 101–133 (2013)
3. Bachmair, L., Ganzinger, H.: On restrictions of ordered paramodulation with simplification. In: Stickel, M.E. (ed.) CADE-10. LNCS, vol. 449, pp. 427–441. Springer, Heidelberg (1990)
4. Bachmair, L., Ganzinger, H.: Resolution theorem proving. In: Robinson, A., Voronkov, A. (eds.) Handbook of Automated Reasoning, ch. 2, vol. I, pp. 19–99. Elsevier (2001)
5. Baumgartner, P., Fuchs, A., Tinelli, C.: Lemma Learning in the Model Evolution Calculus. In: Hermann, M., Voronkov, A. (eds.) LPAR 2006. LNCS (LNAI), vol. 4246, pp. 572–586. Springer, Heidelberg (2006)

6. Biere, A., Heule, M., van Maaren, H., Walsh, T. (eds). Handbook of Satisfiability. IOS Press (2009)
7. Davis, M., Logemann, G., Loveland, D.W.: A machine program for theorem-proving. Communications of the ACM **5**(7), 394–397 (1962)
8. Ganzinger, H., Korovin, K.: New directions in instantiation-based theorem proving. In: LICS 2003, pp. 55–64. IEEE Computer Society (2003)
9. Ganzinger, H., Meyer, C., Weidenbach, C.: Soft typing for ordered resolution. In: McCune, W. (ed.) CADE-14. LNAI, vol. 1249, pp. 321–335, Townsville, Australia. Springer, Heidelberg (1997)
10. Bayardo Jr., R.J., Schrag, R.: Using CSP look-back techniques to solve exception-ally hard SAT instances. In: Freuder, E.C. (ed.) CP. LNCS, vol. 1118, pp. 46–60. Springer, Heidelberg (1996)
11. Moskewicz, M.W., Madigan, C.F., Zhao, Y., Zhang, L., Malik, S.: Chaff: Engineer-ing an efficient SAT solver. In: DAC, pp. 530–535. ACM (2001)
12. Nieuwenhuis, R., Oliveras, A., Tinelli, C.: Solving SAT and SAT modulo theo-ries: From an abstract Davis-Putnam-Logemann-Loveland procedure to DPLL(T). Journal of the ACM **53**, 937–977 (2006)
13. Nieuwenhuis, R., Rubio, A.: Paramodulation-based theorem proving. In: Robin-son, A., Voronkov, A. (eds.) Handbook of Automated Reasoning, ch. 7, vol. I, pp. 371–443. Elsevier (2001)
14. Piskac, R., de Moura, L.M., Bjørner, N.: Deciding effectively propositional logic using DPLL and substitution sets. Journal of Automated Reasoning **44**(4), 401–424 (2010)
15. Robinson, J.A.: A machine-oriented logic based on the resolution principle. Journal of the ACM **12**(1), 23–41 (1965)
16. Silva, J.P.M., Sakallah, K.A.: GRASP - a new search algorithm for satisfiability. In: ICCAD, pp .220–227. IEEE Computer Society Press (1996)
17. Smullyan, R.M.: First-Order Logic. Ergebnisse der Mathematik und ihrer Gren-zgebiete. Springer (1968) (revised republication 1995 by Dover Publications)
18. Teucke, A., Weidenbach, C.: First-order logic theorem proving and model building via approximation and instantiation. In: Lutz, C., Ranise, S. (eds) FroCos 2015. LNAI, vol. 9322. Springer (2015)
19. Weidenbach, C.: Combining superposition, sorts and splitting. In: Robinson, A., Voronkov, A. (eds) Handbook of Automated Reasoning, ch. 27, vol. 2, pp. 1965–2012. Elsevier (2001)

Analysis

Being and Change: Reasoning About Invariance

Frank S. de Boer[1,2]($\boxtimes$) and Stijn de Gouw[1]

[1] Centre Mathematics and Computer Science, Amsterdam, Netherlands
{frb,cdegouw}@cwi.nl
[2] Leiden Advanced Institute of Computer Science, Leiden, Netherlands

Abstract. We introduce a new way of reasoning about invariance in terms of *foot-prints* in a Hoare logic for recursive programs with (unbounded) arrays. A foot-print of a statement is a predicate that describes that part of the state that can be changed by the statement. We define invariance of an assertion with respect to a foot-print by means of a logical operation. This new Hoare logic is applied in a new simpler and modular proof of correctness of the well-known Quicksort sorting algorithm.

1 Introduction

During a visit of Ernst-Ruediger Olderog at the CWI in 2014, together with Krzysztof R. Apt we discussed different alternative proofs of correctness of the well-known Quicksort sorting algorithm [Hoa62]. These discussions resulted in various proof strategies which have been further detailed by Ernst-Ruediger and which form the starting point of this paper.

Proving correctness of (imperative) programs in Hoare logic is in general a challenging task, even for what seem to be relatively simple programs (measured for example in terms of the lines of code). Most of the complications are due to the basic fact that an imperative program specifies what *changes*, whereas an assertion describes what *is*. Consequently, most of the effort in proving correctness goes in specifying and verifying what does *not* change, i.e., what is *invariant*.

Proving correctness of recursive programs in Hoare logic requires special auxiliary rules (axioms) for reasoning about invariance [AdBO09], the so-called *adaptation* rules. These rules are used to adapt a given correctness formula, for example by adding to the pre- and postcondition an invariant. Hoare introduced in [Hoa71] one rule, *the* adaptation rule, which generalizes these rules. In his seminal paper [Old83] Ernst-Ruediger Olderog studied the expressiveness and the completeness of the adaptation rule.

Programs with, for example, *array* variables give rise to *aliasing*, i.e., the phenomenon of two *syntactically* different expressions which refer to the same *memory location*. In the presence of aliasing we cannot determine *statically* anymore general invariant properties, whereas the standard adaptation rules are based on such a static determination, namely checking syntactically whether a given assertion contains occurrences of variables changed in the given program.

R. Meyer et al. (Eds.): Olderog-Festschrift, LNCS 9360, pp. 191–204, 2015.
DOI: 10.1007/978-3-319-23506-6_13

The adaptation rules, including Hoare's rule, therefore are of limited use in proving invariant properties of (recursive) programs with arrays. This limitation in general greatly complicates the correctness proofs because it does not fully support *modularity*: invariant properties in general are verified in terms of the internal control structure of a given program.

For recursive programs with array variables we extend in this paper the standard pre-postcondition specification with a *footprint*. A footprint is a set of predicates indexed by an array name. The arity of the predicate equals that of the array (interpreted as a function). Such a predicate associated with an array describes that subset of the *domain* of the array which can be changed by the program. We show how to extend the syntactic characterization of invariance to these footprints by means of a logical operation. We prove soundness of this logical operation and apply the logic to the verification of the well-known Quicksort sorting algorithm, which results in a simpler and modular proof.

Related Work. A large body of related work focuses on reasoning about invariant properties of object-oriented programs. For example, *dynamic frames* [Wei11] have been introduced as an extension of Hoare logic where invariant properties are specified and verified in terms of an explicit heap representation. Such a representation however in general does not match the abstraction level of object-oriented languages like Java. In *separation logic* [Rey05] invariant properties of object-oriented programs are specified and verified in terms of a logical operation of *separation conjunction* which allows to split the heap into two disjoint parts. The resulting logic however is undecidable for its propositional subset [BK14]. Moreover, in [CYO01] it is shown that validity of the first-order language restricted to so-called "points to" predicate is not recursively enumerable, and as such not axiomatizable. In general, such axiomatizations are needed to prove formally the verification conditions which establish program correctness.

In contrast our approach, which can be extended to object-oriented programs (see Section 5), is based on standard predicate logic which allows established theorem proving techniques/engines. Further it allows reasoning at an abstraction level which coincides with the programming language, i.e., it does not require special predicates like the "points to" predicate.

2   Adaptation Rules

To allow for modular reasoning in Hoare logic, adaptation rules are needed to adapt the specifications in correctness formulas to a specific context. This section discusses four such adaptation rules [AdBO09], abstracting from the programming language: a conjunction rule, an existential introduction rule, an invariance rule, and a substitution rule. The invariance rule provides a basic form to reason about invariance using a simple *syntactic* test. A more precise *semantic* form, introduced in the next section, is needed for arrays. The adaptation rules are amenable to this extension; the next section shows that a modest addition to these rules suffices. In this section we clarify the relation between these adaptation rules and Hoare's single rule of adaptation [Hoa71], which was analyzed by Olderog in [Old83]. The precise definitions of the adaptation rules are given below (for details about the standard logical operations used, like substitution $p[z := x]$ of z for x in p, we refer to [AdBO09]):

RULE A1: CONJUNCTION

$$\frac{\{p_1\}\ S\ \{q_1\}, \{p_2\}\ S\ \{q_2\}}{\{p_1 \wedge p_2\}\ S\ \{q_1 \wedge q_2\}}$$

RULE A2: ∃-INTRODUCTION

$$\frac{\{p\}\ S\ \{q\}}{\{\exists z : p\}\ S\ \{q\}}$$

where z does not occur in S or q.

RULE A3: INVARIANCE

$$\frac{\{p\}\ S\ \{q\}}{\{r \wedge p\}\ S\ \{r \wedge q\}}$$

where S does not assign to the variables in the formula r.

RULE A4: SUBSTITUTION

$$\frac{\{p\}\ S\ \{q\}}{\{p[z := t]\}\ S\ \{q[z := t]\}}$$

where z does not occur in S, and S does not assign the variables in the term t.

The invariance rule above provides a basic way to reason about assertions whose truth remains invariant under execution of S. However, in the presence of arrays, the invariance rule is rather crude. Due to the syntactic check in the side-condition, if *any* array element is assigned to, any assertion that mentions the array cannot be used with the invariance rule, even if the assertion accesses only those indices that are not assigned. The next section shows how to address this problem.

Hoare's rule of adaptation is:

RULE (H) : HOARE-ADAPT

$$\frac{\{p\}\ S\ \{q\}}{\{\exists z : (p \wedge \forall y : (q[x := y] \rightarrow r[x := y]))\}\ S\ \{r\}}$$

where z does not occur in S or r, x is the list of all variables occurring in S and y is a list of fresh variables.

The question arises: what is the relation between Hoare's rule (H) and the other adaptation rules? The next example by Olderog [Old83] shows that they differ in proof strength.

Lemma 1. *Let k be a variable that does not occur in S. From the correctness formula $\{x = k\}\ S\ \{x = k\}$ we can derive $\{x = k + 1\}\ S\ \{x = k + 1\}$ by A4, but not by rule (H).* □

Proof: To derive $\{x = k + 1\}\ S\ \{x = k + 1\}$ simply apply A4, substituting $k + 1$ for k. To see that this is not derivable with (H), note that the precondition $\exists z : x = k \wedge \forall y : y = k \rightarrow y = k + 1$ given by (H) simplifies to *false*. □

Theorem 1. *In the presence of the consequence rule, (H) is derivable by A2, A3.* □

Proof: we show that from $\{p\}\ S\ \{q\}$, we can derive

$$\{\exists z : (p \wedge \forall y : (q[x := y] \rightarrow r[x := y]))\}\ S\ \{r\}$$

where z does not occur in S or r, x is the list of all variables occurring in S and y is a list of fresh variables. Assume

$$\{p\}\ S\ \{q\}$$

From A3:

$$\{p \wedge \forall y : (q[x := y] \rightarrow r[x := y])\}\ S\ \{q \wedge \forall y : (q[x := y] \rightarrow r[x := y])\}$$

Consequence rule:

$$\{p \wedge \forall y : (q[x := y] \rightarrow r[x := y])\}\ S\ \{r\}$$

From A2:

$$\{\exists z : (p \wedge \forall y : (q[x := y] \rightarrow r[x := y]))\}\ S\ \{r\}$$

□

Thus, in the presence of the consequence rule, theorem 1 and lemma 1 show that A1, A2, A3 and A4 are strictly stronger than rule (H).

3   Reasoning About Invariance

The Programming Language. We assume a basic imperative programming language featuring the usual sequential control structures. We distinguish between two kinds of (typed) variables: *simple* variables like $x, y, u, v, \ldots$ which range over elements of the included basic types **integer** or **Boolean**, and *array* variables like $a, b, \ldots$ of a higher type $T_1 \times \ldots \times T_n \rightarrow T$, where the argument types and the result type are basic types. Semantically arrays are functions, e.g., an array of type **integer** $\rightarrow$ **integer** is *unbounded*. Expressions are side-effect free (every operator in the language is semantically interpreted as a *total* function, e.g., division by zero results by definition in, for example, zero). A *subscripted* variable of type T is of the form $a[s_1, \ldots, s_n]$, where a is of some type $T_1 \times \ldots \times T_n \rightarrow T$ and

s_i is an expression of type T_i, for $i = 1, \ldots, n$. For technical convenience only we restrict here to array assignments of the form $a[s_1, \ldots, s_n] := t$, where the argument expressions $s_1, \ldots, s_n$ do not contain subscripted variables. A program consists of a statement S and a set of procedure declarations $P(u_1, \ldots, u_n) ::= S$, with formal parameters $u_1, \ldots, u_n$ of a basic type and body S. A procedure call is of the form $P(t_1, \ldots, t_n)$, where t_i is an expression of a basic type which equals that of the corresponding formal parameter.

Correctness Formulas. Assertions $p, q, \ldots$ are logical formula, defined as usual (as in [AdBO09]) (in contrast to program assignments, in assertions we do allow nested subscripted variables). By

$$F : \{p\}\ S\ \{q\}$$

we denote a *correctness formula* with a footprint F. A footprint F is a (finite) set of uniquely labeled formulas $a : p(x_1, \ldots, x_n)$, where n is the arity of array a. All the formulas of the footprint F are *syntactically invariant* in that they do not contain any program variables which can be affected by an execution of S

The *partial correctness* interpretation of $F : \{p\}\ S\ \{q\}$ (we assume an implicitly given set of procedure declarations D) extends that of $\{p\}\ S\ \{q\}$ with the following clause:

$\sigma \models p$ and
$< S, \sigma > \rightarrow^* < a[s_1, \ldots, s_n] := t; S', \sigma' >$ and $a : r(x_1, \ldots, x_n) \in F$
implies
$\sigma' \models r(s_1, \ldots, s_n)$.

Here $\rightarrow^*$ denotes the reflexive, transitive closure of the transition system for recursive programs (see Section 5.2 in [AdBO09]). In words, the above additional clause states that whenever an assignment to an array is executed the corresponding footprint should hold.

The Hoare Logic of Footprints. The footprint of an array assignment is captured by the following rule:

RULE 1: ARRAY ASSIGNMENT

$$\frac{q[a[s_1, \ldots, s_n] := t] \rightarrow p(s_1, \ldots, s_n)}{\{a : p(x_1, \ldots, x_n)\} : \{q[a[s_1, \ldots, s_n] := t]\}\ a[s_1, \ldots, s_n] := t\ \{q\}}$$

Here the weakest precondition $q[a[s_1, \ldots, s_n] := t]$ is calculated by means of a *substitution* operation $[a[s_1, \ldots, s_n] := t]$ which takes into account aliasing (see Section 2.7 in [AdBO09]). For the soundness of this rule, we refer to [AdBO09] to a proof that $\sigma \models q[a[s_1, \ldots, s_n] := t]$ if and only if $\sigma[a[s_1, \ldots, s_n] := t] \models q$, where $\sigma[a[s_1, \ldots, s_n] := t]$ denotes the result of executing the assignment $a[s_1, \ldots, s_n] := t$ in σ. It remains to show that the above additional clause defining the semantics of a foot-print is valid, i.e., $\sigma \models q[a[s_1, \ldots, s_n] := t]$ implies $\sigma \models p(s_1, \ldots, s_n)$. This follows immediately from the premise.

As a very simple example, it is trivial to derive

$$\{a : x = j\} : \{\textbf{true}\}\ a[j] := 1\ \{\textbf{true}\}$$

In order to reason *semantically* about invariance in terms of footprints we introduce the *restriction* $q \uparrow F$ of a formula q which "talks" only about that part of the state *disjoint* from the footprint. First we transform q in a formula q' in prenex normal form such that its matrix r is in disjunctive normal form. For technical convenience only and without loss of generality we assume that r contains no nested subscripted variables. The formula $q \uparrow F$ then can be obtained from q' by simply adding for each subscripted variable $a[s_1, \ldots, s_n]$ appearing as an argument in a literal the formula $\neg p_a(s_1, \ldots, s_n)$ to the conjunct in which the literal appears, where $p_a(x_1, \ldots, x_n) \in F$. More formally, we replace every literal $l(s_1, \ldots, s_n)$ in q' by $l(s_1, \ldots, s_n) \wedge \bigwedge_i \neg p_{a_i}(\bar{t}_i)$, where i ranges over those indices such that $s_i \equiv a_i(\bar{t}_i)$.

Given the above we can now introduce the following rule:

RULE 2: SEMANTIC INVARIANCE

$$\frac{F : \{p\}\ S\ \{q\}}{F : \{p \wedge r \uparrow F\}\ S\ \{q \wedge r \uparrow F\}}$$

where none of the *simple* variables which appear free in r appear in S at the left-hand-side of an assignment.

Let us illustrate the use of this latter rule by a very simple example. We want to prove

$$\{\forall i : i \neq j \rightarrow a[i] = 0\}\ a[j] := 1\ \{\forall i : i \neq j \rightarrow a[i] = 0\}$$

As already stated above it is trivial to derive from the above array assignment rule that

$$\{a : x = j\} : \{\textbf{true}\}\ a[j] := 1\ \{\textbf{true}\}$$

Calculating next

$$(\forall i : i \neq j \rightarrow a[i] = 0) \uparrow x \neq j$$

yields the formula

$$\forall i : i = j \vee (a[i] = 0 \wedge i \neq j).$$

Clearly this latter formula is logically equivalent to $\forall i : i \neq j \rightarrow a[i] = 0$ itself. So we can apply the above RULE 2 which gives us the desired result.

Soundness of the above proof system derives in a straightforward manner from the following lemma (soundness proofs of the remaining rules are standard, see [AdBO09]).

Lemma 2. *(Soundness): Let a be an array variable that does not appear (free) in the formulas of the footprint F. Further, let $\sigma' = \sigma[a[\bar{s}] := u]$ and $\sigma \models p_a(\bar{s})$. It follows that $\sigma \models r \uparrow F$ iff $\sigma' \models r \uparrow F$.*

Proof: By definition of $r \uparrow F$ it suffices to show the above for any literal l. By definition of σ', we have that $\sigma(t) = \sigma'(t)$, for any term t which does not involve the array variable a. So it suffices to show that $\sigma \models l \uparrow F$ or $\sigma' \models l \uparrow F$ implies that $\sigma(a[\bar{t}]) = \sigma'(a[\bar{t}])$, for any subscripted variable $a[\bar{t}]$ appearing as argument of l. By definition of $l \uparrow F$, we have that $l \uparrow F$ implies $\neg p_a(\bar{t})$. Consequently, since $\sigma \models p_a(\bar{s})$, we have that $\sigma(\bar{s}) \neq \sigma(\bar{t})$, which in turn implies that $\sigma(a[\bar{t}]) = \sigma[a[\bar{t}] := u](a[\bar{t}]) = \sigma'(a[\bar{t}])$. $\qquad\square$

We have the following extension of the consequence rule.

RULE 3: CONSEQUENCE

$$\frac{p \rightarrow p' \quad F : \{p'\} \ S \ \{q'\} \quad q' \rightarrow q}{F : \{p\} \ S \ \{q\}}$$

The following two rules deal with recursion. For technical convenience only we restrict to procedure declarations with read-only formal parameters and procedure calls with actual parameters which are not affected by the call. In Rule 5 "$\vdash$" denotes the derivability in the proof system itself. It allows to introduce assumptions about recursive calls (see [AdBO09]).

RULE 4: INSTANTIATION

$$\frac{F : \{p\} \ P(\bar{u}) \ \{q\}}{F[\bar{u} := \bar{t}] : \{p[\bar{u} := \bar{t}]\} \ P(\bar{t}) \ \{q[\bar{u} := \bar{t}]\}}$$

where $P(\bar{u}) ::= S \in D$ and S does not assign to the variables appearing $\bar{t}$.

RULE 5: RECURSION

$$\frac{\begin{array}{l} F : \{p_1\} \ P_1(\bar{u}_1) \ \{q_1\}, \ldots, F : \{p_n\} \ P_n(\bar{u}_n) \ \{q_n\} \vdash F : \{p\} \ S \ \{q\}, \\ F : \{p_1\} \ P_1(\bar{u}_1) \ \{q_1\}, \ldots, F : \{p_n\} \ P_n(\bar{u}_n) \ \{q_n\} \vdash \\ \qquad F : \{p_i\} \ S_i \ \{q_i\}, \ i \in \{1, \ldots, n\} \end{array}}{F : \{p\} \ S \ \{q\}}$$

where $P_i(\bar{u}_i) :: S_i \in D$ for $i \in \{1, \ldots, n\}$.

To extend the standard auxiliary rules (as discussed in the previous section) to footprints is straightforward (we omit the similar straightforward extensions of the standard axiom and rules, e.g., the axiom for assignments to simple variables, the conditional rule, the while rule, and the rule for sequential composition).

RULE A5: CONJUNCTION

$$\frac{F : \{p_1\} \ S \ \{q_1\}, F : \{p_2\} \ S \ \{q_2\}}{F : \{p_1 \wedge p_2\} \ S \ \{q_1 \wedge q_2\}}$$

RULE A6: ∃-INTRODUCTION

$$\frac{F : \{p\}\ S\ \{q\}}{F : \{\exists z : p\}\ S\ \{q\}}$$

where z does not occur in F, S or q.

RULE A7: SUBSTITUTION

$$\frac{F : \{p\}\ S\ \{q\}}{F[z := t] : \{p[z := t]\}\ S\ \{q[z := t]\}}$$

where z does not occur in S and S does not change any of the variables in the term t.

The following invariance rule additionally allows to adjust the footprint.

RULE A8: INVARIANCE

$$\frac{(r \wedge F) \rightarrow F' \quad F : \{p\}\ S\ \{q\}}{F' : \{p \wedge r\}\ S\ \{q \wedge r\}}$$

where S does not assign to the variables in the formula r. Further, $(r \wedge F) \rightarrow F'$ holds if for every $a : p'(\bar{x}) \in F$ there exists $a : p''(\bar{x}) \in F'$ such that $(r \wedge p'(\bar{x})) \rightarrow p''(\bar{x})$.

4   Case Study: Quicksort

We illustrate the use of footprints in a proof of correctness of the well-known quicksort sorting algorithm [Hoa62, FH71]:

$$QS(l, r) ::$$
$$\quad \textbf{if } l < r$$
$$\quad \textbf{then } P(l, r);$$
$$\quad\quad \textbf{begin}$$
$$\quad\quad \textbf{local } u := m;$$
$$\quad\quad QS(l, u - 1);$$
$$\quad\quad QS(u, r)$$
$$\quad\quad \textbf{end}$$
$$\quad \textbf{fi}$$

Here $P(l, r)$ calls the *partitioning* algorithm which operates on an array $a :$ **integer** $\rightarrow$ **integer** and generates a value for the global integer variable m. The partitioning algorithm satisfies the following contracts.

- $A_1 \equiv x \in [l : r] : \{l < r\}\ P(l, r)\ \{m \in (l : r] \wedge a[l : m - 1] \le a[m : r]\}$
- $A_2 \equiv x \in [l : r] : \{a = a_0\}\ P(l, r)\ \{perm(a, a_0, l, r)\}$

where $m \in (l : r]$ abbreviates $l < m \wedge m \leq r$ and $a[l : m - 1] \leq a[m : r]$ abbreviates $\forall i \in [l : m - 1] : \forall j \in [m : r] : a[i] \leq a[j]$. The postcondition of the first contract thus states that the array segment $a[l : r]$ can be split into two segments $a[l : m-1]$ and $a[m : r]$ such that all numbers in $a[l : m-1]$ are smaller or equal to all numbers in $a[m : r]$. The predicate $perm(a, a_0, l, r)$ states that the array a is a permutation of a_0 on the interval $[l : r]$, which can be expressed by the assertion

$$\exists b : \forall i, j \in [l : r] : \exists k \in [l : r] : (i \neq j \rightarrow b[i] \neq b[j]) \wedge b[i] = k \wedge a[i] = a_0[b[i]]$$

where b is an array of type **integer** $\rightarrow$ **integer**. Finally, the footprint $x \in [l : r]$ of the array a states that the array a is only changed on the interval $[l : r]$ (we thus omit for notational convenience the label "a").

Given these contracts we want to prove the following specifications of Quicksort:

- $B_1 \equiv x \in [l : r] : \{\textbf{true}\} \ QS(l, r) \ \{sorted(a[l : r])\}$
- $B_2 \equiv x \in [l : r] : \{a = a_0\} \ QS(l, r) \ \{perm(a, a_0, l, r)\}$

where $sorted(a[l : r])$ abbreviates the assertion

$$\forall i \in [l : r - 1] : a[i] \leq a[i + 1]$$

Let S_{QS} denote the body of the procedure QS. By the recursion rule it suffices to prove (assuming the contracts A_1 and A_2)

$$A_1, A_2, B_1, B_2 \vdash$$
$$x \in [l : r] : \{\textbf{true}\} \ S_{QS} \ \{sorted(a[l : r])\} \tag{1}$$

and

$$A_1, A_2, B_2 \vdash$$
$$x \in [l : r] : \{a = a_0\} \ S_{QS} \ \{perm(a, a_0, l, r)\} \tag{2}$$

Proof of Obligation (1). By the conditional rule it suffices to prove that

$$x \in [l : r] : \{l \geq r\} \ skip \ \{sorted(a[l : r])\} \tag{3}$$

and

$$A_1, B_1, B_2 \vdash x \in [l : r] : \{l < r\} \ T \ \{sorted(a[l : r])\} \tag{4}$$

where T denotes the then-branch of S.

The first follows directly from a trivial application of the consequence rule. Using A_1, the (standard) assignment axiom, the block [AdBO09] and sequential composition rule, it is straightforward to establish proof obligation (4) from

$$x \in [l : r] : \quad \begin{array}{c} \{u \in (l : r] \wedge a[l : u - 1] \leq a[u : r]\} \\ QS(l, u - 1); QS(u, r) \\ \{sorted(a[l : r])\} \end{array} \tag{5}$$

In order to establish this proof obligation, we first instantiate l and r by u and r, respectively, in B_1:

$$x \in [u : r] : \{\mathbf{true}\} \; QS(u, r) \; \{sorted(a[u : r])\}$$

By instantiating l and r by u and r, respectively, in B_2, and the conjunction rule:

$$x \in [u : r] : \{a = a_0\} \; QS(u, r) \; \{sorted(a[u : r]) \wedge perm(a, a_0, u, r)\}$$

It is straightforward to check that

$$(sorted(a[l : u - 1]) \wedge a[l : u - 1] \le a_0[u : r]) \uparrow x \in [u : r] \leftrightarrow$$
$$(sorted(a[l : u - 1]) \wedge a[l : u - 1] \le a_0[u : r])$$

Thus applying the footprint rule 2:

$$x \in [u : r] : \quad \begin{array}{c} \{a = a_0 \wedge sorted(a[l : u - 1]) \wedge a[l : u - 1] \le a_0[u : r]\} \\ QS(u, r) \\ \{sorted(a[u : r]) \wedge perm(a, a_0, u, r) \wedge sorted(a[l : u - 1]) \wedge a[l : u - 1] \le a_0[u : r]\} \end{array}$$

Next observe that

– $perm(a, a_0, u, r) \wedge a[l : u - 1] \le a_0[u : r]$ implies $a[l : u - 1] \le a[u : r]$, and
– $sorted(a[l : u - 1]) \wedge sorted(a[u : r]) \wedge a[l : u - 1] \le a[u : r]$ implies $sorted(a[l : r])$.

Thus by the consequence rule:

$$x \in [u : r] : \quad \begin{array}{c} \{a = a_0 \wedge sorted(a[l : u - 1]) \wedge a[l : u - 1] \le a_0[u : r]\} \\ QS(u, r) \\ \{sorted(a[l : r])\} \end{array}$$

Existential elimination (of a_0) and consequence rule:

$$x \in [u : r] : \quad \begin{array}{c} \{sorted(a[l : u - 1]) \wedge a[l : u - 1] \le a[u : r]\} \\ QS(u, r) \\ \{sorted(a[l : r])\} \end{array}$$

Invariance rule A8 (adjusting the footprint: $u \in (l : r] \wedge x \in [u : r]$ implies $x \in [l : r]$):

$$x \in [l : r] : \quad \begin{array}{c} \{u \in (l : r] \wedge sorted(a[l : u - 1]) \wedge a[l : u - 1] \le a[u : r]\} \\ QS(u, r) \\ \{u \in (l : r] \wedge sorted(a[l : r])\} \end{array}$$

Consequence rule:

$$x \in [l:r]: \quad \frac{\{u \in (l:r) \wedge sorted(a[l:u-1]) \wedge a[l:u-1] \le a[u:r]\}}{QS(u,r)} \qquad (6)$$
$$\{sorted(a[l:r])\}$$

Following a similar pattern as above, from assumption B_1 we obtain by instantiation

$$x \in [l:u-1]: \{\mathbf{true}\}\ QS(l,u-1)\ \{sorted(a[l:u-1])\}$$

By instantiating assumption B_2 and the conjunction rule:

$$x \in [l:u-1]: \{a=a_0\}\ QS(l,u-1)\ \{sorted(a[l:u-1]) \wedge perm(a,a_0,l,u-1)\}$$

It is straightforward to check that

$$(a_0[l:u-1] \le a[u:r]) \uparrow x \in [l:u-1] \leftrightarrow (a_0[l:u-1] \le a[u:r])$$

Thus applying the footprint rule 2:

$$x \in [l:u-1]: \quad \frac{\{a=a_0 \wedge a_0[l:u-1] \le a[u:r]\}}{QS(l,u-1)}$$
$$\{sorted(a[l:u-1]) \wedge perm(a,a_0,l,u-1) \wedge a_0[l:u-1] \le a[u:r]\}$$

Since

$$perm(a,a_0,l,u-1) \wedge a_0[l:u-1] \le a[u:r]$$

implies $a[l:u-1] \le a[u:r]$, we obtain by the consequence rule:

$$x \in [l:u-1]: \quad \frac{\{a=a_0 \wedge a_0[l:u-1] \le a[u:r]\}}{QS(l,u-1)}$$
$$\{sorted(a[l:u-1]) \wedge a[l:u-1] \le a[u:r]\}$$

Existential elimination (of a_0) and consequence rule:

$$x \in [l:u-1]: \quad \frac{\{a[l:u-1] \le a[u:r]\}}{QS(l,u-1)}$$
$$\{sorted(a[l:u-1]) \wedge a[l:u-1] \le a[u:r]\}$$

Invariance rule A8 (adjusting the footprint: $u \in (l:r) \wedge x \in [l:u-1]$ implies $x \in [l:r]$):

$$x \in [l:r]: \quad \frac{\{u \in (l:r) \wedge a[l:u-1] \le a[u:r]\}}{QS(l,u-1)} \qquad (7)$$
$$\{u \in (l:r) \wedge sorted(a[l:u-1]) \wedge a[l:u-1] \le a[u:r]\}$$

Sequential composition applied to the correctness formulas (6) and (7) finally establishes proof obligation (5).

Proof of Obligation (2). In this proof we use the next lemma.

Lemma 3. *Suppose z does not occur in S and let p be a binary transitive relation. From $F : \{x = z\}\ S\ \{p(x, z)\}$ we can derive $F : \{p(x, z)\}\ S\ \{p(x, z)\}$.*

Proof: Assume $F : \{x = z\}\ S\ \{p(x, z)\}$. Apply the standard invariance rule, where y is a fresh variable:

$$F : \{x = z \wedge p(z, y)\}\ S\ \{p(x, z) \wedge p(z, y)\}$$

Consequence (rule 3):

$$F : \{x = z \wedge p(z, y)\}\ S\ \{p(x, y)\}$$

Existential elimination (rule A6) and consequence (rule 3):

$$F : \{p(x, y)\}\ S\ \{p(x, y)\}$$

Substitution (rule A7):
$$F : \{p(x, z)\}\ S\ \{p(x, z)\}$$

$\square$

We are now ready to establish proof obligation (2). By the conditional rule it suffices to prove

$$x \in [l : r] : \{l \geq r \wedge a = a_0\}\ skip\ \{perm(a, a_0, l, r)\} \tag{8}$$

and

$$B_2 \vdash x \in [l : r] : \{l < r \wedge a = a_0\}\ T\ \{perm(a, a_0, l, r)\} \tag{9}$$

where T denotes the then-branch of S. The first follows trivially by the consequence rule. To establish proof obligation (9), we obtain from assumption B_2 by instantiation

$$x \in [u : r] : \{a = a_0\}\ QS(u, r)\ \{perm(a, a_0, u, r)\}$$

Note that $perm(a, a_0, l, u - 1) \uparrow x \in [u : r]$ is logically equivalent to $perm(a, a_0, l, u - 1)$ itself. Thus applying the footprint rule 2:

$$x \in [u : r] :$$
$$\{a = a_0 \wedge perm(a, a_0, l, u - 1)\}\ QS(u, r)\ \{perm(a, a_0, l, u - 1) \wedge perm(a, a_0, u, r)\}$$

Consequence rule:

$$x \in [u : r] : \{a = a_0\}\ QS(u, r)\ \{perm(a, a_0, l, r)\}$$

Note that $perm$ is a binary, transitive predicate in the first two arguments. Thus lemma 3 gives:

$$x \in [u : r] : \{perm(a, a_0, l, r)\}\ QS(u, r)\ \{perm(a, a_0, l, r)\}$$

Invariance rule A8 (adjusting the footprint) and consequence rule (weakening the postcondition):

$$x \in [l : r] : \{u \in (l : r] \wedge perm(a, a_0, l, r)\} \; QS(u, r) \; \{perm(a, a_0, l, r)\} \quad (10)$$

As above, we can derive

$$x \in [l : u - 1] : \{perm(a, a_0, l, r)\} \; QS(l, u - 1) \; \{perm(a, a_0, l, r)\}$$

Invariance rule A8 (adjusting the footprint again):

$$x \in [l : r] :$$
$$\{u \in (l : r] \wedge perm(a, a_0, l, r)\} \; QS(l, u - 1) \; \{u \in (l : r] \wedge perm(a, a_0, l, r)\}$$
$$(11)$$

Sequential composition (applied to the correctness formulas (10) and (11)):

$$x \in [l : r] :$$
$$\{u \in (l : r] \wedge perm(a, a_0, l, r)\} \; QS(l, u - 1); QS(r, u) \; \{perm(a, a_0, l, r)\}$$

The remainder of the proof follows in a straightforward manner by instantiating A_1 and A_2, and applying the conjunction rule, the block rule (for $u := m$), and the rule for sequential composition.

5   Future Work

First we want to prove (relative) completeness of the Hoare logic extended with footprints. We conjecture that this requires a straightforward extension of the usual Gorelick (relative) completeness proof (see [Apt84]).

It is not difficult to extend the notion of footprints to reasoning about invariant properties of object-oriented programs. The basic idea is simply to include for each *field f* a monadic predicate $p_f :$ **Object** $\rightarrow$ **Boolean** which holds for all those objects which have updated their field f. This extension we want to integrate with our proof theory of *abstract object creation* [AdBG09] which allows specification and verification of dynamic heap structures at an abstraction level that coincides with the Java programming language and which already has been implemented in the KeY theorem prover [BHS07].

References

[AdBG09] Ahrendt, W., de Boer, F.S., Grabe, I.: Abstract Object Creation in Dynamic Logic. In: Cavalcanti, A., Dams, D.R. (eds.) FM 2009. LNCS, vol. 5850, pp. 612–627. Springer, Heidelberg (2009)

[AdBO09] Apt, K.R., de Boer, F.S., Olderog, E.-R.: Verification of Sequential and Concurrent Programs, Texts in Computer Science. Springer (2009)

[Apt84] Apt, K.R.: Ten years of hoare's logic: A survey part II: nondeterminism. Theor. Comput. Sci. **28**, 83–109 (1984)

[BHS07] Beckert, B., Hähnle, R., Schmitt, P.H.: Verification of object-oriented software: The KeY approach. Springer (2007)

[BK14] Brotherston, J., Kanovich, M.I.: Undecidability of propositional separation logic and its neighbours. J. ACM **61**(2), 14 (2014)

[CYO01] Calcagno, C., Yang, H., O'Hearn, P.W.: Computability and Complexity Results for a Spatial Assertion Language for Data Structures. In: Hariharan, R., Mukund, M., Vinay, V. (eds.) FSTTCS 2001. LNCS, vol. 2245, pp. 108–119. Springer, Heidelberg (2001)

[FH71] Foley, M., Hoare, C.A.R.: Proof of a recursive program: Quicksort. Comput. J. **14**(4), 391–395 (1971)

[Hoa62] Hoare, C.A.R.: Quicksort. Comput. J. **5**(1), 10–15 (1962)

[Hoa71] Hoare, C.A.R.: Procedures and parameters: An axiomatic approach. In: Symposium on Semantics of Algorithmic Languages, pp. 102–116 (1971)

[Old83] Olderog, E.-R.: On the notion of expressiveness and the rule of adaption. Theor. Comput. Sci. **24**, 337–347 (1983)

[Rey05] Reynolds, J.C.: An Overview of Separation Logic. In: Meyer, B., Woodcock, J. (eds.) VSTTE 2005. LNCS, vol. 4171, pp. 460–469. Springer, Heidelberg (2008)

[Wei11] Weiß, B.: Deductive Verification of Object-Oriented Software: Dynamic Frames, Dynamic Logic and Predicate Abstraction. PhD thesis, Karlsruhe Institute of Technology (2011)

Toward Compact Abstractions for Processor Pipelines

Sebastian Hahn[1]([✉]), Jan Reineke[1], and Reinhard Wilhelm[1,2]

[1] Informatik, Saarland University, Saarbrücken, Germany
sebastian.hahn@cs.uni-saarland.de
[2] AbsInt Angewandte Informatik GmbH, Saarbrücken, Germany

Abstract. Hard real-time systems require programs to react on time. Static timing analysis derives timing guarantees by analyzing the behavior of programs running on the underlying execution platform. Efficient abstractions have been found for the analysis of caches. Unfortunately, this is not the case for the analysis of processor pipelines. Pipeline analysis typically uses an expensive powerset domain of concrete pipeline states. Therefore, pipeline analysis is the most complex part of timing analysis. We propose a compact abstract domain for pipeline analysis. This pipeline analysis determines the minimal progress of instructions in the program through the pipeline.

We give a concrete semantics for an in-order pipeline, which forms the basis for an abstract semantics. On the way, we found out that in-order pipelines are not guaranteed to be free of timing anomalies, i.e. local worst decisions do not lead to the global worst case. We prove this by giving an example. A major problem is how to find an abstract semantics that guarantees progress on the abstract side. It turns out that monotonicity on the partial progress order is sufficient to guarantee this.

1 Introduction

In state-of-the-art timing analysis, microarchitectural analysis, i.e. the part dealing with the influence of the underlying hardware platform on the execution time behavior, is the most complex part. There are two main reasons for this.

The first one is the complexity of modern microprocessors, which feature (cyclic) interdependencies between components. These interdependencies make it hard to impossible to decompose the analysis into several more efficient sub-analyses. Additionally, they often result in so called timing anomalies (i.e. following local worst cases does not lead to global worst cases) that complicate the analysis.

This work was supported by the Deutsche Forschungsgemeinschaft (DFG) as part of the Transregional Collaborative Research Centre SFB/TR 14 (AVACS) and by the Saarbrücken Graduate School of Computer Science which receives funding from the DFG as part of the Excellence Initiative of the German Federal and State Governments.

© Springer International Publishing Switzerland 2015
R. Meyer et al. (Eds.): Olderog-Festschrift, LNCS 9360, pp. 205–220, 2015.
DOI: 10.1007/978-3-319-23506-6_14

The second, not quite independent, reason is that no compact abstract domain for pipeline analysis has been found, yet, making it necessary to fall-back to a very expensive powerset domain of concrete pipeline states. The resulting state space exploration has to investigate all possible transitions of instructions through the microarchitecture in order not to miss the worst-case behavior. In this context, the aforementioned timing anomalies prevent state space reduction based on local decisions.

There are approaches that tackle the efficiency problem. Wilhelm [11] uses a symbolic representation of the elements of the powerset domain. Other approaches almost exclusively try to overcome the complexity of modern micro-processors. They range from stepping-back towards very simplistic pipeline designs [7] to limiting processor features such that decomposition into more efficient analyses are possible. One example is the PRET architecture [5] that features a thread-interleaved pipeline basically leading to sequential execution w.r.t. a single thread.

Not much research has so far been undertaken to develop compact abstract domain for pipelines. In the following, we present (speculative) ideas and thoughts on how such a compact domain could look like. We also determine sufficient conditions on the concrete pipeline behavior that admit compact domains while leading to precise results. Enforcing these conditions in hardware can lead to degradation of the system's overall performance. Although it cannot be expected that the efficiency problem is ultimately "solved", i.e. compact domains for arbitrarily complex architectures are found, it is a step in this new direction. In any case, it will provide a better understanding of how to model the microarchitectural timing behavior.

2   Background

2.1   Pipelines

We consider a normal RISC-like 5-stage in-order pipeline as depicted in Figure 1, i.e. program instructions are executed in an overlapped fashion, but in the order they occur in the program. First, an instruction is fetched from memory. Second, the instruction is decoded and operands are fetched from the register file. Next, the instruction is executed and potentially a memory address is generated and the corresponding memory access is initiated. In the next stage, pending data memory operations are finished. Last, the results computed by the instruction are written back to the register file. The fetch stage and the memory stage access a common background memory, possibly via separate instruction and data caches. The progress of an instruction in the pipeline is stalled when data dependencies would be violated or when an instruction is waiting for a memory access to be serviced.

Definition 1. *We call a pipeline* in-order *if each stage processes the instructions in the order they occur in the program.*

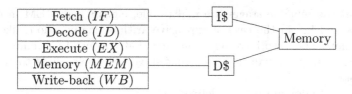

Fig. 1. 5-stage RISC pipeline. Fetch and memory stage access a common background memory through separate caches.

$$
\begin{array}{ll}
\text{IF} & \boxed{(i_8, ml)} \\
\text{ID} & \boxed{} \\
\text{EX} & \boxed{(i_7, 2)} \\
\text{MEM} & \boxed{(i_6, 0)} \\
\text{WB} & \boxed{}
\end{array}
\qquad
\lambda i_j \in \mathcal{I}. \begin{cases}
(pre, 0) & j > 8 \\
(IF, ml) & j = 8 \\
(EX, 2) & j = 7 \\
(MEM, 0) & j = 6 \\
(post, 0) & j < 6
\end{cases}
$$

Fig. 2. A concrete pipeline state in stage-centric and instruction-centric representation. The numbers denote the individual latencies, i.e. the cycles needed for the instruction to become ready in its current stage.

There are more advanced pipelining techniques that feature dynamic scheduling to reorder instruction and execute them out-of-(program)-order, speculation across branches, and additional buffers to decouple the pipeline and the memory hierarchy. We discuss the influence of these features on a compact representation for pipeline analysis in Section 5. For the remainder of this article, we focus on the presented in-order processor architecture.

2.2 Concrete Semantics of an In-Order Pipeline

In the following, we give a concrete semantics of an in-order pipeline. The remainder of this article is based on this concrete semantics.

As depicted in Figure 2, there are two equivalent views of a concrete pipeline state: A stage-centric view describing which stage is occupied by which instruction and an instruction-centric view describing which instruction occupies which stage. We select the second view because the abstract semantics, to be presented later, will represent the guaranteed progress of instructions through the pipeline. We use the first view for visualization purposes only.

Domain. An instruction in the pipeline can occupy one of the stages IF, ID, EX, MEM, and WB. We further distinguish between instructions that have not yet entered the pipeline, which are in the conceptual stage "pre", and instructions that have already left the pipeline, which are in the conceptual stage "post". Together, we obtain the following set of stages:

$$\mathcal{S} := \{pre, IF, ID, EX, MEM, WB, post\}.$$

Some of the pipeline stages are multi-cycle, e.g. IF and MEM in case of a cache miss and EX in the case of expensive arithmetic operations like floating point division. Thus, we introduce counters that capture how many cycles an instruction needs to remain in its current stage until being able to advance to the next stage.

The concrete domain is then defined as

$$Pipe := \mathcal{I} \to \mathcal{S} \times \mathbb{N},$$

where $\mathcal{I}$ denotes the set of instruction instances that form the instruction sequence $i_1, i_2, \ldots, i_n$ occurring during program execution.

Cycle Update. The cycle update $cycle : Pipe \to Pipe$ describes the concrete behavior of the pipeline informally described above, i.e. how a pipeline state changes during the execution of one processor cycle. The structure is quite generic and can be adapted to different pipeline designs: An instruction can advance in the pipeline if the instruction is ready to move to the next pipeline stage and this next pipeline stage would be free in the next cycle. An instruction might not be ready if there are unsatisfied data dependencies or it needs to wait for a memory transfer. In this case the instruction stays in the same stage, but its counter of remaining wait cycles might be decremented. If the next pipeline stage is still occupied in the next cycle, the instruction is stalled and stays unmodified in its current stage.

The next pipeline stage will be free in the next cycle if it is already free or if the instruction occupying it can move on to the next stage. An instruction in the WB stage is considered to always find its (fictive) next stage in the next cycle.

$$cycle(p : Pipe) :=$$

$$\lambda i \in \mathcal{I}. \begin{cases} (stage(i), cnt'(i)) & : \neg ready(i) \\ (stage'(i), latency(stage'(i), i)) & : ready(i) \wedge willbefree(stage'(i)) \\ (stage(i), cnt(i)) & : ready(i) \wedge \neg willbefree(stage'(i)) \end{cases}$$

where $(stage(i), cnt(i)) := p(i)$ and $cnt'(i)$, $stage'(i)$, $ready(i)$, $willbefree(s)$, and $latency(s, i)$ are defined as follows:

$$cnt'(i) := \begin{cases} cnt(i) - 1 & : cnt(i) > 0 \\ 0 & : cnt(i) = 0 \end{cases}$$

$$stage'(i) := \begin{cases} post & : stage(i) = WB \vee (stage(i) = ID \wedge i = nop) \\ WB & : stage(i) = MEM \\ MEM & : stage(i) = EX \\ EX & : stage(i) = ID \\ ID & : stage(i) = IF \\ IF & : stage(i) = pre \wedge tofetch(i) \\ pre & : stage(i) = pre \wedge \neg tofetch(i) \end{cases}$$

$$busfree := \neg\exists i.(stage(i) = IF \vee stage(i) = MEM) \wedge cnt(i) > 0$$
$$ready(i) := (cnt(i) = 0) \wedge (stage(i) = ID \Rightarrow \text{data dependencies satisfied})$$
$$\wedge (stage(i) = EX \wedge i = load/store \Rightarrow (dcachehit \vee busfree))$$
$$\wedge (stage(i) = pre \Rightarrow (icachehit \vee (busfree \wedge \neg dcachemiss)))$$

$$willbefree(s) := (s = post) \vee (\neg\exists i.stage(i) = s)$$
$$\vee (\exists i.stage(i) = s \wedge ready(i) \wedge willbefree(stage'(i)))$$

$$latency(s, i) := \begin{cases} 0 & : s \in \{ID, WB, pre, post\} \\ lat(i) & : s = EX \\ 0 & : (s = IF \wedge icachehit) \vee (s = MEM \wedge dcachehit) \\ ml & : (s = IF \wedge icachemiss) \vee (s = MEM \wedge dcachemiss) \end{cases}$$

where ml denotes the cache miss latency, and $lat(i)$ denotes the execution latency of instruction i that can depend on the operand values (e.g. in case of a division). Some of the values, i.e. lat, $icachehit$, $icachemiss$, $dcachehit$, $dcachemiss$, and $tofetch(i)$, depend on the environment of the pipeline that is known during an actual execution. A latency of 1 cycle is assumed for the access to an L1 cache. We omitted the state of the environment, such as caches, in this formulation for the sake of readability.

2.3 State-of-the-art Pipeline Analysis

Current pipeline analyses rely on expensive powerset domains as abstract domains. An abstract state is a set of concrete pipeline states. The abstract domain is thus given by

$$Pipe_{ps}^{\#} := (2^{Pipe}, \subseteq, \cup).$$

Basically, state-of-the-art pipeline analysis [4,9] computes the so-called collecting semantics of pipeline states, i.e. it computes for each program point the set of concrete pipeline states that reach the program point during execution. The transfer function $f_{ps}^{\#} : Pipe_{ps}^{\#} \rightarrow Pipe_{ps}^{\#}$ can thus be defined using the concrete $cycle$-function applied element-wise. The (abstract) transfer of a single concrete pipeline state can nevertheless lead to several successors due to uncertainty in the environment. E.g. it might be uncertain whether a memory access hits or misses the caches, or what the values of operands are. In these cases, the necessary parts of the environment are concretized resulting in several environments. The $cycle$ of the concrete state is then applied with each such (partially concretized) environment.

3 Are In-Order Pipelines Interesting?
Or: What About Timing Anomalies?

Microarchitectural analysis has to cope with uncertainty about (components of the) execution states. First, it does not generally know the initial execution state,

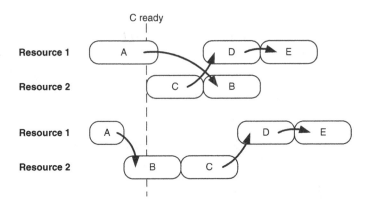

Fig. 3. Timing Anomaly: The local worst case does *not* lead to the global worst case [6].

e.g. the initial occupation of the pipeline stages with old instructions. Secondly, the analysis works with an abstract model of the architecture, potentially omitting some details in a safe way, thereby introducing uncertainty in the pipeline and the environment (e.g. caches). Thirdly, it typically combines information about the execution states resulting from different paths leading to one program point. Uncertainty about some components of the execution state may result in non-deterministic decisions to be made by the analysis based on the powerset domain.

A (timing) anomaly is a scenario where the local worst-case in a non-deterministic decision does not lead to the global worst case. A classical example depicted in Figure 3 is a cache miss (locally worse than a cache hit) that leads to a shorter overall execution time, because in the hit case operations are re-ordered in a way better suiting the subsequent program. This is often referred to as scheduling anomaly which is typically present in architectures featuring out-of-order execution. These anomalies hinder local state space reductions that would simplify the state-exploring pipeline analysis described in the previous section.

In-order pipelines with timing anomalies have been discussed earlier in literature. The anomalous behavior was exclusively triggered by odd cache behavior such as pseudo round-robin replacement [9] or partial cache line fills [3]. In-order pipelines with LRU caches have (implicitly) been considered anomaly-free as they execute instructions in a very regular fashion. As an example, Wenzel et al [10] identified *resource allocation decisions* as necessary condition for timing anomalies. A resource allocation decision describes a situation where a latency variation causes instructions to execute in different orders in the functional units of an superscalar processor.

The pipeline described in Section 2.2 does not allow *resource allocation decisions*, but nevertheless features a timing anomaly. All instructions enter the (sole) functional unit in program order – independent of any latency variation. The definition of a resource allocation decision does not capture, that a latency variation at the beginning of an instruction sequence can have an anomalous impact on

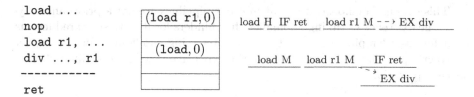

Fig. 4. Left: The program to execute that leads to an anomalous behavior. **Middle**: The pipeline state serving as starting point for the anomaly. **Right**: The ongoing execution of the program demonstrating the anomaly.

the *latency* (not the ordering) of later instructions. We will demonstrate that a timing anomaly is possible anyway and thus the property of in-order execution is not sufficient. However, the definition of Wenzel et al. [10] could be extended to also exclude reorderings of memory accesses.

Let us consider an in-order pipeline with separate caches, but common background memory as described earlier in Figure 1. The anomaly is based on the observation that in-order pipelines still allow the fetch of a instruction to be scheduled before the data access of a preceding instruction. In the following, we discuss the anomaly in more detail.

We make use of the following assumptions on the pipeline behavior that partially extend the formal description in Section 2.2. The pipeline has the ability to eliminate instructions from the pipeline that have no effect, e.g. nop instructions or predicated instructions whose condition is false. Furthermore, the pipeline features a long-lasting instruction such as floating-point division whose latency is at least as long as an L1 cache miss. This seems unrealistic at first glance, however when second-level caches are used to serve first-level misses it might very well be realistic.

Consider the program and (part of) its execution in Figure 3. The division instruction is data-dependent on the second load denoted by the dashed arrow in the right column. The dashed line in the left column denotes the beginning of a new cache line.

The pipeline state depicted in the middle column arises during execution of the program and serves as a starting point for the timing anomaly. We consider the following cache environment: The first load instruction might hit (H) or miss (M) the data cache, resulting in a non-deterministic decision that triggers the anomalous behavior. The second load instruction misses the data cache. The fetch of the return instruction misses the instruction cache. The right column shows the ongoing execution of the program, demonstrating the possibility of a timing anomaly. The stage between the two loads is free resulting from the eliminated nop. The anomaly is due to the fact, that the second load can advance to the *EX* stage while waiting for the first load to complete its miss. Thus it can be started before the fetch of the return instruction which suits the execution of the data-dependent division instruction. In the hit case, the second load becomes ready too late and is blocked by the already ongoing instruction fetch.

This example demonstrates that in-order pipelines are not a priori anomaly-free. The existence of timing anomalies thus hinders local state space reductions even for in-order pipelines. So, they are interesting candidates for a compact abstract representation—getting rid of the expensive powerset domain and state space exploration.

4   Compact Abstract Pipeline Domain Based on Minimal Progress

One idea for a compact representation of pipelines is inspired by our efficient cache analysis [2]. In this cache analysis, must and may analyses are employed to under-/overapproximate the cache content by tracking the maximal/minimal age of a cache block. In analogy, the idea for pipeline analysis is to track minimal/maximal progress of the instructions in the pipeline. The minimal progress metric can be used to guarantee that an instruction has eventually finished execution. One concern with minimal progress is, that – despite being conservative – some progress must be guaranteed for each call of the abstract domain's transfer function. Otherwise no bound can be derived.

4.1   Abstract Domain

First, we consider the pipeline behavior for one fixed instruction sequence $i_1 i_2 \ldots i_n$ with $i_j \in \mathcal{I}$. This eliminates uncertainty about the program's control-flow. We briefly discuss how to handle diverging and merging control-flow later.

Thus, the variation in execution times stems from cache uncertainty and variable-execution-latency instructions.

Minimal Progress. The abstract domain maps each instruction in the sequence to its minimal progress

$$Pipe^{\#} := \mathcal{I} \to \mathcal{P}_{min},$$

where $\mathcal{P}_{min} := \mathcal{S} \times \mathbb{N}$.

Note, that the domain is identical to the concrete domain, however the interpretation of a domain element is different. An element $ap \in Pipe^{\#}$ describes for each instruction its *minimal progress*, i.e. the pipeline stage that the instruction reached *at least*. Thereby, we establish an ordering on concrete pipeline states.

Defining the Partial Order. First, we define a progress order on the stages an instruction can be in. The idea behind the ordering is, that later stages are "better" in the sense that execution should not take longer starting from later stages. The order $\sqsubseteq_{\mathcal{S}}$ is then given by

$$post \sqsubseteq_{\mathcal{S}} WB \sqsubseteq_{\mathcal{S}} MEM \sqsubseteq_{\mathcal{S}} EX \sqsubseteq_{\mathcal{S}} ID \sqsubseteq_{\mathcal{S}} IF \sqsubseteq_{\mathcal{S}} pre.$$

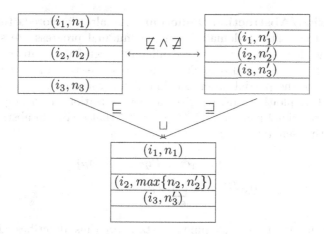

Fig. 5. Example of minimal-progress based join function. Take the minimum of the progress of individual instructions.

Some of the stages are multi-cycle, so we extend this to an ordering on *progress* $\sqsubseteq_{\mathcal{P}_{min}}$ as follows

$$(s, n) \sqsubseteq_{\mathcal{P}_{min}} (s', n') \Leftrightarrow s \sqsubseteq_{\mathcal{S}} s' \vee (s = s' \wedge n \leq n').$$

As the ordering $\sqsubseteq_{\mathcal{P}_{min}}$ is total, the induced join is

$$p \sqcup_{\mathcal{P}_{min}} p' = (p \sqsubseteq_{\mathcal{P}_{min}} p') \; ? \; p' : p.$$

The minimal-progress order on individual pipeline stages can then be extended to whole pipeline states. Two abstract pipeline states are ordered, if the minimal progress of all instructions is ordered:

$$s \sqsubseteq s' \Leftrightarrow \forall i \in \mathcal{I}.s(i) \sqsubseteq_{\mathcal{P}_{min}} s'(i).$$

Note, that the order respects the "has left the pipeline" property, i.e. whether an instruction has left the pipeline and thus finished its execution. Formally, if $s(i) = post$ for $i \in \mathcal{I}$ and $s \in Pipe^{\#}$, then

$$\forall s' \in Pipe^{\#}. \; s' \sqsubseteq s \Rightarrow s'(i) = post.$$

The join function $\sqcup$ is induced by the partial order $\sqsubseteq$ and corresponds to taking the minimal progress for each instruction:

$$s \sqcup s' = \lambda i \in \mathcal{I}.s(i) \sqcup_{\mathcal{P}_{min}} s'(i).$$

As an example consider the illustration in Figure 5.

Concretization/Abstraction Function. As already noted, the concrete domain and the abstract domain based on minimal progress are structurally equivalent, yet their interpretations are different. Therefore, an abstract minimal-progress pipeline state can also be viewed as a concrete state and vice versa. Note, that we can use the partial order and join defined above for concrete pipeline states as well. Exploiting this, we give the concretization function $\gamma : Pipe^\# \to 2^{Pipe}$ and abstraction function $\alpha : 2^{Pipe} \to Pipe^\#$ that relate our abstract domain to the collecting semantics domain and vice versa

$$\gamma(ap) := \{cp \in Pipe \mid cp \sqsubseteq ap\}$$
$$\alpha(CP) := \bigsqcup_{cp \in CP} cp$$

An abstract minimal-progress pipeline state ap thus describes all concrete pipeline states that have *at least* the progress of ap. The concretization and the abstraction function form a Galois connection [1].

4.2   Transfer Function

Before we discuss the transfer function, we present the general correctness criterion.

Definition 2 (Local Consistency and Best Abstract Transformer, [1]). *Let C and A be the concrete and abstract domains, and let $\gamma : A \to C$ and $\alpha : C \to A$ be the concretization and abstraction functions, and let $f : C \to C$ be the concrete transformer. An abstract transformer $f^\# : A \to A$ is* locally consistent *if and only if*

$$\forall a \in A. \ \gamma(f^\#(a)) \sqsupseteq_C f(\gamma(a)).$$

Let $f_{best}^\# = \alpha \circ f \circ \gamma$. If α and γ form a Galois connection, $f_{best}^\#$ is the best *abstract transformer. The best abstract transformer is locally consistent.*

Local consistency implies global consistency of the analysis results, i.e. the correctness of the overall analysis. Thus, it is sufficient to demonstrate local consistency of our abstract transformer as correctness proof.

Next, we try to come up with an abstract transformer for the abstract pipeline domain based on minimal progress.

The Easy Part

The transfer function takes a minimal progress abstract pipeline state from $Pipe^\#$ and computes the effect of the execution for a certain amount of time, e.g. one cycle. To be useful, the transfer function must always be able to guarantee strict progress in the sense of our partial order $\sqsubseteq$ defined in   Section 4.1.

The question to answer is: On what does the progress of an instruction depend? Clearly, the progress of an instruction depends on whether the next

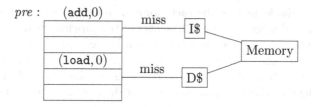

Fig. 6. How can progress be guaranteed during cycle transfer of this minimal-progress abstract state?

stage will be free – which in turn depends on the progress the instructions in these stages will make in the *current* cycle. This can be observed directly as the function *willbefree* in Section 2.2 is recursive. As a consequence, the transfer function should proceed backwards through the pipeline, i.e. the progress of instructions in later stages should be determined first. An instruction at least in stage *write back* will be at least in stage *post* after one cycle – no further dependencies.

Next there are data dependencies that cause hazards, so it also matters how far the dependent instructions have at least advanced in the pipeline (see *ready* in Section 2.2). Most of the data dependencies can be removed by employing techniques like forwarding; however some remain. Consider a `load` instruction followed by an arithmetic operation depending on the loaded value. If the arithmetic operation is *at least* in the decode stage, progress can only be guaranteed if the load is *at least* in the write-back phase. Observe that data dependencies have the same "direction" – upstream instructions depend on downstream instructions – as "resource dependencies" discussed in the previous paragraph.

The Hard Part

Unfortunately, the progress dependencies can be bidirectional in general, i.e. the progress of an instruction may also depend on the progress of an instruction further upstream in the pipeline. As an example, consider an instruction in the memory stage just about to request memory as part of a data access that is blocked by an already ongoing instruction fetch. This is caused by the unified background memory with sequential access.

Why are bidirectional dependencies problematic? A downstream instruction may be stalled by an upstream instruction *and* vice versa. If under the abstraction, it cannot be determined which of the two instructions is progressing and which is stalled, then *no* progress is guaranteed for either of the two.

Consider the example abstract state in Figure 6. Recall that the positions of the instructions represent their *minimal* progress, i.e. instructions could be further down the pipeline in a concrete execution. Considering the data access that might just be about to happen: it could be blocked as the instruction fetch could already have started. On the other hand, consider the instruction access that might just be about to happen: it could be blocked as the data access could already have started. Combining these two arguments, it follows that with the information available in the abstract domain, no progress can be guaranteed out of this state during one cycle. Essentially the pipeline deadlocks under the abstraction and no execution-time bound can be derived at all.

Note, that the uncertainty of whether instruction or data are scheduled for the bus does not exist within the powerset domain since in each concrete state it is always clear whether instruction fetch or data fetch acquire the bus first.

Monotonicity

A sufficient criterion to guarantee progress on each call of the transfer function $cycle^\#$ is monotonicity of the cycle update $cycle$. Monotonicity states that the transfer function preserves the ordering of states:

Definition 3 (Monotonicity). *Let two states $s_1, s_2 \in Pipe$. We call the (concrete) transfer function cycle monotone if and only if*

$$s_1 \sqsubseteq s_2 \Rightarrow cycle(s_1) \sqsubseteq cycle(s_2).$$

Theorem 1. *The abstract transformer $cycle^\# := cycle$ is the best abstract transformer of the minimal-progress abstract domain if cycle is monotone.*

Proof. We have to prove that $cycle^\# = cycle = \alpha \circ cycle \circ \gamma$. By plugging in the definition of γ and α, the claim becomes

$$cycle(ap) = \bigsqcup_{cp \sqsubseteq ap} cycle(cp).$$

Using the monotonicity property of $cycle$ concludes the proof.

Corollary 1. *The abstract transformer $cycle^\# := cycle$ is a sound abstract transfer function of the minimal-progress abstract domain.*

Combined with the property that the order $\sqsubseteq$ respects whether instructions are finished, and given that we start with a correct initial value, it follows that the analysis leads to overall sound results.

However, the transfer function $cycle$ as described in Section 2.2 is *not* monotone. This can be derived from the timing-anomalous behavior we described in Section 3. After one cycle, the state in the load-hit case made more progress compared to the load-miss case. But at the end, the load-miss case leads to a state that has progressed more.

Proposition 1 (Absence of Timing Anomalies). *If the transfer function cycle is monotone, the powerset-domain-based analysis can safely follow local worst-cases.*

The terminology local worst-/best-case suggests that, after one cycle, the state following the local best-case should have at least the progress as the state following the local worst-case has. The monotonicity property guarantees, that further cycling will always preserve this progress ordering. Thus, it is safe to follow the state with the minimal progress (arising from the local worst-case) – if the above characterization of local worst-/best-case is appropriate.

Having completely separate instruction and data memory ensures monotonicity. Then, the progress of an instruction solely depends monotonically on the progress of instructions further down in the pipeline. However, this scenario is unrealistic – applications could rely on self-modifying code or need to load data from the instruction memory (e.g. constant pools).

Another "hardware" attempt to ensure monotonicity is to never start a memory request upon an instruction-cache miss as long as an instruction, potentially accessing data memory, could be blocked by this. This way, we enforce a stronger property than in-order execution as we defined it above. In-order execution for example still allows that the fetch (memory access) of a later instruction can be scheduled before the data access of an earlier execution. In some sense, the execution is not in-order w.r.t. to externally visible events such as the acquisition of the memory bus.

Definition 4 (Strictly In-Order). *We call a pipeline* strictly in-order *if each resource processes the instructions in the order they occur in the program.*

These resources include the pipeline stages as well as the common background memory. The definition enforces, that all memory accesses of one instruction (i.e. the instruction fetch and potential data accesses) happen before any memory access of a later instruction.

Recall the definition of the concrete pipeline semantics in   Section 2.2   that is not strictly in-order. We modify the underlying pipeline such that it becomes strictly in-order as follows:

$$ready(i) := \dots \wedge (stage(i) = pre \Rightarrow (icachehit \vee$$
$$(busfree \wedge \forall pr \in prev(i).\ (pr \neq ld/str \vee stage(pr) \notin \{IF, ID, EX\}))).$$

An instruction miss that could block the bus for earlier data memory accessing instructions is delayed until no such instructions are in the "critical area" any more. In the case of an instruction cache hit, no such actions are necessary as the caches are separated.

Proposition 2. *The strictly in-order pipeline just described is monotone in the sense of Definition 3.*

The detailed proof of this proposition is quite lengthy and therefore we only present a sketch here. Given two pipeline states s_1, s_2 such that $s_1 \sqsubseteq s_2$. The proof uses a case distinction of the progress of an instruction in s_2. The cases should be considered in a bottom-up fashion starting with *post* and ending with *pre*. This represents the progress dependencies on the progress of instructions further down the pipeline. Then, we exploit that each instruction has at least the same progress in s_1. Using the definition of the concrete semantics, it follows that $cycle(s_1) \sqsubseteq cycle(s_2)$.

Note that in general, even a strictly in-order pipeline may feature timing anomalies, e.g., if it contains multiple incomparable functional units as described by Wenzel et al [10].

Outlook: Enriched Abstraction

An alternative to enforcing monotonicity of the concrete behavior of the pipeline by hardware modifications is to come up with more expressive abstractions. The idea is to enrich the abstraction with further (instrumented) properties about the time that has been spent at least in a specific stage.

An analogous idea has been successfully used in shape analysis via three-valued logic [8]. Additionally introduced *instrumentation predicates* made it possible to establish and preserve complex statements about heap-allocated data structures.

Transferred to our domain: To ensure that *some* progress is always made in the abstraction, one can instrument the semantics, so that every instruction tracks the number of cycles it has spent in a pipeline stage. Independently of the executed instruction, there is an upper bound on the time needed to pass a stage that can be determined from the concrete microarchitectural behavior. As soon as an instruction exceeds this bound, it is guaranteed to have advanced to the next stage.

Another possibility is to employ additional *relational* information. The example in Figure 6 shows, that progress cannot be guaranteed individually neither for the *add* nor the *load* instruction. However, we know that at least one of the two instructions makes progress in each cycle. Thus, an abstraction that tracks the progress of both instructions in a relational manner is eventually able to guarantee that both instructions have progressed to the next stage.

4.3   Diverging and Joining Control Flow

So far, we considered instruction sequences without branching and joining of control flow. The actual static analysis is performed on a control-flow graph with branches, control joins, and loops. A detailed and formal explanation of the consequences is out of the scope of this article, however we want to give a rough idea of how to extend the domain.

The problem with branching/joining is that one abstract state would need to talk about the behavior of different instruction sequences coming from/going to different branches of the control-flow graph. The obvious solution is to keep several abstract states – namely one per branch/different instruction sequence. After several abstract transformer cycles, the differing instructions will finally leave the pipeline and allow to join the remaining abstract states according to their minimal progress.

An efficient representation could be based on directed, acyclic graphs. Nodes in the graph are the instructions currently processed in the pipeline associated with their minimal progress. An edge points to the preceding instructions. In case of branch, several instructions have the same preceding instruction. In case of a control-flow join, the first common instruction has several preceding instructions and thus several outgoing edges.

4.4   Why Maximal Progress is not so Important

In this article, we focused on the minimal progress of instructions within a pipeline which is sufficient to derive an upper bound on the execution times of a program. In analogy to must-/may-cache analyses, an abstract domain tracking the *maximal* progress can be defined. This is needed e.g. for the calculations of lower bounds on the execution times of a program. Furthermore, in the case of non-monotone transformers it might be useful to prune some cases as infeasible.

5   Open Problems

So far, we examined in-order pipelines with separate caches which are well-suited candidates for compact abstractions due to their regular behavior. Modern processors, however, invest far more complexity to cleverly predict and optimize the executed instruction sequence. Their behavior is sensitive even to small local changes – leading to large global changes. This complicates the search for compact representations.

Branch Prediction and Speculation. Speculation techniques allow to execute instructions although it is unclear whether they should be executed at all (e.g. due to an unresolved branch). Speculative execution is known to cause timing anomalies [6] and is also problematic from the point of view of guaranteed progress. Corresponding concrete transformers are non-monotone: Speculatively executed instructions that progress further can turn out to be detrimental for the overall progress. Besides direct effects such as unnecessary and expensive memory accesses (see [6] for an example), speculation can pollute the cache leading to indirect effects due to reloads later.

Buffers such as Store Buffers. Additional buffers in the pipeline allow to further decouple the pipeline and the memory. As an example, stores complete to a store buffer such that the pipeline can continue execution while – in parallel – the store is actually performed in memory. Such behavior introduces additional dependencies, e.g. instruction fetches, data loads, and stores compete for the exclusive bus resource.

Out-Of-Order Execution. Data dependencies can hinder the execution of the current instruction in a program. Out-of-order execution allows to reorder instructions and thus to execute subsequent instructions whose dependencies are already satisfied. This complicates the dependencies of an instruction's progress – it might depend on the progress of instructions later in the program. Out-of-order execution is also known to cause timing anomalies [6].

6 Summary and Conclusion

We introduced design principles for pipelines with compact abstractions. We focus on an abstraction that is based on minimal progress of instructions through the pipeline. Any useful abstract transformer should guarantee some progress in each abstract transition. Otherwise, no execution-time bounds can be derived.

We showed that in-order pipelines are not automatically free of timing anomalies. Further, we found that monotonicity of the concrete transformer is sufficient for the absence of timing anomalies. Then, we defined strictly in-order pipelines and showed that these provide for monotone concrete transformers and thus for compact and effective abstractions.

References

1. Cousot, P., Cousot, R.: Systematic design of program analysis frameworks. In: Aho, A.V., Zilles, S.N., Rosen, B.K. (eds.) Conference Record of the Sixth Annual ACM Symposium on Principles of Programming Languages, pp. 269–282. ACM Press (1979)
2. Ferdinand, C., Wilhelm, R.: Efficient and precise cache behavior prediction for real-time systems. Real-Time Systems 17(2–3), 131–181 (1999)
3. Gebhard, G.: Timing anomalies reloaded. In: Lisper, B. (ed.) 10th International Workshop on Worst-Case Execution Time Analysis, WCET. OASICS, vol. 15, pp. 1–10. Schloss Dagstuhl - Leibniz-Zentrum für Informatik, Germany (2010)
4. Langenbach, M., Thesing, S., Heckmann, R.: Pipeline modeling for timing analysis. In: Hermenegildo, M.V., Puebla, G. (eds.) SAS 2002. LNCS, vol. 2477, pp. 294–309. Springer, Heidelberg (2002)
5. Liu, I., Reineke, J., Lee, E.A.: A PRET architecture supporting concurrent programs with composable timing properties. In: Conference Record of the Forty Fourth Asilomar Conference on Signals, Systems and Computers, pp. 2111–2115. IEEE (2010)
6. Reineke, J., Wachter, B., Thesing, S., Wilhelm, R., Polian, I., Eisinger, J., Becker, B.: A definition and classification of timing anomalies. In: Mueller, F. (ed.) 6th Intl. Workshop on Worst-Case Execution Time (WCET) Analysis. OASICS, vol. 4. Internationales Begegnungs- und Forschungszentrum fuer Informatik (IBFI), Schloss Dagstuhl, Germany (2006)
7. Rochange, C., Sainrat, P.: A time-predictable execution mode for superscalar pipelines with instruction prescheduling. In: Bagherzadeh, N., Valero, M., Ramírez, A. (eds.) Proceedings of the Second Conference on Computing Frontiers, pp. 307–314. ACM (2005)
8. Sagiv, M., Reps, T., Wilhelm, R.: Parametric shape analysis via 3-valued logic. In: Proceedings of the 26th ACM SIGPLAN-SIGACT Symposium on Principles of Programming Languages. POPL 1999, pp. 105–118. ACM, New York (1999)
9. Thesing, S.: Safe and precise WCET determination by abstract interpretation of pipeline models. Ph.D. thesis, Saarland University (2005)
10. Wenzel, I., Kirner, R., Puschner, P., Rieder, B.: Principles of timing anomalies in superscalar processors. In: Fifth International Conference on Quality Software (QSIC 2005). IEEE (2005)
11. Wilhelm, S.: Efficient analysis of pipeline models for WCET computation. In: Wilhelm, R. (ed.) 5th Intl. Workshop on Worst-Case Execution Time (WCET) Analysis. OASICS, vol. 1. Internationales Begegnungs- und Forschungszentrum für Informatik (IBFI), Schloss Dagstuhl, Germany (2005)

Synthesis

Bounded Synthesis for Petri Games

Bernd Finkbeiner[(✉)]

Universität des Saarlandes, Saarbrücken, Germany
finkbeiner@cs.uni-saarland.de

Abstract. Petri games, introduced in recent joint work with Ernst-Rüdiger Olderog, are an extension of Petri nets for the causality-based synthesis of distributed systems. In a Petri game, each token is a player in a multiplayer game, played between the "environment" and "system" teams. In this paper, we propose a new technique for finding winning strategies for the system players based on the bounded synthesis approach. In bounded synthesis, we limit the size of the strategy. By incrementally increasing the bound, we can focus the search towards small solutions while still eventually finding every finite winning strategy.

1 Introduction

The ambition to translate formal *specifications* into executable *programs*, and to do so *automatically*, without a human programmer, dates back to the early beginnings of computer science [1,2,10,19]. In the area of *reactive systems*, which includes hardware circuits, communication protocols, and generally all systems that interact continuously with their environment, the first formalization of the problem is generally attributed to Alonzo Church at the Summer Institute of Symbolic Logic in 1957 at Cornell University:

> Given a requirement which a circuit is to satisfy [...]. The synthesis problem is then to find recursion equivalences representing a circuit that satisfies the given requirement (or alternatively, to determine that there is no such circuit) [2].

Synthesis algorithms have the potential to dramatically simplify the development of complex systems. Instead of manually writing a program, one only needs to specify the actions available to the system and the objective, or *winning condition*, that one would like the system to guarantee against all possible behaviors of the system's environment. A strategy that achieves the winning condition by reacting to the actions of the environment with appropriate system actions is then constructed automatically.

Over the years, Church's synthesis problem has been studied in many variations. Of particular interest for the purposes of this paper is the problem of synthesizing *distributed* systems [6–9,13,15,16,18,20]. Many modern reactive

This research was partially supported by the German Research Council (DFG) in the Transregional Collaborative Research Center SFB/TR 14 AVACS.

© Springer International Publishing Switzerland 2015
R. Meyer et al. (Eds.): Olderog-Festschrift, LNCS 9360, pp. 223–237, 2015.
DOI: 10.1007/978-3-319-23506-6_15

systems are distributed in the sense that they consist of multiple processes with individual inputs, which they may or may not share with other processes. A key challenge in the design of such systems is to decide how the processes should interact so that each process obtains the information needed to carry out its functionality.

Computationally, the synthesis of distributed systems is a very hard problem, often with nonelementary lower bounds on the complexity [9]. An encouraging observation is, however, that practical specifications often have reasonably small implementations. This observation is exploited by the *bounded synthesis* approach [7,20], which restricts the space of potential solutions by some (iteratively growing) bound. Despite the uncomfortable lower bounds on the worst-case behavior of the synthesis algorithms, bounded synthesis is often able to find an implementation in reasonable time, as long as the solution is small with respect to a suitable parameter such as the number of states of the implementation.

In this paper, we propose a first bounded synthesis approach for Petri games. *Petri games* [5], introduced in recent joint work with Ernst-Rüdiger Olderog, are an extension of Petri nets for the causality-based synthesis of distributed systems. In a Petri game, each token is a player in a multiplayer game, played between the "environment" and the "system" team. In the tradition of Zielonka's automata [21], Petri games model distributed systems with *causal memory*, i.e., the processes memorize their causal histories and learn about each other's histories when synchronizing. The environment tokens represent independent sources of input, such as different users of the system. The system players represent the processes of the system. Each system player is only allowed to act on information it actually knows, either through direct interaction with the environment, or indirectly, through synchronization with other system players. Since the different system players have the same objective but different knowledge about the system state, a winning strategy usually involves an active synchronization between the system players to ensure that every player has the knowledge needed to win the game. A good example for the type of system that can be constructed with Petri games is the distributed burglar alarm system discussed in detail in [5]: a break-in may occur at one of several locations, and the alarm system at that location must inform the other distributed components about this, so that the alarm can be activated in all locations. Petri games have also been used to synthesize controllers for robots in production plants [4], where the Petri net is used to capture the concurrency, usage constraints, and uncertain availability of machines in the plant. The winning condition is to accomplish certain tasks, such as to process a certain number of orders on a certain number of machines, despite the actions of the hostile environment, which may declare a subset of the machines to be defect.

Strategies in a Petri game are, in general, infinite objects, because they are defined in terms of the (infinite) unfolding of the net. The reason for this construction is that different places in the unfolding reflect different causal histories; this enables the strategy to act depending on the individual history of a player. In practice we are, however, mostly interested in winning strategies that can be represented by a finite net. For the special case of a single environment token,

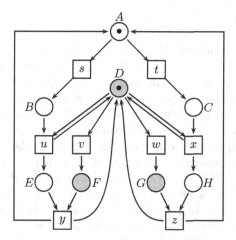

Fig. 1. Example Petri game. Places belonging to the system player are shown in gray, all other places belong to the environment player. The set of bad markings consists of the markings $\{B, G\}$ and $\{C, F\}$, where the net has reached a deadlock. The system player can win the game by waiting in place D until a synchronization with the environment token in places B or C becomes possible. The move after the synchronization then depends on whether the synchronization was via transition u or via transition x. In case of u, the system player takes transition v, thus avoiding $\{B, G\}$; in case of x, the system player takes w, thus avoiding $\{C, F\}$.

the existence of a winning strategy can be decided, and a finite representation of the winning strategy can be constructed in single-exponential time via a reduction to two-player games over finite graphs [5]. This synthesis algorithm has been implemented as a BDD-based fixed point iteration in the tool ADAM [4]. The bounded synthesis approach of the present paper complements the symbolic algorithm of ADAM with a satisfiability-based approach. We bound the size of the solutions of interest by setting a bound on the number of instances of each place of the Petri game. The existence of a winning strategy is then encoded as a quantified boolean formula, where the choices of the strategy appear as existentially quantified variables. We use a QBF solver to extract a satisfying assignment to these variables, which defines a winning strategy that meets the specified bounds.

The remainder of the paper is structured as follows. We begin in Section 2 with a review of the main notions for Petri games from [5]. In Section 3, we define *bounded* strategies. Section 4 then presents an encoding of the existence of winning strategies as a quantified boolean formula. In Section 5 we discuss the analysis of trade-offs, in particular the trade-off between the memory allocated to different players, and the trade-off between memory and proof complexity.

2 Petri Games

Petri games were introduced in [5] as an extension of Petri nets. In the following we briefly review the main definitions. To simplify the presentation, we consider in this paper only *safe* nets, where each place can be marked with at most one token. The definitions for the general case are given in [5].

Figure 1 shows a simple Petri game, which will serve as our running example in the following. If we ignore the distinction between gray and white places for the moment, then the net shown in the figure is a standard Petri net. A *Petri net* $\mathcal{N} = (\mathcal{P}, \mathcal{T}, \mathcal{F}, In)$ consists of disjoint sets $\mathcal{P}$ of *places* and $\mathcal{T}$ of *transitions*, a *flow relation* $\mathcal{F} \subseteq (\mathcal{P} \times \mathcal{T}) \cup (\mathcal{T} \times \mathcal{P})$, and an *initial marking* $In \subseteq \mathcal{P}$. We depict places as circles and transitions as rectangles. Note that the flow relation defines a bipartite graph, i.e., the flow relation connects places with transitions and transitions with places. Places and transitions are generically called *nodes*. A *finite* Petri net is a Petri net with finitely many nodes. For nodes x, y we write $x \mathcal{F} y$ for $(x, y) \in \mathcal{F}$.

The presence of a flow between two nodes models a causal dependency. The *preset* of a node x, denoted by $^{\bullet}x$, is the set $\{y \mid y \mathcal{F} x\}$. The *postset* of x, denoted by $x^{\bullet}$, is the set $\{y \mid x \mathcal{F} y\}$.

The behavior of a Petri net is defined in terms of its *markings*, which are subsets of the places. A transition t is *enabled* in a marking $M \subseteq \mathcal{P}$ if $^{\bullet}t \subseteq M$. A marking M' is *reachable* from a marking M *in one step*, denoted by $M \to M'$ if there is a transition t that is enabled in M and $M' = (M \smallsetminus {}^{\bullet}t) \cup t^{\bullet}$. A sequence $M_1 M_2 M_3 \dots M_n$ such that $M_i \to M_{i+1}$ for all $i \in \{1, \dots, n-1\}$ is a *firing sequence* of $\mathcal{N}$. The set of *reachable markings* is defined as $\mathcal{R}(\mathcal{N}) = \{M' \mid In \to^* M'\}$ where $\to^*$ is the reflexive and transitive closure of $\to$.

In the example of Fig. 1, the initial marking $\{A, D\}$ (which is depicted by the black dots on places A and D) has four enabled transitions, namely s, t, v and w, which result in the successor markings $\{B, D\}, \{C, D\}, \{A, F\}$ and $\{A, G\}$, respectively. An example for a firing sequence is the infinite repetition of the sequence $\{A, D\} \{B, D\} \{E, D\} \{E, F\} \{A, D\}$. Note that there is no reachable marking that contains both B and C. Places B and C are in conflict: transition s, which adds a token to B, also removes the token from A, which is needed to enable t. We say that two nodes x and y are *in conflict*, denoted by $x \sharp y$, if there exists a place $p \in \mathcal{P}$, different from x and y, from which one can reach x and y via the transitive closure $\mathcal{F}^+$ of $\mathcal{F}$, exiting p by different transitions.

A *Petri game* $\mathcal{G} = (\mathcal{P}_S, \mathcal{P}_E, \mathcal{T}, \mathcal{F}, In, \mathcal{B})$ is a finite Petri net where the set of places has been partitioned into a subset $\mathcal{P}_S$ belonging to the *system* and a subset $\mathcal{P}_E$ belonging to the *environment*; additionally, the Petri game identifies a set $\mathcal{B} \subseteq 2^{\mathcal{P}}$ of *bad* markings (from the point of view of the system), which indicate a victory for the environment[1]. We call the Petri net $\mathcal{N} = (\mathcal{P}, \mathcal{T}, \mathcal{F}, In)$ with $\mathcal{P} = \mathcal{P}_S \cup \mathcal{P}_E$ the *underlying* Petri net of $\mathcal{G}$.

---

[1] In [5], the bad markings are given as a set of bad places that must be avoided by the system.

We view each token as a player in a multiplayer game. Informally, the tokens on the environment places show the complete, unrestricted behavior, while the tokens on the system places restrict the behavior strategically by forbidding a subset of the transitions in the postset of the currently occupied place.

In the example game of Fig. 1, places belonging to the system are shown in gray, all other places belong to the environment. Let the set of bad markings consist of the markings $\{B, G\}$ and $\{C, F\}$, where the net has reached a deadlock. The example has two tokens, one that moves on environment places, one on system places. We will refer to these two tokens as the system player and the environment player, respectively. Note that, in general, there may be more than one token in a team, which means that these players have the same objective, but not necessarily the same knowledge about decisions that have been made by the players of the other team.

Initially, in place D, the system player must choose a subset of the transitions u, v, w and x. To avoid the bad markings, it is crucial that the system player stays in place D until it is clear whether the environment player has played transition s or t. If the system player proceeds with, for example, transition v, and the environment player turns out to be in place C, then the bad marking $\{C, F\}$ is reached. To win the game, the system player must forbid transitions v and w and thus stay in place D until a synchronization with the environment token in places B or C becomes possible. After the synchronization via u or x, the system player knows whether the environment player is in place E or H and can, correspondingly, enable v if the synchronization was via transition u or w if the synchronization was via transition x. Transition v thus leads to the marking $\{E, F\}$, transition w to the marking $\{G, H\}$ and the bad markings $\{B, G\}$ and $\{C, F\}$ are avoided.

The example illustrates that information about other players can be obtained through synchronization. The system player does *not* know whether the environment player went from place A via s to B or via t to C *until* the players synchronize via transitions u or x. The formalization of this idea is based on the notions of occurrence nets, branching processes, and unfoldings.

Informally, the unfolding of a Petri net is constructed by splitting any places with multiple incoming flows into separate copies that each only have a single incoming flow. Loops are unrolled into an infinite structure. Figure 2 shows the unfolding of the example game from Fig. 1. Note that every place in the unfolding has a unique causal history. Each place thus captures precisely the knowledge of the player when the token reaches the place in a play of the game. This correspondence between nodes and knowledge is exploited in the definition of strategies: a strategy fixes for each place in the unfolding the set of transitions that are not forbidden by any of the players. In Fig. 2, the strategy discussed above is depicted with thick lines. Note that, because there are multiple instances of place D, the system token in D can choose to forbid different transitions depending on whether transition u, transition x, or neither u or x has occurred in the present round of the game.

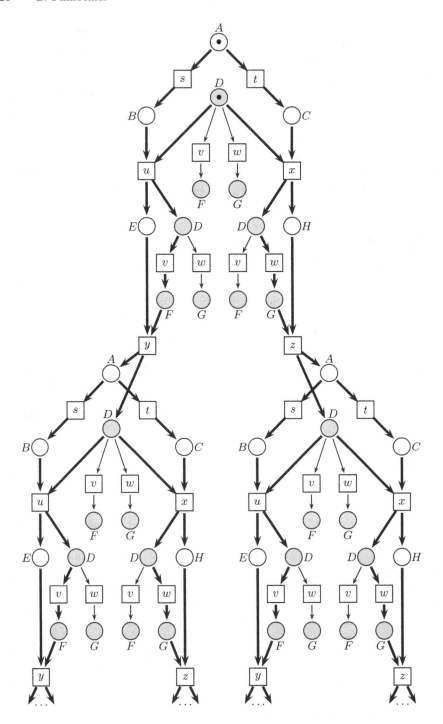

Fig. 2. Unfolding and winning strategy of the Petri game from Fig. 1. The winning strategy is shown with thick lines.

Formally, an *occurrence net* is a Petri net where (ON1) each place has at most one incoming transition; (ON2) the inverse flow relation $\mathcal{F}^{-1}$ is well-founded, i.e., starting from any node of $\mathcal{N}$ there does not exist an infinite path following the flow relation backwards; (ON3) no transition $t \in \mathcal{T}$ is in *self-conflict*, i.e., $t\sharp t$ does not hold for any transition t, and (ON4) the initial marking is the set of places without incoming transitions.

Two nodes x, y of an occurrence net are *causally related* if $x \mathcal{F}^* y$ or $y \mathcal{F}^* x$, where $\mathcal{F}^*$ denotes the reflexive and transitive closure of $\mathcal{F}$. They are *concurrent* if they are neither causally related nor in conflict. A *homomorphism* from a Petri net $\mathcal{N}_1$ to a Petri net $\mathcal{N}_2$ is a mapping $h : \mathcal{P}_1 \cup \mathcal{T}_1 \rightarrow \mathcal{P}_2 \cup \mathcal{T}_2$ that preserves the type of the elements, i.e., $h(\mathcal{P}_1) \subseteq \mathcal{P}_2$ and $h(\mathcal{T}_1) \subseteq \mathcal{T}_2$, and the pre- and postsets of the transitions, i.e., for all transitions $t \in \mathcal{T}_1 : h \downarrow {}^\bullet t$ is a bijection from ${}^\bullet t$ onto ${}^\bullet h(t)$ and $h \downarrow t^\bullet$ is a bijection from $t^\bullet$ onto $h(t)^\bullet$, where $h \downarrow D$ denotes the restriction of h to the domain D. If additionally the restriction $h \downarrow In_1$ is a bijection from In_1 onto In_2, then h is called *initial*.

A *branching process* of a net $\mathcal{N}$ is a pair $\beta = (\mathcal{N}^B, \lambda^B)$, where $\mathcal{N}^B$ is an occurrence net and λ^B is a homomorphism from $\mathcal{N}^B$ to $\mathcal{N}$ that is injective on transitions with the same preset, i.e., for all transitions t_1 and t_2 of the branching process, if ${}^\bullet t_1 = {}^\bullet t_2$ and $\lambda^B(t_1) = \lambda^B(t_2)$, then t_1 and t_2 must be the same transition. If λ^B is initial, β is called an *initial branching process*.

The *unfolding of a net* $\mathcal{N}$ is an initial branching process $\beta_U = (\mathcal{N}^U, \lambda^U)$ that is *complete* in the sense that every transition of the net is recorded in the unfolding, i.e., for every transition t and every set C of concurrent places, if $\lambda^U \downarrow C$ is a bijection from C onto ${}^\bullet t$, then there exists a transition $t^U \in \mathcal{T}^U$ such that ${}^\bullet t^U = C$ and $\lambda^U(t^U) = t$. The *unfolding of a game* $\mathcal{G}$ is the unfolding of the underlying net $\mathcal{N}$.

A branching process β_1 is a *subprocess* of a branching process β_2 if the identity on the nodes of β_1 is an initial homomorphism from β_1 to β_2. A *strategy* for the system players is a subprocess $\sigma = (\mathcal{N}^\sigma, \lambda^\sigma)$ of the unfolding $\beta_U = (\mathcal{N}^U, \lambda)$ of $\mathcal{N}$ subject to the following conditions: (S1) σ is deterministic in all system places, i.e., for all reachable markings $M \in \mathcal{R}(\mathcal{N}^\sigma)$ and all system places p, there is at most one transition $t \in \mathcal{T}^\sigma$ such that $p \in {}^\bullet t$ and ${}^\bullet t \subseteq M$; (S2) the strategy does not restrict local transitions of the environment, i.e., if, for a transition $t \in \mathcal{T}^U$, ${}^\bullet t \subseteq \mathcal{P}_E^\sigma$, then $t \in \mathcal{T}^\sigma$; and (S3) if an instance t of a transition is forbidden by σ there exists a place $p \in {}^\bullet t$ where σ uniformly forbids all instances t' of this transition.

A strategy σ for the system players is *winning* if the bad markings are unreachable in the strategy, i.e., $\mathcal{B} \cap \mathcal{R}(\mathcal{N}^\sigma) = \emptyset$. For example, the strategy shown in Fig. 2 is a winning strategy for the system player of the Petri game in Fig. 1. To avoid trivial solutions, we look for strategies σ that are *deadlock avoiding* in the sense that for all reachable markings $M \in \mathcal{R}(\mathcal{N}^\sigma)$, if there exists an enabled transition in the unfolding, i.e., $t \in \mathcal{T}^U$ with ${}^\bullet t \subseteq M$, then there exists an enabled transition in the strategy as well, i.e., $t \in \mathcal{T}^\sigma$ with ${}^\bullet t \subseteq M$. Note that we allow the strategy to produce a deadlock if the deadlock was already present in the game. In such situations we say that the game has *terminated*. If

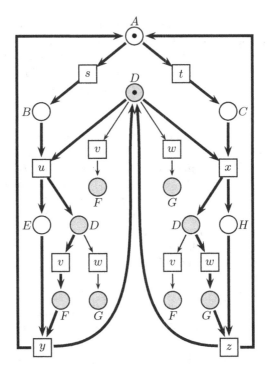

Fig. 3. Bounded unfolding und winning strategy of the Petri game from Fig. 1 for a bound b that allows for three instances of place D: $b(p) = 1$ for $p \in \{A, B, C, E, H\}$ and $b(p) = 3$ for $p \in \{D, F, G\}$. The b-bounded unfolding admits the winning strategy shown with thick lines.

termination is undesired (as in the example of Fig. 1), such markings must be explicitly included in the set of bad markings.

3   Bounded Strategies

Since strategies are subprocesses of the unfolding, they are in general infinite objects, even if the state space of an actual controller implementing the strategy turns out to be finite. A strategy $\sigma = (\mathcal{N}^\sigma, \lambda^\sigma)$ is *finitely generated* if there exists a finite net $\mathcal{N}^f$ and a homomorphism λ from $\mathcal{N}^\sigma$ to $\mathcal{N}^f$ such that $(\mathcal{N}^\sigma, \lambda)$ is an unfolding of $\mathcal{N}^f$. We say that σ is *finitely generated* by $\mathcal{N}^f$.

We search for finitely generated strategies by considering bounded unfoldings of the game. A *bound* $b : \mathcal{P} \to \mathbb{N}$ assigns to each place of the game a natural number. A pair $(\mathcal{N}^b, \lambda^b)$ consisting of a finite net $\mathcal{N}^b$ and a homomorphism λ^b from $\mathcal{N}^b$ to $\mathcal{N}$ is a *b-bounded unfolding* of the game $\mathcal{G}$ if there exists a homomorphism λ from the net $\mathcal{N}^U$ of the unfolding $(\mathcal{N}^U, \lambda^U)$ of $\mathcal{G}$ to $\mathcal{N}^b$ such that $\lambda^U(p) = \lambda^b(\lambda(p))$ for all nodes p of $\mathcal{N}^U$, and, furthermore, each place p of $\mathcal{G}$ occurs at most $b(p)$ times in $\mathcal{N}^b$, i.e., $|(\lambda^b)^{-1}(p)| \le b(n)$ for every $p \in \mathcal{P}$. Figure 3

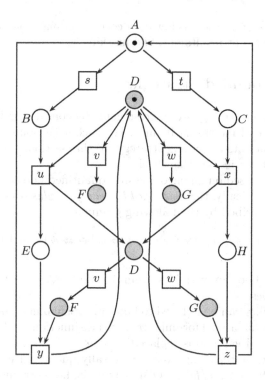

Fig. 4. Bounded unfolding of the Petri game from Fig. 1 for a bound b' that allows only two instances of place D: $b'(p) = 1$ for $p \in \{A, B, C, E, H\}$ and $b'(p) = 2$ for $p \in \{D, F, G\}$. The b'-bounded unfolding does not admit a winning strategy.

shows a b-bounded unfolding of the Petri game from Fig. 1 for the bound b with $b(p) = 1$ for $p \in \{A, B, C, E, H\}$ and $b(p) = 3$ for $p \in \{D, F, G\}$.

We find bounded strategies by restricting the flow of a bounded unfolding. A *b-bounded strategy* is a finite net $\mathcal{N}^f$ such that there is a b-bounded unfolding $\mathcal{N}^b$ of G with $\mathcal{P}^f = \mathcal{P}^b$, $\mathcal{T}^f = \mathcal{T}^b$, $In^f = In^b$ and $\mathcal{F}^f \subseteq \mathcal{F}^b$, and there is a strategy σ that is finitely generated by $\mathcal{N}^f$. We say that $\mathcal{N}^b$ *admits* the bounded strategy $\mathcal{N}^f$. The bounded strategy $\mathcal{N}^f$ is winning iff σ is winning.

In Fig. 3, a bounded strategy is depicted as part of the bounded unfolding. The thick lines indicate the flow that is preserved by the strategy. Note that the strategy from Fig. 2 is finitely generated by this bounded strategy. The bounded strategy is thus winning. Obviously, not every bounded unfolding admits a winning strategy. Consider, for example, the b'-bounded unfolding of our game in Fig. 4. The bound b' with $b'(p) = 1$ for $p \in \{A, B, C, E, H\}$ and $b'(p) = 2$ for $p \in \{D, F, G\}$ only allows two instances of D. The transitions u and x thus lead to the same instance of place D. As a result, the information whether the environment token is in place E or H is not available in the place reached by the two transitions. The strategy cannot forbid both outgoing transitions v and w, because this would result in a deadlock; however, no matter if the strategy chooses to enable v or w, there is always a corresponding choice of s vs. t for the environment player that results in a bad marking.

4   Finding Bounded Strategies

We look for winnning strategies of a given game by considering bounded unfoldings for a sequence of increasing bounds. For each such bounded unfolding, we check whether it admits a winning strategy. In this section, we describe an efficient method that carries out this check.

Our method is based on an encoding into quantified boolean formula (QBF) satisfiability. Syntactically, the *quantified boolean formulas* over a set of boolean variables $\mathcal{V}$ are described by the following grammar:

$$\phi ::= x \mid \phi \wedge \phi \mid \phi \vee \phi \mid \neg\phi \mid \phi \Rightarrow \phi \mid \phi \Leftrightarrow \phi \mid \textit{true} \mid \textit{false}$$
$$\Phi ::= \phi \mid \Phi \wedge \Phi \mid \Phi \vee \Phi \mid \exists x.\ \Phi \mid \forall x.\ \Phi$$

where x denotes boolean variables from $\mathcal{V}$ and $\wedge, \vee, \neg, \Rightarrow, \Leftrightarrow$ are the usual boolean connectives.

QBF satisfiability can be reduced to boolean satisfiability (SAT) by replacing every existentially quantified formula $\exists x.\phi$ by a disjunction $\phi[x/\textit{true}] \vee \phi[x/\textit{false}]$ where the quantified variable is replaced by *true* and *false*, respectively, in the two disjuncts, and by replacing every universally quantified formula $\forall x.\phi$ by a conjunction $\phi[x/\textit{true}] \wedge \phi[x/\textit{false}]$. QBF solving is, however, more difficult than SAT, both in theory (PSPACE-complete vs. NP-complete), and in practice [17]. Nevertheless, QBF is increasingly being used for practical applications and several powerful QBF solvers are available (cf. [14]).

Let $\mathcal{N}^b$ be a bounded unfolding. We encode the existence of a winning strategy that is admitted by $\mathcal{N}^b$ as a formula

$$\Phi_n = \exists V_S.\ \forall V_{T,n}.\ \phi_n,$$

where V_S is a set of boolean variables that encode which transitions are chosen by the strategy, $V_{T,n}$ is a set of boolean variables that encode a sequence of transitions, and ϕ_n is a boolean formula that expresses that if the choices of V_S and $V_{T,n}$ result in a firing sequence in $\mathcal{N}^b$, then the sequence is won by the system players. The index $n \in \mathbb{N}$ is a natural number that indicates the length of the firing sequence to be considered. We consider firing sequences that end in a repeated marking. Since there are only $2^{|\mathcal{P}^b|}$ many different markings, such a repetition must occur after at most exponentially many steps, and it suffices to set $n = 2^{|\mathcal{P}^b|} + 1$. However, we leave n as a parameter, which allows us to restrict the encoding to shorter firing sequences.

Our encoding of the existence of a winning strategy as a quantified boolean formula resembles the reduction of the bounded *model checking* problem to SAT [11]. The main difference is that in model checking, one is interested in finding a *single* firing sequence that leads from the initial marking to a bad marking, while in synthesis one must ensure that *all* firing sequences are correct.

We accomplish this by quantifying universally over the variables in $V_{T,n}$, which select the sequence of transitions, and by requiring that every firing sequence has a loop (unless it ends in a deadlock).

Proposition 1. *Let $\mathcal{N}^b$ be a bounded unfolding. There is a family of quantified boolean formulas Φ_n such that Φ_n is satisfiable for some $n \leq 2^{|\mathcal{P}^b|} + 1$ iff $\mathcal{N}^b$ admits a deadlock-avoiding winning strategy for the system players.*

Proof. We define $\Phi_n = \exists V_S.\ \forall V_{T,n}.\ \phi_n$ as follows. The set $V_S = \{(p, \lambda^b(t)) \mid p \in \mathcal{P}_S^b, t \in T^b, p \in {}^\bullet t\}$ consists of boolean variables encoding the system strategy. There is a variable for each pair of a system place and a transition where the preset of some instance of the transition contains the system place[2]. The set $V_{T,n} = \{(p, i) \mid p \in \mathcal{P}^b, i \in \{1, \ldots n\}\}$ contains one boolean variable for each place and index position between 1 and n, representing a sequence of markings of length n. The formula ϕ_n expresses that every sequence up to length n is winning for the system players. If the sequence reaches the full length n (i.e., there is no previous deadlock) it must be a loop:

$$\phi_n = \left(\bigwedge\nolimits_{i \in \{1, \ldots, n\}} sequence_i \Rightarrow winning_i \right) \wedge (sequence_n \Rightarrow loop_n)$$

Condition $sequence_n$ describes that the sequence of markings encoded by $V_{T,n}$ is indeed a firing sequence:

$$sequence_i = initial \wedge \neg deadlock_1 \wedge flow_1 \wedge \neg deadlock_2 \wedge flow_2 \wedge$$
$$\ldots \wedge \neg deadlock_{i-1} \wedge flow_{i-1}$$

where $initial$, $deadlock_i$, and $flow_i$ encode that the first marking is the initial marking, the occurrence of deadlock in the ith marking, and the satisfaction of the flow relation from the ith to the $(i+1)$st marking, respectively:

$$initial = \left(\bigwedge\nolimits_{p \in In^b} (p, 1) \right) \wedge \left(\bigwedge\nolimits_{p \in \mathcal{P}^b \smallsetminus In^b} \neg(p, 1) \right)$$
$$deadlock_i = \bigwedge\nolimits_{t \in T^b} \left(\bigvee\nolimits_{p \in {}^\bullet t} \neg(p, i) \right) \vee \left(\bigvee\nolimits_{p \in {}^\bullet t \cap \mathcal{P}_S^b} \neg(p, \lambda^b(t)) \right)$$
$$flow_i = \bigvee\nolimits_{t \in T^b} \left(\bigwedge\nolimits_{p \in {}^\bullet t} (p, i) \right) \wedge \left(\bigwedge\nolimits_{p \in {}^\bullet t \cap \mathcal{P}_S^b} (p, \lambda^b(t)) \right) \wedge \left(\bigwedge\nolimits_{p \in t^\bullet} (p, i+1) \right)$$
$$\wedge \left(\bigwedge\nolimits_{p \in \mathcal{P}^b \smallsetminus ({}^\bullet t \cup t^\bullet)} (p, i) \Leftrightarrow (p, i+1) \right) \wedge \left(\bigwedge\nolimits_{p \in {}^\bullet t \smallsetminus t^\bullet} \neg(p, i+1) \right)$$

Condition $winning_i$ ensures that there are no bad markings, that all deadlocks are terminating markings (i.e., the deadlock was already present in the net), and that the strategy is deterministic in all markings (as required by condition (S1)):

---

[2] The variable refers to $\lambda^b(t) \in T$ instead of $t \in T^b$ because of condition (S3) on strategies, which requires that all instances of a transition must be uniformly forbidden.

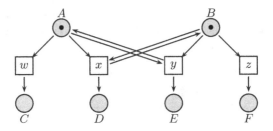

Fig. 5. Trade-off between the memory requirements of two players. The Petri game has a single bad marking $\{C, F\}$. Both the b-bounded unfolding with $b(A) = 1$ and $b(B) = 2$ and the b'-bounded unfolding with $b'(A) = 2$ and $b'(B) = 1$ admit a winning strategy, but the b''-bounded unfolding with $b''(A) = 1$ and $b''(B) = 1$ does not admit a winning strategy.

$$winning_i = nobadmarking_i \wedge deadlocksterm_i \wedge deterministic_i$$
$$nobadmarking_i = \bigwedge_{M \in \mathcal{B}} \left(\bigvee_{p \in M} \neg(p, i) \right) \vee \left(\bigvee_{p \in \mathcal{P}^b \setminus M} (p, i) \right)$$
$$deadlocksterm_i = deadlock_i \Rightarrow terminating_i$$
$$terminating_i = \bigwedge_{t \in \mathcal{T}^b} \bigvee_{p \in {}^\bullet t} \neg(p, i)$$
$$deterministic_i = \bigwedge_{t_1, t_2 \in \mathcal{T}^b, t_1 \neq t_2, {}^\bullet t_1 \cap {}^\bullet t_2 \cap \mathcal{P}^b_S \neq \emptyset}$$
$$\left(\bigvee_{p \in {}^\bullet t_1 \cup {}^\bullet t_2} \neg(p, i) \right) \vee \left(\bigvee_{p \in ({}^\bullet t_1 \cup {}^\bullet t_2) \cap \mathcal{P}^b_S} \neg(p, \lambda^b(t)) \right)$$

Finally, condition $loop_n$ ensures that the nth marking is a repetition of an earlier marking:

$$loop_n = \bigvee_{i \in \{1, \ldots n-1\}} \left(\bigwedge_{p \in \mathcal{P}^b} (p, i) \Leftrightarrow (p, n) \right)$$

$\square$

For each choice of n, the formula Φ_n can be translated into conjunctive normal form by tools like the Boolean circuit tool package (BCpackage) [12] and then solved by a standard QBF solver like DepQBF [14].

5 Trade-Offs

An interesting type of analysis made possible by the bounded synthesis approach is to trade different bounds against each other. The Petri game in Fig. 5 illustrates a trade-off between the memory needed in different processes of a system. The bad marking of the game is the combination of the places C of the first player and F of the second player. The bad marking is reached if the two players choose their local transitions w and z. To avoid the bad marking, one of the players must *wait* for the other player before taking the local transition. For example, the token in A could initially enable transition y, and, once the token in B has moved via y to E, safely take w to C. This winning strategy requires two instances of A and only a single instance of B. Symmetrically, the token in B could first enable x and then z, provided that two instances of B are available.

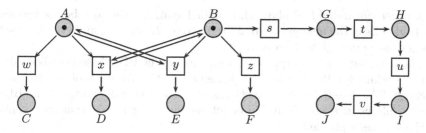

Fig. 6. Trade-off between memory and proof complexity. The Petri game has a single bad marking $\{C, F\}$. The b-bounded unfolding with $b(A) = 1$ and $b(B) = 2$, the b'-bounded unfolding with $b'(A) = 2$ and $b'(B) = 1$, and also the b''-bounded unfolding with $b''(A) = 1$ and $b''(B) = 1$ admit a winning strategy. However, the synthesis of the winning strategy in the b''-bounded unfolding requires a firing sequence of length six, while the firing sequences of the winning strategies in the b-bounded and b'-bounded unfoldings only have length three.

A winning strategy thus requires either two instances of A or two instances of B. Both the b-bounded unfolding with $b(A) = 1$ and $b(B) = 2$ and the b'-bounded unfolding with $b'(A) = 2$ and $b'(B) = 1$ admit a winning strategy, indicating that the memories allocated to the two players can be traded against each other, while the b''-bounded unfolding with $b''(A) = 1$ and $b''(B) = 1$ does not admit a winning strategy.

Since the encoding of the bounded synthesis problem in Section 4 is parametric in the length of the firing sequences, i.e., in the complexity of the *correctness proof*, another interesting trade-off to be analyzed is between memory and proof complexity. This is illustrated by the Petri game in Fig. 6, where, similarly to the previous example, the goal is to avoid the bad marking in places C and F. This time, however, the players can avoid the synchronization, if the token on place A takes the local transition to C and the token on B moves along the long chain towards the right. While this solution can be implemented within bound $b''(A) = b''(B) = 1$, the length of the firing sequence corresponds to the length of the chain plus the firing of w, resulting in length six, while the b-bounded unfolding with $b(A) = 1, b(B) = 2$ (and, analogously, the b'-bounded unfolding with $b'(A) = 2, b'(B) = 1$) admits a winning strategy where the firing sequence has only three markings.

6   Conclusions

We have presented a bounded synthesis method for Petri games. Similarly to the bounded synthesis approach for the synthesis of distributed systems from temporal logic [7,20], our approach limits the size of the solution and therefore finds small solutions fast. Petri games appear to be particularly well-suited for bounded synthesis because the net typically provides more structure than a logical specification. Because specific bounds can be set for individual places (and, hence, for individual players), bounded synthesis can also be used to analyze trade-offs in the memory needed for different the players, and even trade-offs between memory and proof complexity.

The bounded synthesis approach complements the BDD-based symbolic decision procedure for Petri games implemented in the ADAM tool [4]. In model checking, SAT-based bounded methods often dramatically outperform BDD-based symbolic methods [3]. It will be interesting to see if the situation is similar for the synthesis problem.

It is important to note, however, that the two methods are not directly comparable. While the symbolic decision procedure is limited to games with a single environment token, the bounded approach is universally applicable. On the other hand, the symbolic decision procedure can prove the absence of a strategy (of arbitrary size), while the bounded approach is currently limited to proving the existence of a strategy. Combining the two approaches is an interesting topic for future work.

Acknowledgments. I am deeply grateful to Ernst-Rüdiger Olderog for our productive and most enjoyable collaboration on Petri games. Thanks are also due to the anonymous referees for their helpful comments on this paper.

References

1. Büchi, J.R., Landweber, L.H.: Solving sequential conditions by finite-state strategies. Transactions of the American Mathematical Society **138** (1969)
2. Church, A.: Applications of recursive arithmetic to the problem of circuit synthesis. In: Summaries of the Summer Institute of Symbolic Logic, vol. 1, pp. 3–50. Cornell Univ., Ithaca (1957)
3. Copty, F., Fix, L., Fraer, R., Giunchiglia, E., Kamhi, G., Tacchella, A., Vardi, M.Y.: Benefits of bounded model checking at an industrial setting. In: Berry, G., Comon, H., Finkel, A. (eds.) CAV 2001. LNCS, vol. 2102, p. 436. Springer, Heidelberg (2001)
4. Finkbeiner, B., Gieseking, M., Olderog, E.-R.: ADAM: causality-based synthesis of distributed systems. In: Kroening, D., Păsăreanu, C.S. (eds.) CAV 2015. LNCS, vol. 9206, pp. 433–439. Springer, Heidelberg (2015)
5. Finkbeiner, B., Olderog, E.: Petri games: synthesis of distributed systems with causal memory. In: Peron, A., Piazza, C. (eds.) Proc. Fifth Intern. Symp. on Games, Automata, Logics and Formal Verification (GandALF). EPTCS, vol. 161, pp. 217–230 (2014). http://dx.doi.org/10.4204/EPTCS.161.19
6. Finkbeiner, B., Schewe, S.: Uniform distributed synthesis. In: Proc. LICS, pp. 321–330. IEEE Computer Society Press (2005)
7. Finkbeiner, B., Schewe, S.: Bounded synthesis. International Journal on Software Tools for Technology Transfer **15**(5–6), 519–539 (2013). http://dx.doi.org/10.1007/s10009-012-0228-z
8. Gastin, P., Lerman, B., Zeitoun, M.: Distributed games with causal memory are decidable for series-parallel systems. In: Lodaya, K., Mahajan, M. (eds.) FSTTCS 2004. LNCS, vol. 3328, pp. 275–286. Springer, Heidelberg (2004)

9. Genest, B., Gimbert, H., Muscholl, A., Walukiewicz, I.: Asynchronous games over tree architectures. In: Fomin, F.V., Freivalds, R., Kwiatkowska, M., Peleg, D. (eds.) ICALP 2013, Part II. LNCS, vol. 7966, pp. 275–286. Springer, Heidelberg (2013)
10. Green, C.: Application of theorem proving to problem solving. In: Proceedings of the 1st International Joint Conference on Artificial Intelligence. IJCAI 1969, pp. 219–239. Morgan Kaufmann Publishers Inc., San Francisco (1969). http://dl.acm.org/citation.cfm?id=1624562.1624585
11. Heljanko, K.: Bounded reachability checking with process semantics. In: Larsen, K.G., Nielsen, M. (eds.) CONCUR 2001. LNCS, vol. 2154, pp. 218–232. Springer, Heidelberg (2001)
12. Junttila, T.A., Niemelä, I.: Towards an efficient tableau method for boolean circuit satisfiability checking. In: Palamidessi, C., Moniz Pereira, L., Lloyd, J.W., Dahl, V., Furbach, U., Kerber, M., Lau, K.-K., Sagiv, Y., Stuckey, P.J. (eds.) CL 2000. LNCS (LNAI), vol. 1861, pp. 553–567. Springer, Heidelberg (2000)
13. Kupferman, O., Vardi, M.Y.: Synthesizing distributed systems. In: Proc. LICS, pp. 389–398. IEEE Computer Society Press (2001)
14. Lonsing, F., Biere, A.: DepQBF: A dependency-aware QBF solver. JSAT 7(2–3), 71–76 (2010)
15. Madhusudan, P., Thiagarajan, P.S., Yang, S.: The MSO theory of connectedly communicating processes. In: Sarukkai, S., Sen, S. (eds.) FSTTCS 2005. LNCS, vol. 3821, pp. 201–212. Springer, Heidelberg (2005)
16. Madhusudan, P., Thiagarajan, P.S.: Distributed controller synthesis for local specifications. In: Orejas, F., Spirakis, P.G., van Leeuwen, J. (eds.) ICALP 2001. LNCS, vol. 2076, p. 396. Springer, Heidelberg (2001)
17. Mangassarian, H.: QBF-based formal verification: Experience and perspectives. JSAT 133–191
18. Pnueli, A., Rosner, R.: Distributed reactive systems are hard to synthesize. In: Proc. FOCS 1990, pp. 746–757 (1990)
19. Rabin, M.O.: Automata on Infinite Objects and Church's Problem, Regional Conference Series in Mathematics, vol. 13. Amer. Math. Soc. (1972)
20. Schewe, S., Finkbeiner, B.: Bounded synthesis. In: Namjoshi, K.S., Yoneda, T., Higashino, T., Okamura, Y. (eds.) ATVA 2007. LNCS, vol. 4762, pp. 474–488. Springer, Heidelberg (2007)
21. Zielonka, W.: Asynchronous automata. In: Rozenberg, G., Diekert, V. (eds.) Book of Traces, pp. 205–248. World Scientific (1995)

Mediator Synthesis in a Component Algebra with Data

Lukáš Holík[1][(✉)], Malte Isberner[2], and Bengt Jonsson[3]

[1] Brno University of Technology, Brno, Czech Republic
holik@fit.vutbr.cz
[2] Technical University of Dortmund, Dortmund, Germany
malte.isberner@cs.uni-dortmund.de
[3] Uppsala University, Uppsala, Sweden
bengt@it.uu.se

Abstract. We formulate a compositional specification theory for components that interact by directed synchronous communication actions. The theory is an extension of interface automata which is also able to capture both absence of deadlock as well as constraints on data parameters in interactions. We define refinement, parallel composition, and quotient. The quotient is an adjoint of parallel composition, and produces the most general component that makes the components cooperate to satisfy a given system specification. We show how these operations can be used to synthesize mediators that allow components in networked systems to interoperate. This is illustrated by application to the synthesis of mediators in e-commerce applications.

1   Introduction

Modern software-intensive systems are increasingly composed of independently developed and network-connected software components. In many cases, these components exhibit heterogeneous behaviour, e.g., employing different protocols, which prevents them from cooperating to achieve user-level goals. Such cases call for the synthesis of *mediators*, which are intermediary software entities that allow software components to interact by coordinating their behaviours. Mediator synthesis has many different applications, including protocol converters [22], web service composition [4], and driver synthesis [20].

Component-based development, including mediator synthesis, should be performed within a specification theory. The theory should express how specifications capture the requirements for a component to function in an intended system context, while operators and refinement relations allow the composition and comparison of specifications, in analogy with how components are composed and refined towards an overall system design. Several such theories have been proposed, one of the earliest by Olderog and Hoare for Hoare-style specifications of Communicating Sequential Processes [14]. A more recent theory is that of interface automata of de Alfaro and Henzinger [7], in which components are assumed to communicate by synchronisation of input and output (I/O) actions, with the understanding that outputs are non-blocking. If an output is issued when another component

© Springer International Publishing Switzerland 2015
R. Meyer et al. (Eds.): Olderog-Festschrift, LNCS 9360, pp. 238–259, 2015.
DOI: 10.1007/978-3-319-23506-6_16

is unwilling to receive it, a communication mismatch is said to occur. This allows to capture assumptions on the behaviour of a component's environment.

The theory of interface automata is formulated in a finite-state, finite-alphabet setting. This falls short of adequately capturing the behavior of realistic systems, where communicating components exchange messages containing data values (e.g., header information or payload) ranging over large or infinite domains, and the overall control flow depends crucially on these values. Furthermore, the theory does not have any facilities for modeling progress or deadlock properties.

In this paper, we address the aforementioned issues by extending the formalism of interface automata to model both progress and constraints on data parameters. The constraints on data parameters are restricted to equalities and negated equalities between data values; this still allows to use the formalism in many interesting applications. We define refinement, *parallel composition* for structural composition of components, and *quotient* for synthesizing new components to satisfy partial requirements. The quotient is an adjoint of parallel composition, and produces the most general component that makes the system components cooperate to satisfy a given system specification.

Thereafter, we demonstrate how our specification theory and its operators can be used to automatically synthesize mediators in component-based systems. Several previous approaches to mediator synthesis require the user to supply a specification of the overall system functionality. However, the facilities of our theory allow to automatically generate such a system specification in many situations. More precisely, the ability to model constraints on data allows to capture data flow properties of a system, which in turn induces constraints on the control flow that result in a specification. We further show how the ability to specify progress properties results in the pruning of unproductive behavior from the resulting mediator. We have implemented the computation of quotients, and show the applicability of our approach by synthesizing mediators in e-commerce applications.

Related Work. Our component algebra is rooted in the theory of interface automata, which was developed for finite-state specifications, without covering data or progress [7]. The extension to progress was developed in our previous work [6]. Another related theory is the trace theory of Dill [8], which is based on the same distinction between inputs and outputs, and a notion of progress based on infinite traces. Another variation of this theory is the modal interfaces by Raclet and others [15–18], and the modal specifications by Larsen et al. [12], where *must* and *may* transitions play the role of input and output transitions. Yet another related theory is the ioco theory for model based testing due to Tretmans [21], which has a notion of deadlock that is very similar to ours.

For finite-state specifications, several works have presented a construction for computing quotients. Bhaduri and Ramesh [5] showed that for finite-state interface automata, which do not capture progress properties, quotient can be reformulated by combining parallel composition and renaming of interface actions:

this reformulation does not hold when deadlock and progress are considered. In our previous work [2], we used finite-state quotient to synthesize mediators. In the absence of facilities for modeling data, it was necessary to use an ontology for correlating different actions. In the current paper, the ability to specify constraints on data makes the use of ontologies unnecessary.

All works mentioned in the preceding paragraph are developed for the finite-state case, not covering data. There is one work that extends the theory of modal specifications with data [1]. It is based on a different model of interaction between processes, where data values are communicated via shared variables rather than as parameters in synchronization primitives. Moreover, the theory of modal specifications is based on a semantic foundation different from ours, in which specifications are compared in terms of the sets of transition systems that implement them. Our theory is based on notions of simulation between specifications, which allows mainstream techniques from controller synthesis to be adapted to the extension with data.

The synthesis of mediators has been addressed in works that target web service composition, e.g., in [3,4,9,10]. The authors of these works model components, representing web services, using automata extended with data. Mediators are synthesized using planning techniques, typically by a forward search in the space of possible mediators. The synthesized mediators are not required to be "best" or "most general" in a sense provided by some speficiation theory. In [4], loop-free controllers are synthesized that guarantee that the composed system reaches a target state while avoiding unsafe states. In contrast, our specification is a combination of a safety condition and a deadlock-freeness condition; these two kinds of specifications are of incomparable expressive power. Further difference include that our synthesized controllers may contain loops, and [4] under-approximates infinite data domains by finite ones.

2   Processes

In this section, we define our model of processs with data.

We assume an unbounded domain $\mathcal{D}$ of *data values* and a set of *actions*. Each action has a certain *arity*, which is the number of data parameters it takes. A *symbol* is a term of the form $\alpha(d_1, \ldots, d_n)$, where α is an action of arity n, and $d_1, \ldots, d_n$ are data values in $\mathcal{D}$. We further assume a set of *variables*, ranged over by x, y, z, etc., and a set of *formal parameters* ranged over by p_i for $i \in \mathbb{N}$. A *parameterized symbol* is of form $\alpha(p_1, \ldots, p_n)$, where α is an action of arity n, and $p_1, \ldots, p_n$ are distinct formal parameters. A *guard* is a conjunction of equalities and negated equalities over variables and formal parameters.

Definition 1. A *process* $\mathcal{P}$ is a tuple $\mathcal{P} = (I_\mathcal{P}, O_\mathcal{P}, L_\mathcal{P}, \ell_\mathcal{P}^0, X_\mathcal{P}, \sigma_\mathcal{P}^0, \delta_\mathcal{P}, Quie_\mathcal{P})$, where

- $I_\mathcal{P}$ is a finite set of *input actions*,
- $O_\mathcal{P}$ is a finite set of *output actions*,
- $L_\mathcal{P}$ is a finite set of *locations*,

- $\ell_{\mathcal{P}}^0 \in L_{\mathcal{P}}$ is the *initial location*,
- $X_{\mathcal{P}}$ is a tuple $x_1, \ldots, x_k$ of *variables*,
- $\sigma_{\mathcal{P}}^0 \colon X_{\mathcal{P}} \to \mathcal{D}$ is the *initial valuation*, assigning to each variable in $X_{\mathcal{P}}$ an initial value,
- $\delta_{\mathcal{P}}$ is a finite set of *transitions*, each of which is of form $\langle \ell, stmt, \ell' \rangle$ where $\ell, \ell' \in L_{\mathcal{P}}$ are locations, and $stmt$ is a statement of form

$$\alpha(p_1, \ldots, p_n) \; ; \; g \; ; \; x_1, \ldots, x_k := e_1, \ldots, e_k \quad ,$$

  where

  - $\alpha(p_1, \ldots, p_n)$ is a parameterized symbol with $\alpha \in (I_{\mathcal{P}} \cup O_{\mathcal{P}})$,
  - g is a guard over $X_{\mathcal{P}}$ and $p_1, \ldots, p_n$,
  - $x_1, \ldots, x_k := e_1, \ldots, e_k$ assigns to each variable $x_i \in X_{\mathcal{P}}$ an expression e_i over variables in $X_{\mathcal{P}}$ and formal parameters in $p_1, \ldots, p_n$.

- $Quie_{\mathcal{P}}$ maps each location in $L_{\mathcal{P}}$ to a predicate over $X_{\mathcal{P}}$. $\square$

We write $\bar{p}$ for $p_1, \ldots, p_n$, $\bar{d}$ for $d_1, \ldots, d_n$, $\bar{x}$ for $x_1, \ldots, x_k$, and similarly for $\bar{e}$. We write $A_{\mathcal{P}}$ for $I_{\mathcal{P}} \cup O_{\mathcal{P}}$. We use the term α-*transition* for a transition $\langle \ell, \; \alpha(\bar{p}); g; \bar{x} := \bar{e} \; , \ell' \rangle$ with the action α. For an action α and a location $\ell \in L_P$, let g_ℓ^α denote the disjunction of the guards of the outgoing α-transitions from ℓ; if there is no such statement, then g_ℓ^α is defined as *false*.

Intuitively, a process is at any point in time in a state, given by a location and a valuation of its variables. The process can change its state by performing a transition $\langle \ell, \; \alpha(\bar{p}); g; \bar{x} := \bar{e} \; , \ell' \rangle$, provided that its current location is ℓ, and that there is some $\alpha(\bar{d})$ that makes the guard $g[\bar{d}/\bar{p}]$ evaluate to true under the current valuation. It synchronizes via the symbol $\alpha(\bar{d})$, binds the formal parameters $\bar{p}$ to data values $\bar{d}$, and simultaneously assigns new values to the variables according to $\bar{x} := \bar{e}$.

The distinction between input and output actions is interpreted as follows. When in a state $\langle \ell, \sigma \rangle$, a process $\mathcal{P}$ is willing to receive any input $\alpha(\bar{d})$ that is enabled in the current state. Some input symbols may not be enabled: this expresses the assumption that the environment never generates these inputs when $\mathcal{P}$ is in $\langle \ell, \sigma \rangle$. This is in contrast to I/O-automata [11,13], which in every state must be prepared to receive any input. A process $\mathcal{P}$ can also emit any output symbol that is enabled. The predicate $Quie_{\mathcal{P}}(\ell)$ constrains when $\mathcal{P}$ can be *quiescent*, in the sense that if no input is received, the system *must* eventually emit some output symbol, *unless* the predicate $Quie_{\mathcal{P}}(\ell)$ is satisfied under the current valuation σ. Quiescence can be used to model termination or deadlock. We require that any process $\mathcal{P}$ is *consistent* with $Quie_{\mathcal{P}}$ in the sense that for any location ℓ of $\mathcal{P}$, the formula

$$Quie_{\mathcal{P}}(\ell) \vee \bigvee_{\alpha \in O_{\mathcal{P}}} \exists \bar{p}. \; g_\ell^\alpha$$

is universally valid. Here, $\bar{p}$ refers to the formal parameters of the respective action α.

Determinism. We assume that processes are *deterministic*, i.e., if g_1 and g_2 are the guards of two outgoing α-statements from a location ℓ, then g_1 and g_2 are mutually exclusive. Note that this still allows to express non-determinism on the level of data values. Consider, for instance, a process which nondeterministically selects an identifier, assigns it to a variable id and then transmits it in an action $showid(p)$, where p is id. We can represent this behavior by a single transition labeled by the statement

$$showid(p) \ ; \ true \ ; \ id := p \ .$$

Since this transition has the guard *true*, the parameter of *showid* can be an arbitrary identifier. We have just slightly remodeled the process so that id is assigned in connection with the transition, rather than before the transition; the external behavior remains the same. We use this modeling idiom in several of the examples in Section 6.

Underlying Theory. In this paper, we have restricted the relations in guards and predicates to equalities and negated equalities, and avoided functions in expressions (i.e., each e_i in an assignment $\overline{x} := \overline{e}$ is either a formal parameter or a variable). Under these restrictions, the elimination of existential quantifiers is fairly simple (as described at the end of this section). Our solution to the synthesis problem does not inherently rely on these restrictions. In principle, the set of relations and functions could be extended to include other theories as well. Our constructions would still work under restrictions that guarantee termination of concerned algorithms. We leave the precise formulation of such restrictions as future work.

Semantics of Processes. For a set X of variables, an *X-valuation* is a partial mapping from X to $\mathcal{D}$. For an X-valuation σ and a guard g over X, we let $\sigma \models g$ denote that g evaluates to true under the valuation σ. A *state* of $\mathcal{P}$ is a pair $\langle \ell, \sigma \rangle$ where $\ell \in L_{\mathcal{P}}$ and σ is an $X_{\mathcal{P}}$-valuation. The *initial state* of $\mathcal{P}$ is $\langle \ell_{\mathcal{P}}^0, \sigma_{\mathcal{P}}^0 \rangle$, where $\ell_{\mathcal{P}}^0$ is the initial location and $\sigma_{\mathcal{P}}^0$ is the initial valuation. A *step* of $\mathcal{P}$, denoted by $\langle \ell, \sigma \rangle \xrightarrow{\alpha(\bar{d})}_{\mathcal{P}} \langle \ell', \sigma' \rangle$, transfers $\mathcal{P}$ from $\langle \ell, \sigma \rangle$ to $\langle \ell', \sigma' \rangle$ while performing the symbol $\alpha(\bar{d})$. According to whether α is in $I_{\mathcal{P}}$ or $O_{\mathcal{P}}$, we call $\langle \ell', \sigma' \rangle$ an *input* or *output* successor of $\langle \ell, \sigma \rangle$. It is derived from a transition $\langle \ell, stmt, \ell' \rangle \in \delta_{\mathcal{P}}$, with stmt of the form $\alpha(p_1, \ldots, p_n); g; x_1, \ldots, x_k := e_1, \ldots, e_k$, such that $\sigma \models g[\bar{d}/\bar{p}]$, and for each $x_i \in X_{\mathcal{P}}$, we have

1. $\sigma'(x_i) = \sigma(e_i)$ when $e_i \in X_{\mathcal{P}}$, and
2. $\sigma'(x_i) = d_j$ when e_i is formal parameter p_j.

For a sequence of symbols $w = \alpha_1(\bar{d}_1) \cdots \alpha_m(\bar{d}_m)$, we write $\langle \ell, \sigma \rangle \xRightarrow{w}_{\mathcal{P}} \langle \ell', \sigma' \rangle$ to denote that there is a sequence of steps $\langle \ell, \sigma \rangle \xrightarrow{\alpha_1(\bar{d}_1)}_{\mathcal{P}} \langle \ell_1, \sigma_1 \rangle \cdots \langle \ell_{m-1}, \sigma_{m-1} \rangle \xrightarrow{\alpha_m(\bar{d}_m)}_{\mathcal{P}} \langle \ell', \sigma' \rangle$. We write $\langle \ell, \sigma \rangle \Rightarrow_{\mathcal{P}} \langle \ell', \sigma' \rangle$ to express that there exists a w such that $\langle \ell, \sigma \rangle \xRightarrow{w}_{\mathcal{P}} \langle \ell', \sigma' \rangle$. A state $\langle \ell, \sigma \rangle$ is *reachable* if $\langle \ell_{\mathcal{P}}^0, \sigma_{\mathcal{P}}^0 \rangle \stackrel{w}{\Longrightarrow}_{\mathcal{P}} \langle \ell, \sigma \rangle$ for some w. A *trace* of $\mathcal{P}$ is a sequence w such that $\langle \ell_{\mathcal{P}}^0, \sigma_{\mathcal{P}}^0 \rangle \stackrel{w}{\Longrightarrow}_{\mathcal{P}} \langle \ell, \sigma \rangle$ for some $\langle \ell, \sigma \rangle$. The trace is *quiescent* if $\sigma \models Quie_{\mathcal{P}}(\ell)$. We let $T_{\mathcal{P}}$ denote the set of traces of $\mathcal{P}$, and let $Q_{\mathcal{P}}$ denote the set of quiescent traces of $\mathcal{P}$.

A symbol $\alpha(\bar{d})$ is *enabled* in a state $\langle \ell, \sigma \rangle$ if there is a step $\langle \ell, \sigma \rangle \xrightarrow{\alpha(\bar{d})}_{\mathcal{P}} \langle \ell', \sigma' \rangle$ for some state $\langle \ell', \sigma' \rangle$. Intuitively, $\sigma \models g_\ell^\alpha[\bar{d}/\bar{p}]$ iff $\alpha(\bar{d})$ is enabled in $\langle \ell, \sigma \rangle$.

Computing Pre- and Postconditions. The algorithms in later sections use, as a basic building block, the computation of pre- and postconditions of statements. Let us first consider postconditions. Let φ be a formula and let g be a guard. Let $\bar{x}'$ be a vector of the same length as $\bar{x}$, containing fresh variables. The postcondition with respect to an assignment and a guard, respectively, are computed in the standard way as

$$post(\bar{x} := \bar{e} \ ; \ \varphi) := \exists \bar{x}'. \ (\varphi[\bar{x}'/\bar{x}] \wedge \bar{x} = \bar{e}[\bar{x}'/\bar{x}]) \quad \text{and} \quad post(g \ ; \ \varphi) := \varphi \wedge g.$$

Putting the above rules together, we derive the postcondition of a statement as follows.

$$post(\alpha(\bar{p}); g; \bar{x} := \bar{e}; \ \varphi) \ := \ \exists \bar{p}. \ \exists \bar{x}'. \ ((g \wedge \varphi)[\bar{x}'/\bar{x}] \wedge \bar{x} = \bar{e}[\bar{x}'/\bar{x}]).$$

Let us next consider preconditions in the analogous way. The precondition of an assignment and a guard, respectively, is defined as

$$pre(\bar{x} := \bar{e}; \varphi) := \varphi[\bar{e}/\bar{x}] \quad \text{and} \quad pre(g; \varphi) := g \wedge \varphi,$$

and a precondition of a whole statement is obtained by putting these together:

$$pre(\alpha(\bar{p}); g; \bar{x} := \bar{e} \ ; \ \varphi) \ = \ \exists \bar{p}. \ (g \wedge \varphi[\bar{e}/\bar{x}]) \ .$$

The existential quantifiers that arise in both the post- and precondition computation can, in the equality domain, be eliminated by a procedure involving two steps, *saturation* and *elimination*. Assume that φ is a formula in disjunctive normal form (DNF) over equalities and negated equalities over a set of variables. Formula $\exists (x_1, \ldots, x_n). \ \varphi$ is obtained by transforming every clause of φ in the following way:

1. *Saturation:* compute the reflexive, transitive and symmetric closure of the equality relation $=$ as partially given by the respective clause. This induces a partition on the set of variables. Let $[x]_=$ denote the class of this partition containing x. Then, transform φ by replacing every conjunct $x = y$ with $\bigwedge_{x' \in [x]_=, y' \in [y]_=} x' = y'$, and $x \neq y$ with $\bigwedge_{x' \in [x]_=, y' \in [y]_=} x' \neq y'$.
2. *Elimination:* from the saturated predicate, remove all conjuncts involving some of $x_1, \ldots, x_n$. Thus, the conjunct $x = y$ ($x \neq y$) is removed (or replaced with *true*) if x or y (or both) are in $\{x_1, \ldots, x_n\}$.

Example. Let φ be the predicate $x = y \wedge y \neq z$. For eliminating the existential quantifier in $\exists (y, z). \ \varphi$, we first saturate φ, which yields $x = y \wedge y \neq z \wedge x \neq z$. Since every conjunct in the saturated formula involves either y or z (or both), elimination leaves us with the empty conjunction, which is equivalent to *true*.

3   Refinement

We adapt the refinement relation between interface automata [7] to our processes with data, and at the same time we extend the definition to include quiescence.

Definition 2. Let $\mathcal{S}$ and $\mathcal{P}$ be processes. We say that $\mathcal{S}$ is *refined* by $\mathcal{P}$, denoted $\mathcal{P} \sqsubseteq \mathcal{S}$, if $I_{\mathcal{S}} \subseteq I_{\mathcal{P}}$, $O_{\mathcal{P}} \subseteq O_{\mathcal{S}}$, and whenever t is a trace of both $\mathcal{S}$ and $\mathcal{P}$ (i.e., $t \in T_{\mathcal{S}} \cap T_{\mathcal{P}}$), then

1. for any input symbol i, if $ti \in T_{\mathcal{S}}$ then $ti \in T_{\mathcal{P}}$,
2. for any output symbol o, if $to \in T_{\mathcal{P}}$ then $to \in T_{\mathcal{S}}$,
3. if t is quiescent in $\mathcal{P}$, then it is also quiescent in $\mathcal{S}$ (i.e., $Q_{\mathcal{P}} \cap T_{\mathcal{S}} \subseteq Q_{\mathcal{S}}$).   □

Condition 1 reflects the assumption that the environment does not supply unenabled input symbols. This assumption must not be strengthened by refinement; hence $\mathcal{P}$ must be prepared to accept any input that $\mathcal{S}$ can accept. Condition 2 reflects that the set of enabled output symbols constrains what the process may produce, which must not be weakened by refinement: hence $\mathcal{P}$ may at most produce the outputs that may be produced by $\mathcal{S}$. Condition 3 similarly reflects the view that allowed quiescence is viewed as a constraint on a process.

Given two processes, refinement can be checked by first computing the set of pairs of states of $\mathcal{S}$ and $\mathcal{P}$ which can be reached by the same trace, and thereafter checking the conditions in Definition 2 on these states. As the state space is infinite due to the unboundedness of $\mathcal{D}$, the set of reachable states has to be computed symbolically.

Let us assume (without loss of generality) that $X_{\mathcal{S}} \cap X_{\mathcal{P}} = \emptyset$. To check whether $\mathcal{P} \sqsubseteq \mathcal{S}$, we compute for each pair $\langle \ell_{\mathcal{S}}, \ell_{\mathcal{P}} \rangle$ of locations a predicate $Reach(\langle \ell_{\mathcal{S}}, \ell_{\mathcal{P}} \rangle)$ over $X_{\mathcal{S}} \cup X_{\mathcal{P}}$ such that for an $X_{\mathcal{S}}$-valuation $\sigma_{\mathcal{S}}$ and $X_{\mathcal{P}}$-valuation $\sigma_{\mathcal{P}}$, we have $\sigma_{\mathcal{S}} \cup \sigma_{\mathcal{P}} \models Reach(\langle \ell_{\mathcal{S}}, \ell_{\mathcal{P}} \rangle)$ iff there is a trace $w \in T_{\mathcal{S}} \cap T_{\mathcal{P}}$ such that $\langle \ell_{\mathcal{S}}^0, \sigma_{\mathcal{S}}^0 \rangle \overset{w}{\Longrightarrow}_{\mathcal{S}} \langle \ell_{\mathcal{S}}, \sigma_{\mathcal{S}} \rangle$ and $\langle \ell_{\mathcal{P}}^0, \sigma_{\mathcal{P}}^0 \rangle \overset{w}{\Longrightarrow}_{\mathcal{P}} \langle \ell_{\mathcal{P}}, \sigma_{\mathcal{P}} \rangle$.

We compute values of $Reach$ by a repeated postcondition computation, starting from the strongest condition that holds for the pair of initial states of $\mathcal{S}$ and $\mathcal{P}$. Let φ_0 be the strongest predicate satisfied by the union of the initial valuations $\sigma_{\mathcal{S}}^0 \cup \sigma_{\mathcal{P}}^0$ of $\mathcal{S}$ and $\mathcal{P}$. This predicate expresses precisely that all variables have their initial values. The computation of $Reach$ is then carried out by a standard fixpoint procedure, shown in Algorithm 1.

It is initialized by letting $Reach(\langle \ell_{\mathcal{S}}^0, \ell_{\mathcal{P}}^0 \rangle)$ be φ_0, and letting any other $Reach(\langle \ell_{\mathcal{S}}, \ell_{\mathcal{P}} \rangle)$ be *false*. Therafter the predicates $Reach(\langle \ell_{\mathcal{S}}, \ell_{\mathcal{P}} \rangle)$ are iteratively extended: for each pair of transitions $\langle \ell_{\mathcal{S}}, stmt_{\mathcal{S}}, \ell'_{\mathcal{S}} \rangle$ and $\langle \ell_{\mathcal{P}}, stmt_{\mathcal{P}}, \ell'_{\mathcal{P}} \rangle$ with the same action, we calculate the postcondition ϕ of the joint transition, in which $\mathcal{S}$ and $\mathcal{P}$ synchronize on a common symbol $\alpha(\overline{d})$. More precisely, letting $stmt_{\mathcal{S}}$ be $\alpha(\overline{p}); g_{\mathcal{S}}; \overline{x}_{\mathcal{S}} := \overline{e}_{\mathcal{S}}$, letting $stmt_{\mathcal{P}}$ be $\alpha(\overline{p}); g_{\mathcal{P}}; \overline{x}_{\mathcal{P}} := \overline{e}_{\mathcal{P}}$, and letting φ be a formula over $\overline{x}_{\mathcal{S}}$ and $\overline{x}_{\mathcal{P}}$, we define the *joint postcondition* of $stmt_{\mathcal{S}}$ and $stmt_{\mathcal{P}}$ wrp. to φ, denoted $jointpost(stmt_{\mathcal{S}}, stmt_{\mathcal{P}}; \varphi)$ as

$$jointpost(stmt_{\mathcal{S}}, stmt_{\mathcal{P}}; \varphi) = \exists \overline{p}.\, \exists \overline{x}'_{\mathcal{S}}, \overline{x}'_{\mathcal{P}}. \begin{bmatrix} (g_{\mathcal{S}} \wedge g_{\mathcal{P}} \wedge \varphi)[\overline{x}'_{\mathcal{S}}, \overline{x}'_{\mathcal{P}}/\overline{x}_{\mathcal{S}}, \overline{x}_{\mathcal{P}}] \\ \wedge\, \overline{x}_{\mathcal{S}} = \overline{e}_{\mathcal{S}}[\overline{x}'_{\mathcal{S}}/\overline{x}_{\mathcal{S}}] \\ \wedge\, \overline{x}_{\mathcal{P}} = \overline{e}_{\mathcal{P}}[\overline{x}'_{\mathcal{P}}/\overline{x}_{\mathcal{P}}] \end{bmatrix}$$

Algorithm 1. Computing $Reach(\langle \ell_S, \ell_P \rangle)$ for all $\ell_S \in L_S$ and $\ell_P \in L_P$

---

1 Initialise $Reach$ as φ_0 for $\langle \ell_S^0, \ell_P^0 \rangle$ and $false$ otherwise;
2 **repeat**
3 $\quad$ $changed := false$;
4 $\quad$ **forall the** $\langle \ell_S, stmt_S, \ell_S' \rangle \in \delta_S$ *and* $\langle \ell_P, stmt_P, \ell_P' \rangle \in \delta_P$ *with equal actions*
$\quad$ **do**
5 $\quad\quad$ $\phi := jointpost(stmt_S, stmt_P; Reach(\langle \ell_S, \ell_P \rangle))$;
6 $\quad\quad$ **if** $IsSat(\phi \wedge \neg Reach(\langle \ell_S', \ell_P' \rangle))$ **then**
7 $\quad\quad\quad$ $Reach(\langle \ell_S', \ell_P' \rangle) := Reach(\langle \ell_S', \ell_P' \rangle) \vee \phi$;
8 $\quad\quad\quad$ $changed := true$;
9 **until** $\neg changed$;

---

where $\overline{x}_S'$ is a tuple containing fresh variables, of the same length as $\overline{x}_S$, and similarly for $\overline{x}_P'$. In the algorithm, the predicates $Reach(\langle \ell_S, \ell_P \rangle)$ are iteratively extended: for each pair of transitions $\langle \ell_S, stmt_S, \ell_S' \rangle$ and $\langle \ell_P, stmt_P, \ell_P' \rangle$ with the same action, we calculate their joint postcondition in $\phi := jointpost(stmt_S, stmt_P; Reach(\langle \ell_S, \ell_P \rangle))$. As long as $Reach(\langle \ell_S', \ell_P' \rangle)$ is not weaker than ϕ, it is weakened by updating it to the disjunction of its previous value and ϕ. This process is repeated until convergence.

The following proposition states that, after $Reach$ has been computed, checking refinement $\mathcal{P} \sqsubseteq \mathcal{S}$ amounts to checking validity of three conditions for each pair $\langle \ell_S, \ell_P \rangle$ of locations.

Proposition 1. $\mathcal{P} \sqsubseteq \mathcal{S}$ *if and only if* $I_S \subseteq I_P$, $O_P \subseteq O_S$, *and for each pair* $\langle \ell_S, \ell_P \rangle$ *of locations of* $\mathcal{S}$ *and* $\mathcal{P}$ *it holds that*

1. $Reach(\langle \ell_S, \ell_P \rangle) \Rightarrow (g_{\ell_S}^\alpha \Rightarrow g_{\ell_P}^\alpha)$ *for all input actions* $\alpha \in I_S$,
2. $Reach(\langle \ell_S, \ell_P \rangle) \Rightarrow (g_{\ell_P}^\alpha \Rightarrow g_{\ell_S}^\alpha)$ *for all output actions* $\alpha \in O_P$, *and*
3. $Reach(\langle \ell_S, \ell_P \rangle) \Rightarrow (Quie_P(\ell_P) \Rightarrow Quie_S(\ell_S))$. $\qquad\qquad\square$

Proof. First, assume that $\mathcal{P} \sqsubseteq \mathcal{S}$. We exemplify the proof idea focusing on the first condition only. Let $\sigma_S \cup \sigma_P \models Reach(\langle \ell_S, \ell_P \rangle)$, i.e., there exists a trace $w \in T_S \cap T_P$ such that $\langle \ell_S^0, \sigma_S^0 \rangle \xrightarrow{w}_S \langle \ell_S, \sigma_S \rangle$ and $\langle \ell_P^0, \sigma_P^0 \rangle \xrightarrow{w}_P \langle \ell_P, \sigma_P \rangle$. Furthermore, let $i = \alpha(\overline{d})$, $\alpha \in I_S$, be any input symbol such that $\sigma_S \models g_{\ell_S}^\alpha[\overline{d}/\overline{p}]$, i.e., $\alpha(\overline{d})$ is enabled in $\langle \ell_S, \sigma_S \rangle$ and hence $wi \in T_S$. Since $\mathcal{P} \sqsubseteq \mathcal{S}$ implies that $wi \in T_P$, $\alpha(\overline{d})$ is enabled in $\langle \ell_P, \sigma_P \rangle$ as well, thus $\sigma_P \models g_{\ell_P}^\alpha[\overline{d}/\overline{p}]$, rendering the implication $g_{\ell_S}^\alpha \Rightarrow g_{\ell_P}^\alpha$ true.

For the converse, assume that $\mathcal{P} \not\sqsubseteq \mathcal{S}$. Then, at least one of the conditions of Definition 2 is violated. Again, we exemplify the proof idea by looking at the third condition only, i.e., assuming that there exists a trace $w \in T_S \cap T_P$ such that w is quiescent in $\mathcal{P}$ but not in $\mathcal{S}$. Let $\langle \ell_S^0, \sigma_S^0 \rangle \xrightarrow{w}_S \langle \ell_S, \sigma_S \rangle$ and $\langle \ell_P^0, \sigma_P^0 \rangle \xrightarrow{w}_P \langle \ell_P, \sigma_P \rangle$ (thus, $\sigma_S \cup \sigma_P \models Reach(\langle \ell_S, \ell_P \rangle)$). Since w is quiescent in $\mathcal{P}$, we have $\sigma_P \models Quie_P(\ell_P)$. Conversely, since w is not quiescent in $\mathcal{S}$, we have $\sigma_S \not\models Quie_S(\ell_S)$, and thus $\sigma_S \cup \sigma_P \not\models (Quie_P(\ell_P) \Rightarrow Quie_S(\ell_S))$. $\qquad\square$

4   Parallel Composition

In this section, we generalize parallel composition of interface automata [7] to our processes with data and quiescence. Intuitively, the parallel composition operator yields the combined effect of its operands running asynchronously, but synchronizing on common actions. We use a *broadcast* model of communication in which an output from a component can be received by multiple components. An input $?a(\overline{d})$ and output $!a(\overline{d})$ combine to form an output $!a(\overline{d})$. Here, the attributes ? and ! on actions (as in $!a(\overline{d})$) are not part of the actions, they serve only to remind that the action in question is an input or output in the considered context.

Product Operation. Before defining parallel composition of processes, we will as an auxiliary building block define the *product* of two processes as the process obtained by letting them run in parallel, while synchronizing on common actions and ignoring communication mismatches.

Let us define the parallel composition of two statements $stmt_1 = \alpha(\overline{p}); g_1; \overline{x}_1 := \overline{e}_1$ and $stmt_2 = \alpha(\overline{p}); g_2; \overline{x}_2 := \overline{e}_2$ with the same action α,[1] in two processes with disjoint sets of variables, as $stmt_1 \| stmt_2 = \alpha(\overline{p}); g_1 \wedge g_2; \overline{x}_1, \overline{x}_2 := \overline{e}_1, \overline{e}_2$.

We can now define product of two processes.

Definition 3. Let $\mathcal{P}$ and $\mathcal{Q}$ be two processes. Then define $\delta_{\mathcal{P} \otimes \mathcal{Q}}$ as the set of transitions between product locations in $L_{\mathcal{P}} \times L_{\mathcal{Q}}$, which is obtained from $\delta_{\mathcal{P}}$ and $\delta_{\mathcal{Q}}$ as follows.

- If $\langle \ell_{\mathcal{P}}, stmt, \ell'_{\mathcal{P}} \rangle \in \delta_{\mathcal{P}}$ has an action which is not an action of $\mathcal{Q}$, (i.e., it is non-synchronizing), then $\langle \langle \ell_{\mathcal{P}}, \ell_{\mathcal{Q}} \rangle, stmt, \langle \ell'_{\mathcal{P}}, \ell_{\mathcal{Q}} \rangle \rangle \in \delta_{\mathcal{P} \otimes \mathcal{Q}}$ for any location $\ell_{\mathcal{Q}} \in L_{\mathcal{Q}}$.
- Symmetrically, if $\langle \ell_{\mathcal{Q}}, stmt, \ell'_{\mathcal{Q}} \rangle \in \delta_{\mathcal{Q}}$ has an action which is not an action of $\mathcal{P}$, then $\langle \langle \ell_{\mathcal{P}}, \ell_{\mathcal{Q}} \rangle, stmt, \langle \ell_{\mathcal{P}}, \ell'_{\mathcal{Q}} \rangle \rangle \in \delta_{\mathcal{P} \otimes \mathcal{Q}}$ for any location $\ell_{\mathcal{P}} \in L_{\mathcal{P}}$.
- If $\langle \ell_{\mathcal{P}}, stmt_{\mathcal{P}}, \ell'_{\mathcal{P}} \rangle \in \delta_{\mathcal{P}}$ and $\langle \ell_{\mathcal{Q}}, stmt_{\mathcal{Q}}, \ell'_{\mathcal{Q}} \rangle \in \delta_{\mathcal{Q}}$ have the same action, i.e., they synchronize, then $\langle \langle \ell_{\mathcal{P}}, \ell_{\mathcal{Q}} \rangle, stmt_{\mathcal{P}} \| stmt_{\mathcal{Q}}, \langle \ell'_{\mathcal{P}}, \ell'_{\mathcal{Q}} \rangle \rangle \in \delta_{\mathcal{P} \otimes \mathcal{Q}}$. □

Note that while we restrict ourselves to defining a product transition relation $\delta_{\mathcal{P} \otimes \mathcal{Q}}$ instead of a complete product operation $\mathcal{P} \otimes \mathcal{Q}$ (mostly because there is no reasonable choice for $Quie_{\mathcal{P} \otimes \mathcal{Q}}$), we nonetheless adapt the notation $\overset{w}{\Longrightarrow}_{\mathcal{P} \otimes \mathcal{Q}}$ for sequences of steps, as established in Section 2.

Parallel Composition. We can now define parallel composition of processes. Note that there does not always exist a parallel composition of two processes. A necessary (but not sufficient, cf. Definition 5) precondition that has to be met is stated in the following definition.

---

[1] Similarly to many programming languages, we assume that actions only have positional arguments, i.e., their formal parameters are identified solely by their order of occurrence in $\overline{p}$, not their name (if any).

Definition 4. *Two processes* $\mathcal{P}$ *and* $\mathcal{Q}$ *are composable if* $O_\mathcal{P} \cap O_\mathcal{Q} = \emptyset$.[2]

Intuitively, the parallel composition will be obtained from the product by restricting input transitions so that the product cannot reach an *illegal* state. A state of the product is illegal if one of the processes can generate an output symbol in their joint set of symbols which the other cannot receive. We say that a state of the product is *unsafe* if the product can reach an illegal state by a sequence of output steps. The parallel composition is now obtained by restricting input transitions so that they do not reach unsafe states.

Let us now formalize this intuition. First, we define a mapping, denoted $Illegal_{\mathcal{P}\|\mathcal{Q}}$ from pairs of locations of $\mathcal{P}$ and $\mathcal{Q}$ to the formula

$$Illegal_{\mathcal{P}\|\mathcal{Q}}(\langle \ell_\mathcal{P}, \ell_\mathcal{Q} \rangle) := \bigvee_{\alpha \in O_\mathcal{P} \cap I_\mathcal{Q}} \exists \overline{p}.\ (g^\alpha_{\ell_\mathcal{P}} \wedge \neg g^\alpha_{\ell_\mathcal{Q}}) \vee \bigvee_{\alpha \in O_\mathcal{Q} \cap I_\mathcal{P}} \exists \overline{p}.\ (g^\alpha_{\ell_\mathcal{Q}} \wedge \neg g^\alpha_{\ell_\mathcal{P}})$$

Intuitively, $Illegal_{\mathcal{P}\|\mathcal{Q}}(\langle \ell_\mathcal{P}, \ell_\mathcal{Q} \rangle)$ is true if $\mathcal{P}$ in location $\ell_\mathcal{P}$ can produce a synchronizing output symbol for which $\mathcal{Q}$ does not have a matching input step, or vice versa. Thereafter, we perform the pruning process, by defining the mapping $Unsafe_{\mathcal{P}\|\mathcal{Q}}$ from pairs of locations of $\mathcal{P}$ and $\mathcal{Q}$ to formulas over $X_\mathcal{P} \cup X_\mathcal{Q}$ such that $\sigma_\mathcal{P} \cup \sigma_\mathcal{Q} \models Unsafe_{\mathcal{P}\|\mathcal{Q}}(\langle \ell_\mathcal{P}, \ell_\mathcal{Q} \rangle)$ iff there exists a sequence of symbols $w \in (O_\mathcal{P} \cup O_\mathcal{Q})^*$ such that $\langle \langle \ell_\mathcal{P}, \ell_\mathcal{Q} \rangle, \sigma_\mathcal{P} \cup \sigma_\mathcal{Q} \rangle \stackrel{w}{\Longrightarrow}_{\mathcal{P} \otimes \mathcal{Q}} \langle \langle \ell'_\mathcal{P}, \ell'_\mathcal{Q} \rangle, \sigma'_\mathcal{P} \cup \sigma'_\mathcal{Q} \rangle$ and $\sigma'_\mathcal{P} \cup \sigma'_\mathcal{Q} \models Illegal_{\mathcal{P}\|\mathcal{Q}}(\langle \ell'_\mathcal{P}, \ell'_\mathcal{Q} \rangle)$. The mapping $Unsafe_{\mathcal{P}\|\mathcal{Q}}$ can be computed in a fashion similar to the computation of *Reach*, as illustrated in Algorithm 2.

Algorithm 2. Computing $Unsafe_{\mathcal{P}\|\mathcal{Q}}$ for product of $\mathcal{P}$ and $\mathcal{Q}$

---

1 $Unsafe_{\mathcal{P}\|\mathcal{Q}} := Illegal_{\mathcal{P}\|\mathcal{Q}}$;
2 **repeat**
3      $changed := false$;
4   **forall the** *output transitions* $\langle \ell, stmt, \ell' \rangle$ *of* $\delta_{\mathcal{P} \otimes \mathcal{Q}}$ **do**
5          $\phi := pre(stmt;\ Unsafe_{\mathcal{P}\|\mathcal{Q}}(\ell'))$;
6   **if** $IsSat(\phi \wedge \neg Unsafe_{\mathcal{P}\|\mathcal{Q}}(\ell))$ **then**
7              $Unsafe_{\mathcal{P}\|\mathcal{Q}}(\ell) := Unsafe_{\mathcal{P}\|\mathcal{Q}}(\ell) \vee \phi$;
8              $changed := true$;
9 **until** $\neg changed$;

Definition 5. Let $\mathcal{P}$ and $\mathcal{Q}$ be processes. The parallel composition of $\mathcal{P}$ and $\mathcal{Q}$ exists if and only if 1. $\mathcal{P}$ and $\mathcal{Q}$ are composable, and 2. $\sigma^0_\mathcal{P} \cup \sigma^0_\mathcal{Q} \not\models Unsafe_{\mathcal{P}\|\mathcal{Q}}(\langle \ell^0_\mathcal{P}, \ell^0_\mathcal{Q} \rangle)$. In this case, it is the process $\mathcal{P}\|\mathcal{Q}$, obtained as $\mathcal{P}\|\mathcal{Q} = (I_{\mathcal{P}\|\mathcal{Q}}, O_{\mathcal{P}\|\mathcal{Q}}, L_{\mathcal{P}\|\mathcal{Q}}, \ell^0_{\mathcal{P}\|\mathcal{Q}}, X_{\mathcal{P}\|\mathcal{Q}}, \sigma^0_{\mathcal{P}\|\mathcal{Q}}, \delta_{\mathcal{P}\|\mathcal{Q}}, Quie_{\mathcal{P}\|\mathcal{Q}})$, where

---

[2] Formally, composing $\mathcal{P}$ and $\mathcal{Q}$ also requires $X_\mathcal{P}$ and $X_\mathcal{Q}$ to be disjoint. However, this can be assumed without loss of generality (as remarked in Section 3), as renaming variables does not change the behaviour of a process.

- $I_{\mathcal{P}\|\mathcal{Q}} = (I_{\mathcal{P}} \cup I_{\mathcal{Q}}) \setminus O_{\mathcal{P}\|\mathcal{Q}}$,
- $O_{\mathcal{P}\|\mathcal{Q}} = O_{\mathcal{P}} \cup O_{\mathcal{Q}}$,
- $L_{\mathcal{P}\|\mathcal{Q}} = L_{\mathcal{P}} \times L_{\mathcal{Q}}$,
- $\ell^0_{\mathcal{P}\|\mathcal{Q}} = \langle \ell^0_{\mathcal{P}}, \ell^0_{\mathcal{Q}} \rangle$,
- $X_{\mathcal{P}\|\mathcal{Q}} = X_{\mathcal{P}} \cup X_{\mathcal{Q}}$,
- $\sigma^0_{\mathcal{P}\|\mathcal{Q}} = \sigma^0_{\mathcal{P}} \cup \sigma^0_{\mathcal{Q}}$,
- $\delta_{\mathcal{P}\|\mathcal{Q}}$ is obtained from $\delta_{\mathcal{P}\otimes\mathcal{Q}}$ by strengthening every guard g of every input transitions of form $\langle \langle \ell_{\mathcal{P}}, \ell_{\mathcal{Q}} \rangle, \ \alpha(\overline{p}); g; \overline{x} := \overline{e}, \ \langle \ell'_{\mathcal{P}}, \ell'_{\mathcal{Q}} \rangle \rangle$ in $\delta_{\mathcal{P}\otimes\mathcal{Q}}$ to

$$g \wedge pre(\overline{x} := \overline{e}; \neg Unsafe_{\mathcal{P}\|\mathcal{Q}}(\langle \ell'_{\mathcal{P}}, \ell'_{\mathcal{Q}} \rangle)) \ ,$$

- $Quie_{\mathcal{P}\|\mathcal{Q}}(\langle \ell_{\mathcal{P}}, \ell_{\mathcal{Q}} \rangle) = Illegal_{\mathcal{P}\|\mathcal{Q}}(\langle \ell_{\mathcal{P}}, \ell_{\mathcal{Q}} \rangle) \vee (Quie_{\mathcal{P}}(\ell_{\mathcal{P}}) \wedge Quie_{\mathcal{Q}}(\ell_{\mathcal{Q}}))$. □

An important observation is that parallel composition preserves the consistency requirement introduced in Section 2.

Proposition 2. *Let $\mathcal{P}$ and $\mathcal{Q}$ be processes such that the parallel composition $\mathcal{P} \| \mathcal{Q}$ exists, and that $\mathcal{P}$ and $\mathcal{Q}$ are consistent with $Quie_{\mathcal{P}}$ and $Quie_{\mathcal{Q}}$, respectively. Then, $\mathcal{P} \| \mathcal{Q}$ is consistent with $Quie_{\mathcal{P}\|\mathcal{Q}}$, i.e., for any location $\ell_{\mathcal{P}\|\mathcal{Q}} \in L_{\mathcal{P}\|\mathcal{Q}}$,*

$$Quie_{\mathcal{P}\|\mathcal{Q}}(\ell_{\mathcal{P}\|\mathcal{Q}}) \vee \bigvee_{\alpha \in O_{\mathcal{P}\|\mathcal{Q}}} \exists \overline{p}. \ g^\alpha_{\ell_{\mathcal{P}\|\mathcal{Q}}}$$

is universally valid.

Proof. Let $\ell_{\mathcal{P}\|\mathcal{Q}} = \langle \ell_{\mathcal{P}}, \ell_{\mathcal{Q}} \rangle$, and let $\sigma_{\mathcal{P}\|\mathcal{Q}}$ be a valuation such that $\sigma_{\mathcal{P}\|\mathcal{Q}} \not\models Quie_{\mathcal{P}\|\mathcal{Q}}(\langle \ell_{\mathcal{P}}, \ell_{\mathcal{Q}} \rangle)$, i.e., we have $\sigma_{\mathcal{P}\|\mathcal{Q}} \not\models Illegal_{\mathcal{P}\|\mathcal{Q}}(\langle \ell_{\mathcal{P}}, \ell_{\mathcal{Q}} \rangle)$ and either of $\sigma_{\mathcal{P}\|\mathcal{Q}} \not\models Quie_{\mathcal{P}}(\ell_{\mathcal{P}})$ or $\sigma_{\mathcal{P}\|\mathcal{Q}} \not\models Quie_{\mathcal{Q}}(\ell_{\mathcal{Q}})$ (or both). Without loss of generality, we only consider the first case. Due to the consistency of $\mathcal{P}$, there exists an output action $\alpha \in O_{\mathcal{P}}$ such that $\sigma_{\mathcal{P}\|\mathcal{Q}} \models \exists \overline{p}. \ g^\alpha_{\ell_{\mathcal{P}}}$. If α is not an (input) action of $\mathcal{Q}$, then, by definition of the parallel composition, we have $g^\alpha_{\ell_{\mathcal{P}\|\mathcal{Q}}} \equiv g^\alpha_{\ell_{\mathcal{P}}}$ (note that only guards of input transitions are strengthened in Definition 5) and thus $\sigma_{\mathcal{P}\|\mathcal{Q}} \models \exists \overline{p}. \ g^\alpha_{\ell_{\mathcal{P}\|\mathcal{Q}}}$. Otherwise, we have $g^\alpha_{\ell_{\mathcal{P}\|\mathcal{Q}}} \equiv g^\alpha_{\ell_{\mathcal{P}}} \wedge g^\alpha_{\ell_{\mathcal{Q}}}$. From $\sigma_{\mathcal{P}\|\mathcal{Q}} \not\models Illegal_{\mathcal{P}\|\mathcal{Q}}(\langle \ell_{\mathcal{P}}, \ell_{\mathcal{Q}} \rangle)$ we can conclude that $\sigma_{\mathcal{P}\|\mathcal{Q}} \models \forall \overline{p}. \ (g^\alpha_{\ell_{\mathcal{P}}} \Rightarrow g^\alpha_{\ell_{\mathcal{Q}}})$. Combining this with $\sigma_{\mathcal{P}\|\mathcal{Q}} \models \exists \overline{p}. \ g^\alpha_{\ell_{\mathcal{P}}}$ yields $\sigma_{\mathcal{P}\|\mathcal{Q}} \models \exists \overline{p}. \ g^\alpha_{\ell_{\mathcal{Q}}}$, and thus $\sigma_{\mathcal{P}\|\mathcal{Q}} \models \exists \overline{p}. \ g^\alpha_{\ell_{\mathcal{P}\|\mathcal{Q}}}$. □

The following proposition establishes the important property that refinement is preserved by parallel composition.

Proposition 3. *Let $\mathcal{P}$, $\mathcal{Q}$, $\mathcal{S}$ and $\mathcal{T}$ be processes with $\mathcal{P} \sqsubseteq \mathcal{S}$ and $\mathcal{Q} \sqsubseteq \mathcal{T}$, such that the parallel composition $\mathcal{S} \| \mathcal{T}$ exists. Then the parallel composition $\mathcal{P} \| \mathcal{Q}$ exists and $\mathcal{P} \| \mathcal{Q} \sqsubseteq \mathcal{S} \| \mathcal{T}$.* □

5   Quotient

In this section, we introduce the *quotient* operation, which can be seen as an "inverse" of parallel composition. It is connected to the synthesis problem in the following way: given a specification for a system $\mathcal{R}$, together with a component $\mathcal{P}$ implementing part of $\mathcal{R}$, the quotient, denoted $\mathcal{R} \setminus \mathcal{P}$, yields the *least refined* (in the sense of $\sqsubseteq$) process for the remaining part of $\mathcal{R}$, i.e., such that $\mathcal{P} \| (\mathcal{R} \setminus \mathcal{P}) \sqsubseteq \mathcal{R}$. Therefore, quotient can be thought of as an adjoint of parallel composition.

Looking at the treatment of sets of actions in Definition 5, we see that a necessary requirement for the existence of a quotient is that $O_{\mathcal{P}} \subseteq O_{\mathcal{R}}$. Then, the set of output actions of the quotient must be $O_{\mathcal{R}} \setminus O_{\mathcal{P}}$. However, there is some freedom for the set $I_{\mathcal{R} \setminus \mathcal{P}}$ of input actions of the quotient. From Definition 5, we take as a natural choice $I_{\mathcal{R} \setminus \mathcal{P}}$ to be $(O_{\mathcal{P}} \cup I_{\mathcal{R}})$, since a quotient with these actions will always exist if there is one with a smaller set, and since the subsequently presented technique to produce a quotient will not have to consider the difficulties that come with actions that are not visible to the quotient.

Computing the Quotient. Let us provide an algorithmic construction of the quotient. The structure of our construction is the following. We first construct the product of $\mathcal{P}$ and $\mathcal{R}$ using the product construction of Definition 3. We thereafter construct $\mathcal{R} \setminus \mathcal{P}$ so that it ensures that the combination of $\mathcal{P}$ and $\mathcal{R}$ does not reach an undesired state. An undesired state occurs if 1. $\mathcal{P}$ can produce an output which cannot be produced by $\mathcal{R}$, or 2. $\mathcal{R}$ can receive an input which cannot be received by $\mathcal{P}$, or 3. the state of $\mathcal{R}$ is not quiescent, but it is a *deadlock*, that is, no output transition of $\mathcal{R}$ can be taken without $\mathcal{R} \setminus \mathcal{P}$ losing control (i.e., $\mathcal{R} \setminus \mathcal{P}$ will no longer be able to ensure that an undesired state will not eventually be reached).

To enforce the above criteria, we define a mapping *Bad* from pairs of locations of $\mathcal{P}$ and $\mathcal{R}$ to predicates over $X_{\mathcal{P}} \cup X_{\mathcal{R}}$ such that $\sigma \models Bad(\langle \ell_{\mathcal{P}}, \ell_{\mathcal{R}} \rangle)$ iff there exists a sequence of symbols $w \in O_{\mathcal{P}}^*$ such that $\langle \langle \ell_{\mathcal{P}}, \ell_{\mathcal{R}} \rangle, \sigma \rangle \overset{w}{\Longrightarrow}_{\mathcal{P} \otimes \mathcal{R}}$ $\langle \langle \ell'_{\mathcal{P}}, \ell'_{\mathcal{R}} \rangle, \sigma' \rangle$ and $\langle \langle \ell'_{\mathcal{P}}, \ell'_{\mathcal{R}} \rangle, \sigma' \rangle$ is an undesired state according to any (or all) of the above criteria. In the following, we describe how *Bad* can be computed algorithmically. As a first step, we define a mapping $NotRefine_{\mathcal{R} \setminus \mathcal{P}}$ from pairs of locations of $\mathcal{P}$ and $\mathcal{R}$ to predicates such that $NotRefine_{\mathcal{R} \setminus \mathcal{P}}(\langle \ell_{\mathcal{P}}, \ell_{\mathcal{R}} \rangle)$ is true in the situations that should be avoided according to Criterion 1 or 2 above. We can represent $NotRefine_{\mathcal{R} \setminus \mathcal{P}}(\langle \ell_{\mathcal{P}}, \ell_{\mathcal{R}} \rangle)$ as the formula

$$\bigvee_{\alpha \in I_{\mathcal{R}} \cap I_{\mathcal{P}}} (g_{\ell_{\mathcal{R}}}^{\alpha} \wedge \neg g_{\ell_{\mathcal{P}}}^{\alpha}) \; \vee \; \bigvee_{\alpha \in O_{\mathcal{P}}} (g_{\ell_{\mathcal{P}}}^{\alpha} \wedge \neg g_{\ell_{\mathcal{R}}}^{\alpha})$$

To formalize Criterion 3, we first define deadlock as a predicate parameterized by the set of states which are uncontrollable (represented by the predicate *Bad*), i.e., from which $\mathcal{R} \setminus \mathcal{P}$ cannot guarantee avoiding undesired states. We define $Deadlock^{Bad}(\langle \ell_{\mathcal{P}}, \ell_{\mathcal{R}} \rangle)$ as

$$\neg \bigvee_{\langle \langle \ell_{\mathcal{P}}, \ell_{\mathcal{R}} \rangle, stmt, \langle \ell'_{\mathcal{P}}, \ell'_{\mathcal{R}} \rangle \rangle \in \delta_{\mathcal{P} \otimes \mathcal{R}}^{O_{\mathcal{R}}}} pre(stmt; \; \neg Bad(\langle \ell'_{\mathcal{P}}, \ell'_{\mathcal{R}} \rangle))$$

Algorithm 3. Computing *Bad*

---

1 $Bad := NotRefine_{\mathcal{R} \setminus \mathcal{P}}$;
2 **repeat**
3 | *changed* := *false*;
4 | **forall the** $O_{\mathcal{P}}$-transitions $\langle \ell, stmt, \ell' \rangle$ of $\delta_{\mathcal{P} \otimes \mathcal{R}}$ **do**
5 | | $\phi := pre(stmt; Bad(\ell'))$;
6 | | **if** $IsSat(\phi \wedge \neg Bad(\ell))$ **then**
7 | | | $Bad(\ell) := Bad(\ell) \vee \phi$;
8 | | | *changed* := *true*;
9 | | **forall the** $\langle \ell_{\mathcal{P}}, \ell_{\mathcal{R}} \rangle \in L_{\mathcal{P}} \times L_{\mathcal{R}}$ **do**
10 | | | $\phi := Deadlock^{Bad}(\langle \ell_{\mathcal{P}}, \ell_{\mathcal{R}} \rangle) \wedge \neg Quie_{\mathcal{R}}(\ell_{\mathcal{R}})$;
11 | | | **if** $IsSat(\phi \wedge \neg Bad(\langle \ell_{\mathcal{P}}, \ell_{\mathcal{R}} \rangle))$ **then**
12 | | | | $Bad(\langle \ell_{\mathcal{P}}, \ell_{\mathcal{R}} \rangle) := Bad(\langle \ell_{\mathcal{P}}, \ell_{\mathcal{R}} \rangle) \vee \phi$;
13 | | | | *changed* := *true*;
14 **until** $\neg changed$;

---

where $\delta^{O_{\mathcal{R}}}_{\mathcal{P} \otimes \mathcal{R}}$ are the $O_{\mathcal{R}}$-transitions of $\delta_{\mathcal{P} \otimes \mathcal{R}}$. Notice that $Deadlock^{Bad}(\langle \ell_{\mathcal{P}}, \ell_{\mathcal{R}} \rangle)$ is automatically *true* if $\langle \ell_{\mathcal{P}}, \ell_{\mathcal{R}} \rangle$ has no output transitions. A state $\langle \langle \ell_{\mathcal{R}}, \ell_{\mathcal{P}} \rangle, \sigma \rangle$ is undesired according to Criterion 3 above if

$$\sigma \models Deadlock^{Bad}(\langle \ell_{\mathcal{P}}, \ell_{\mathcal{R}} \rangle) \wedge \neg Quie_{\mathcal{R}}(\ell_{\mathcal{R}}) .$$

The complete computation of *Bad* is illustrated in Algorithm 3.

After *Bad* is computed, we compute $\mathcal{R} \setminus \mathcal{P}$ as the process
$\mathcal{R} \setminus \mathcal{P} = (I_{\mathcal{R} \setminus \mathcal{P}}, O_{\mathcal{R} \setminus \mathcal{P}}, L_{\mathcal{R} \setminus \mathcal{P}}, \ell^0_{\mathcal{R} \setminus \mathcal{P}}, X_{\mathcal{R} \setminus \mathcal{P}}, \sigma^0_{\mathcal{R} \setminus \mathcal{P}}, \delta_{\mathcal{R} \setminus \mathcal{P}}, Quie_{\mathcal{R} \setminus \mathcal{P}})$ where

- $I_{\mathcal{R} \setminus \mathcal{P}} = O_{\mathcal{P}} \cup I_{\mathcal{R}}$,
- $O_{\mathcal{R} \setminus \mathcal{P}} = O_{\mathcal{R}} \setminus O_{\mathcal{P}}$,
- $L_{\mathcal{R} \setminus \mathcal{P}} = L_{\mathcal{P}} \times L_{\mathcal{R}}$,
- $\ell^0_{\mathcal{R} \setminus \mathcal{P}} = \langle \ell^0_{\mathcal{P}}, \ell^0_{\mathcal{R}} \rangle$,
- $X_{\mathcal{R} \setminus \mathcal{P}} = X_{\mathcal{P}} \cup X_{\mathcal{R}}$,
- $\sigma^0_{\mathcal{R} \setminus \mathcal{P}} = \sigma^0_{\mathcal{P}} \cup \sigma^0_{\mathcal{Q}}$,
- $\delta_{\mathcal{R} \setminus \mathcal{P}}$ is obtained from $\delta_{\mathcal{P} \otimes \mathcal{R}}$ by strengthening every guard g of every $O_{\mathcal{R} \setminus \mathcal{P}}$-transition of the form $\langle \langle \ell_{\mathcal{P}}, \ell_{\mathcal{R}} \rangle, \alpha(\overline{p}); g; \overline{x} := \overline{e}, \langle \ell'_{\mathcal{P}}, \ell'_{\mathcal{R}} \rangle \rangle$ in $\delta_{\mathcal{P} \otimes \mathcal{R}}$ to

$$g \wedge pre(\overline{x} := \overline{e}; \neg Bad(\langle \ell'_{\mathcal{P}}, \ell'_{\mathcal{R}} \rangle))$$

- $Quie_{\mathcal{R} \setminus \mathcal{P}} = Quie_{\mathcal{P}}(\ell_{\mathcal{P}}) \Rightarrow Quie_{\mathcal{R}}(\ell_{\mathcal{R}})$

The input transitions of $\mathcal{R} \setminus \mathcal{P}$ are the same as the input $I_{\mathcal{R} \setminus \mathcal{P}}$-transitions of $\delta_{\mathcal{P} \otimes \mathcal{R}}$. Output transitions of $\mathcal{R} \setminus \mathcal{P}$ will be obtained from $O_{\mathcal{R} \setminus \mathcal{P}}$-transitions of $\delta_{\mathcal{P} \otimes \mathcal{R}}$ by equipping them with additional guards which keep computations of $(\mathcal{R} \setminus \mathcal{P}) \| \mathcal{P}$ outside *Bad* and hence also outside undesired states.

The following proposition states that the quotient, if it exists, is indeed the most general component that can cooperate with $\mathcal{P}$ to satisfy $\mathcal{R}$.

Proposition 4. *Let $\mathcal{P}$ and $\mathcal{R}$ be such that $O_\mathcal{P} \subseteq O_\mathcal{R}$ and $I_\mathcal{P} \subseteq (I_\mathcal{R} \cup O_\mathcal{R})$. If $\mathcal{R} \setminus \mathcal{P}$ as computed in this section exists, then*

- *$\mathcal{P} \parallel (\mathcal{R} \setminus \mathcal{P}) \sqsubseteq R$, and*
- *for any $\mathcal{Q}$ with $O_\mathcal{Q} = O_{\mathcal{R} \setminus \mathcal{P}}$ and $I_\mathcal{Q} = I_{\mathcal{R} \setminus \mathcal{P}}$ such that $\mathcal{P} \parallel \mathcal{Q} \sqsubseteq R$, we have $\mathcal{Q} \sqsubseteq (\mathcal{R} \setminus \mathcal{P})$.* □

Pruning the Quotient. The quotient obtained by the above method may contain a significant amount of redundancy. Particularly, 1. some of its states may never be reached, 2. some transitions may never be taken, and 3. some parts of conditions within guards may be true for every computation reaching the source location of the transition.

To obtain a more compact solution, we will prune the redundant parts of $\mathcal{R} \setminus \mathcal{P}$. Using a procedure analogous to Algorithm 1 in Section 3, we compute for every location $\langle \ell_\mathcal{P}, \ell_{\mathcal{R} \setminus \mathcal{P}} \rangle$ of $\mathcal{P} \parallel (\mathcal{R} \setminus \mathcal{P})$ a predicate $Reach(\langle \ell_\mathcal{P}, \ell_{\mathcal{R} \setminus \mathcal{P}} \rangle)$ such that a state $\langle \langle \ell_\mathcal{P}, \ell_{\mathcal{R} \setminus \mathcal{P}} \rangle, \sigma \rangle$ is reachable in $\mathcal{P} \parallel (\mathcal{R} \setminus \mathcal{P})$ iff σ satisfies $Reach(\langle \ell_\mathcal{P}, \ell_{\mathcal{R} \setminus \mathcal{P}} \rangle)$. As before, we assume (w.l.o.g.) that $X_{\mathcal{R} \setminus \mathcal{P}} \cap X_\mathcal{P} = \emptyset$; further, let us assume that $X_\mathcal{P}$ is the tuple $x_1^\mathcal{P}, \ldots, x_k^\mathcal{P}$. Then, for every location $\ell_{\mathcal{R} \setminus \mathcal{P}}$ of $\mathcal{R} \setminus \mathcal{P}$, we will compute

$$Reach(\ell_{\mathcal{R} \setminus \mathcal{P}}) = \exists x_1^\mathcal{P}, \ldots, x_k^\mathcal{P}. \bigvee_{\ell_\mathcal{P} \in L_\mathcal{P}} Reach(\langle \ell_\mathcal{P}, \ell_{\mathcal{R} \setminus \mathcal{P}} \rangle)$$

which characterizes all possible valuations of variables of $\mathcal{R} \setminus \mathcal{P}$ that can appear in a computation of $\mathcal{P} \parallel (\mathcal{R} \setminus \mathcal{P})$ together with $\ell_{\mathcal{R} \setminus \mathcal{P}}$. We then prune the redundancies of the types 1–3 above from $\mathcal{R} \setminus \mathcal{P}$ as follows:

1. Remove locations $\ell_{\mathcal{R} \setminus \mathcal{P}}$ where $IsSat(Reach(\ell_{\mathcal{R} \setminus \mathcal{P}}))$ is *false*.
2. Remove transitions $\langle \ell_{\mathcal{R} \setminus \mathcal{P}}, \alpha(\overline{p}); g; \overline{x} := \overline{e}, \ell'_{\mathcal{R} \setminus \mathcal{P}} \rangle$ where $IsSat(Reach(\ell_{\mathcal{R} \setminus \mathcal{P}}) \wedge g)$ is *false*.
3. For each remaining transition $\langle \ell_{\mathcal{R} \setminus \mathcal{P}}, \alpha(\overline{p}); g; \overline{x} := \overline{e}, \ell'_{\mathcal{R} \setminus \mathcal{P}} \rangle$, we weaken the guard g to $Reach(\ell_{\mathcal{R} \setminus \mathcal{P}}) \Rightarrow g$, possibly enabling further simplification of the formula (e.g., by transforming it to DNF and removing redundant literals and clauses).

6   Applications to the Synthesis of Mediators

We demonstrate our framework and our implementation on examples from mediator synthesis. A *mediator* is a process that mediates communication between several parties with incompatible interfaces, ensuring that they interact to achieve a certain aim, while preventing any communication mismatches. We demonstrate how this task can be specified as a problem of computing a quotient for a specification $\mathcal{R}$ which can be automatically generated from this aim. We have implemented the computation of quotient, and illustrate its application on mediator synthesis in e-commerce applications.

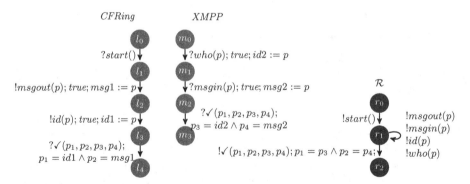

Fig. 1. Messaging clients *CFRing* and *XMPP*, on the left. The synthesis problem is specified by the process $\mathcal{R}$ on the right.

Messaging Protocol. In the first example, the scenario consists of two incompatible messaging protocols, *CFRing* and *XMPP*, both of which need to communicate with one another through a mediator that we must construct. The two messaging clients are represented by the two processes shown on the left in Figure 1. In the description, we omit guards of transitions that are *true*. Variables are initialized by a special value $\perp$.

- The process *CFRing* has input actions *start*, output actions *msgout* and *id*, variables $msg1$ and $id1$, and locations $\{l_0, l_1, l_2, l_3, l_4\}$.
- The process *XMPP* has input actions *msgin* and *who*, variables $msg2$ and $id2$, and locations $\{m_0, m_1, m_2, m_3\}$.

The goal of the communication is that the two clients agree on the message and the id. That is, the values of variables $msg1$ and $id1$ of *CFRing* should on termination be equal to the values of $msg2$ and $id2$ of *XMPP*. This goal can be captured by adding a special action $\checkmark$, which is performed at the end, and which reveals the values of the relevant variables of *CFRing* and *XMPP*. The specification $\mathcal{R}$ allows the components to perform any sequence of actions, but requires that quiescence is reached only after jointly performing the $\checkmark$ action, where the revealed variables satisfy the desired goal constraints. In other words, the communication parties are forced to reveal the values of their variables within action $\checkmark$, and $\mathcal{R}$ checks that they correspond. $\mathcal{R}$ does not specify what exactly happens before, it only states that the trace should begin with *start* and may continue by any sequence of actions *msgout*, *msgin*, *id*, and *who*. To ensure that $\checkmark$ will indeed be performed eventually, that is, that both processes reach their terminal states, $Quie_{\mathcal{R}}$ is defined to be *false* for r_0, r_1 and *true* for r_2.

We follow the scheme of the previous sections, and compute first the process $\mathcal{P} = CFRing \parallel XMPP$ shown in the upper part of Figure 2, which inherits all the transitions of *CFRing* and *XMPP* unsynchronized.

We are looking for the least refined process such that its composition with $\mathcal{P}$ refines $\mathcal{R}$, that is, we are looking for the quotient $\mathcal{R} \setminus \mathcal{P}$. The next step is the construction of the transition relation $\delta_{\mathcal{P} \otimes \mathcal{R}}$ which is shown in Figure 2 (lower part). We proceed by computing the mapping *Bad* that characterizes

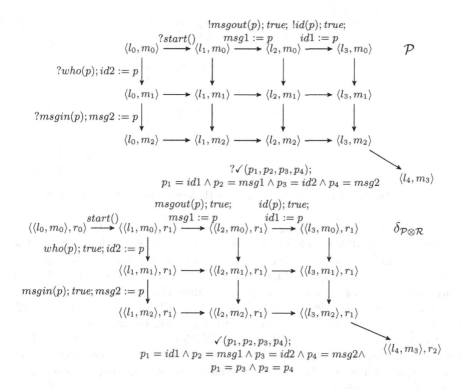

Fig. 2. The process $\mathcal{P} = CFRing \parallel XMPP$ and the transition relation $\delta_{\mathcal{P} \otimes \mathcal{R}}$. The statements of the transitions marked by vertical arrows are defined by the labels on the left margin of the figure. The statements of the transitions marked by horizontal arrows are defined by the labels on the top margin of the figure.

uncontrollable states of $\mathcal{R} \setminus \mathcal{P}$ (from where $\mathcal{R} \setminus \mathcal{P}$ cannot guarantee that no undesired state is reached). This is done by running Algorithm 3.

Let us discuss some steps of the computation of *Bad* in detail. The undesired states here are only those which deadlock and are not quiescent according to $\mathcal{R}$, that is, the location of $\mathcal{R}$ is different from r_2. Since $Bad(\ell)$ is initially *false* for all locations of $\mathcal{R} \setminus \mathcal{P}$, the computation starts by evaluating $Deadlock^{Bad}$. Let us pick the location $\ell = \langle \langle l_3, m_2 \rangle, r_1 \rangle$. $Deadlock^{Bad}(\ell)$ evaluates initially to the disjunction $\varphi \equiv id1 \neq id2 \vee msg1 \neq msg2$, which falsifies the guard of the only output transition of ℓ (leading to $\langle \langle l_4, m_3 \rangle, r_2 \rangle$). We set $Bad(\ell)$ to φ and we then propagate $Bad(\ell) = \varphi$ backwards via output transitions of $\mathcal{R}$. The precondition of φ with respect to the transition from $\ell' = \langle \langle l_2, m_2 \rangle, r_1 \rangle$ to ℓ is evaluated as *true* (particularly, the disjunct $id2 \neq id1$ is first transformed into $id2 = p$ which is then turned into *true* by the existential quantification over p). Hence, we set $Bad(\ell')$ to *true*; therefore, $\mathcal{R} \setminus \mathcal{P}$ must avoid visiting ℓ' in all situations. Note that it is indeed the case that if $\mathcal{R} \setminus \mathcal{P}$ allows *XMPP* to reach state m_2 (where the value of $msg2$ is already fixed) before $msg1$ is generated by *CFRing*, then $\mathcal{R} \setminus \mathcal{P}$

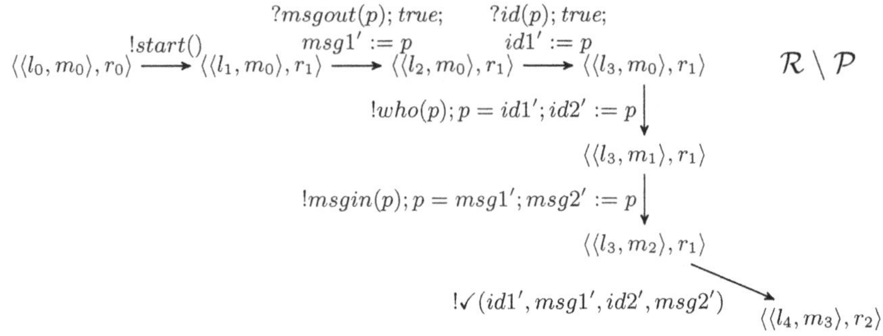

$$\langle\langle l_0, m_0\rangle, r_0\rangle \xrightarrow{\substack{!start()}} \langle\langle l_1, m_0\rangle, r_1\rangle \xrightarrow{\substack{?msgout(p);\,true;\\ msg1':=p}} \langle\langle l_2, m_0\rangle, r_1\rangle \xrightarrow{\substack{?id(p);\,true;\\ id1':=p}} \langle\langle l_3, m_0\rangle, r_1\rangle \qquad \mathcal{R}\setminus\mathcal{P}$$

$$\downarrow \scriptstyle{!who(p);\,p=id1';\,id2':=p}$$

$$\langle\langle l_3, m_1\rangle, r_1\rangle$$

$$\downarrow \scriptstyle{!msgin(p);\,p=msg1';\,msg2':=p}$$

$$\langle\langle l_3, m_2\rangle, r_1\rangle$$

$$\xrightarrow{!\checkmark(id1',msg1',id2',msg2')} \langle\langle l_4, m_3\rangle, r_2\rangle$$

Fig. 3. The resulting mediator $\mathcal{R}\setminus\mathcal{P}$ for the application with messaging.

cannot guarantee $msg1 = msg2$ at the end of the communication which may lead to an undesired deadlock. Let us now look at the precondition of φ wrt. the transition from $\langle\langle l_3, m_1\rangle, r_1\rangle$. Since ℓ is its successor and we have set $Bad(\ell) = \varphi$, $Deadlock^{Bad}(\langle\langle l_3, m_1\rangle, r_1\rangle)$ evaluates to $id2 \neq id1$ (particularly, the disjunct $msg1 \neq msg2$ turns into $msg1 \neq p$ which after the universal quantification over p becomes *false*). (Particularly, $\neg Bad(\ell) \equiv id1 = id2 \land msg1 = msg2$, $pre(\neg\varphi, msgin(p); msg2 := p)$ evaluates to $id1 = id2$ which is then negated.) Intuitively, this reflects the fact that when *CFRing* is in state l_3 and *XMPP* in state m_1, the values of $id1$ and $id2$ should be already equal since they cannot become equal otherwise. The rest of the symbolic backward computation of *Bad* is carried out analogously.

The construction of $\mathcal{R}\setminus\mathcal{P}$ continues by adding guards to the transitions of $\delta_{\mathcal{P}\otimes\mathcal{R}}$ to guarantee that any computation stays outside of *Bad*. Finally, we prune unreachable states, useless transitions, and redundant guard conditions, as described in Section 14. The resulting quotient $\mathcal{R}\setminus\mathcal{P}$ is shown in Figure 3.

E-Commerce. The next example is a more realistic and larger system. Due to its size, we only explain the specification and the functionality of the system and the synthesized mediator. We are given both a client and a customer service application, shown in Figure 4, that together are supposed to realize an e-commerce workflow, but are incompatible. The client *Blue* starts by sending a *StartOrder* message containing its id and expects to receive an id of a new order. It then orders a number of items in some quantities using the *AddToOrder* action, provides its payment information via the *PlaceOrder* action, and expects all items together with their quantities to be confirmed by the customer service. It blocks in case that it does not receive the right confirmation. *Blue* then announces that it is ready to quit the transaction and expects to receive the result of the payment transaction, *Result* (indicating whether or not the payment transaction was successful).

The customer service *Moon* expects to receive the client's id, then it sends a confirmation and sends an id of a new order, together with a client verification. It is then prepared to repeat a loop in which it 1) receives an order of an item, 2)

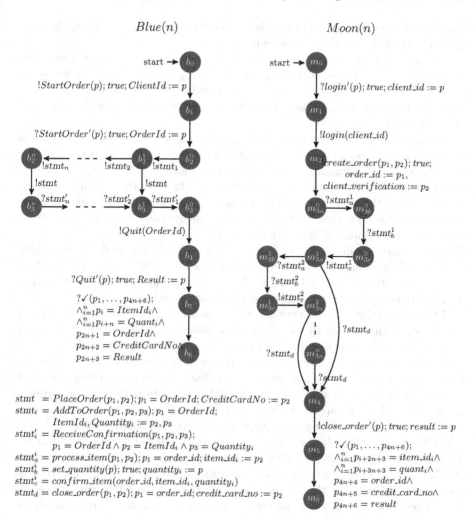

Fig. 4. Client *Blue* and service *Moon*, parameterized by maximum number of ordered items n.

receives a quantity in which the item is ordered, and 3) confirms that the item in the given quantity is ordered. After that, it receives payment information, arranges the payment via a third-party service (which is invisible to the client and not modelled here), and sends the result of the payment transaction.

Ideally, we would like to model the scenario where the client can order any number of items. However, our modelling mechanism allows only processes with a finite number of variables. This scenario would require processes with an unbounded number of variables, both for the *Blue* and *Moon* services and for the specification, as well as for mediator. We therefore restrict ourselves to the case where the number of ordered items is bounded by a constant n, which becomes a parameter of the synthesis problem.

The specification (parameterized by the maximum number of ordered items n) is given by the process $\mathcal{R}(n)$ in Figure 5. $Quie_{\mathcal{R}}$ is defined as *true* for r_2 and *false* otherwise. Similar to the previous examples, $\mathcal{R}$ specifies that both sides finish the transaction and that at the end of the transaction, both sides agree on all the important values.

A mediator for a fixed number of ordered items n can be synthesized analogously to the previous example. We construct the process $\mathcal{P} = Blue(n) \,\|\, Moon(n)$ and synthesize the mediator in the form of the quotient $\mathcal{R} \setminus \mathcal{P}$ by: 1. computing the predicate *Bad* characterizing uncontrollable states of the product of $\mathcal{P}$ and $\mathcal{R}$, using Algorithm 3, 2. strengthening guards in $\delta_{\mathcal{P} \otimes \mathcal{R}}$ so that no uncontrollable state (and hence no undesirable state) can be reached, and 3. pruning the useless states and transitions, utilizing Algorithm 1.

The synthesized mediator for the case $n = 2$ is shown in Figure 6. For simplicity, the figure displays only the $\mathcal{P}$-component of locations of $\mathcal{R} \setminus \mathcal{P}$, i.e., the locations of $Blue(2) \,\|\, Moon(2)$. The location $\langle b_0, m_0 \rangle$ is coupled with r_0, the location $\langle b_6, m_6 \rangle$ with r_2, and all the other displayed locations with r_1.

The functionality of the synthesized mediator can be explained as follows. It first brings the system to the point when *Blue* has ordered its first item (the state $\langle b_2^1, m_{3a}^0 \rangle$). It can now decide to start forwarding the first order to *Moon* (vertical transitions). At the same time, it has to be ready to receive either the second order or the credit card details from *Blue* (the horizontal transitions). If *Blue* ordered also the second item, the mediator will forward the second order to *Moon* after it has forwarded the first one. The mediator will send the payment credit card details to *Moon* only after it has received it from *Blue* and after it has forwarded all orders of *Blue* (this is taken care of by the guard of the statement $stmt_d^*$). It waits for *Blue* to confirm all orders. Sending confirmations to *Blue* is independent from receiving confirmations from *Moon* since to send the right confirmations, the mediator only needs to know what was ordered by *Blue*. The mediator can therefore choose from many variants of interleaving the communication with *Blue* and *Moon*.

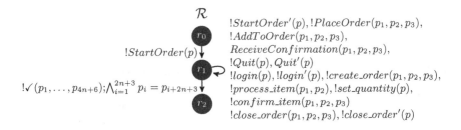

Fig. 5. Specification of the synthesis problem where the number of ordered items is bounded by n is given by the process $\mathcal{R}(n)$.

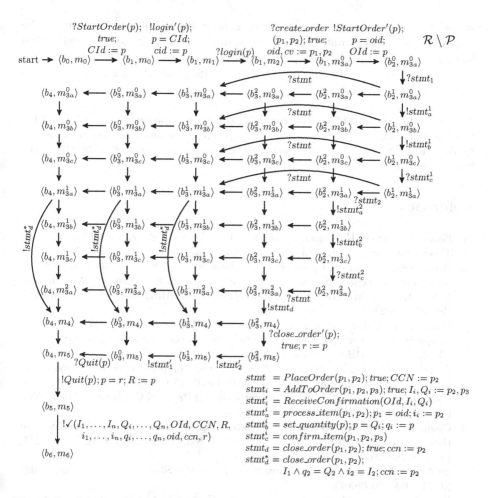

Fig. 6. Resulting mediator $\mathcal{R} \setminus \mathcal{P}$ for the e-commerce service with $n = 2$. The labels of the vertical edges in a row agree with the label of the right-most edge in the row, the labels of the horizontal edges in a column agree with the label of the bottom edge of the column.

7 Summary and Future Work

We have extended the theory of interface automata [7] with data and progress, using mechanisms that have been naturally adapted from similar other works in the literature. The resulting theory allows to capture data-flow behaviour at the modelling level, and can also be used to formulate the mediator synthesis problem. Further extensions can be done along several dimensions. One is to extend the theory to richer sets of relations and functions over the data domain. Another dimension is to handle non-deterministic processes, which brings the problem of uncertainty – the mediator cannot be completely sure about the state of the systems it controls. A further dimension is to extend the specification framework to cover liveness properties. Such extensions have been considered for the finite-state case in the literature (see, e.g., [19]), but their adaption to handle data appears nontrivial.

Acknowledgments. This work is carried out as part of the European FP7 ICT FET CONNECT project (http://connect-forever.eu/). The last author was supported in part by the UPMARC centre of excellence, the first author was supported by the Czech Science Foundation project 202/13/37876P.

References

1. Bauer, S., Larsen, K., Legay, A., Nyman, U., Wasowski, A.: A Modal Specification Theory for Components with Data. Sci. Comput. Program. **83**, 106–128 (2014)
2. Bennaceur, A., Chilton, C., Isberner, M., Jonsson, B.: Automated mediator synthesis: combining behavioural and ontological reasoning. In: Hierons, R.M., Merayo, M.G., Bravetti, M. (eds.) SEFM 2013. LNCS, vol. 8137, pp. 274–288. Springer, Heidelberg (2013)
3. Berardi, D., Calvanese, D., De Giacomo, G., Hull, R., Mecella, M.: Automatic composition of transition-based semantic web services with messaging. In: Proc. of VLDB 2005. VLDB Endowment (2005)
4. Bertoli, P., Pistore, M., Traverso, P.: Automated Composition of Web Services via Planning in Asynchronous Domains. Artif. Intell. **174**(3–4) (2010)
5. Bhaduri, P., Ramesh, S.: Interface Synthesis and Protocol Conversion. Form. Asp. Comput. **20**(2), 205–224 (2008)
6. Chen, T., Chilton, C., Jonsson, B., Kwiatkowska, M.: A compositional specification theory for component behaviours. In: Seidl, H. (ed.) ESOP 2012. LNCS, vol. 7211, pp. 148–168. Springer, Heidelberg (2012)
7. de Alfaro, L., Henzinger, T.A.: Interface Automata. SIGSOFT Softw. Eng. Notes **26**(5), 109–120 (2001)
8. Dill, D.: Trace Theory for Automatic Hierarchical Verification of Speed-independent Circuits. PhD thesis, Carnegie Mellon University (1988)
9. Guermouche, N., Perrin, O., Ringeissen, C.: A mediator based approach for services composition. In: SERA 2008. IEEE (2008)
10. Hull, R., Benedikt, M., Christophides, V., Su, J.: E-services: a look behind the curtain. In: Proc. 22nd ACM Symp. on Principles of Database Systems, pp. 1–14. ACM (2003)
11. Jonsson, B.: Compositional specification and verification of distributed systems. ACM Trans. on Programming Languages and Systems **16**(2), 259–303 (1994)
12. Larsen, K.G., Nyman, U., Wasowski, A.: Modal I/O automata for interface and product line theories. In: De Nicola, R. (ed.) ESOP 2007. LNCS, vol. 4421, pp. 64–79. Springer, Heidelberg (2007)
13. Lynch, N.A., Tuttle, M.R.: Hierarchical correctness proofs for distributed algorithms. In: Proc. 6th ACM Symp. on Principles of Distributed Computing, pp. 137–151 (1987)
14. Olderog, E., Hoare, C.: Specification-oriented semantics for communicating processes. Acta Informatica **23**(1), 9–66 (1986)

15. Raclet, J.-B.: Residual for component specifications. Electronic Notes in Theoretical Computer Science **215**, 93–110 (2008)
16. Raclet, J.-B., Badouel, E., Benveniste, A., Caillaud, B., Legay, A., Passerone, R.: Modal interfaces: unifying interface automata and modal specifications. In: Proc. of EMSOFT 2009, pp. 87–96. ACM (2009)
17. Raclet, J.-B., Badouel, E., Benveniste, A., Caillaud, B., Legay, A., Passerone, R.: A Modal Interface Theory for Component-based Design. Fundamenta Informaticae **108**(1–2), 119–149 (2011)
18. Raclet, J.-B., Badouel, E., Benveniste, A., Caillaud, B., Passerone, R.: Why are modalities good for interface theories? In: ACSD 2009, pp. 119–127. IEEE (2009)
19. Raskin, J.-F., Chatterjee, K., Doyen, L., Henzinger, T.A.: Algorithms for Omega-Regular Games with Imperfect Information. Logical Methods in CS **3**(3) (2007)
20. Ryzhyk, L., Chubb, P., Kuz, I., Sueur, E.L., Heiser, G.: Automatic device driver synthesis with termite. In: SOSP 2009, pp. 73–86. ACM (2009)
21. Tretmans, J.: Model-based testing and some steps towards test-based modelling. In: Bernardo, M., Issarny, V. (eds.) SFM 2011. LNCS, vol. 6659, pp. 297–326. Springer, Heidelberg (2011)
22. Yellin, D.M., Strom, R.E.: Protocol Specifications and Component Adaptors. ACM Trans. Program. Lang. Syst. **19**(2), 292–333 (1997)

Safe and Optimal Adaptive Cruise Control

Kim Guldstrand Larsen[✉], Marius Mikučionis, and Jakob Haahr Taankvist

Department of Computer Science, Aalborg University,
Selma Lagerlöfs Vej 300, 9220 Aalborg Øst, Denmark
kgl@cs.aau.dk

Abstract. In a series of contributions Olderog et al. have formulated and verified safety controllers for a number of lane-maneuvers on multi-lane roads. Their work is characterized by great clarity and elegance partly due to the introduction of a special-purpose Multi-Lane Spatial Logic. In this paper, we want to illustrate the potential of current model-checking technology for automatic synthesis of optimal yet safe (collision-free) controllers. We demonstrate this potential on an Adaptive Cruise Control problem, being a small part of the overall safety problem considered by Olderog[1].

1 Introduction

These days the Google Self-Driving car is about to become a reality: legislation has been passed in several U.S. states allowing driverless cars, in April 2014, Google announced that their vehicles had been logging nearly 1.1 million km, and it is forecast that Google's self-driving cars will hit the roads this summer[1].

Also in Europe driverless cars have been actively pursued, both by the automotive industry itself and within a number of national and European research projects (e.g. FP7 and Horizon2020). With more and more traffic, European roads are becoming increasingly congested, polluted and unsafe. One potential solution to this growing problem is seen to be the use of small, automated, low-polluting vehicles for driverless transport in (and between) cities. Within the last decade, a number of European projects have been launched for making transport systems capable of fully automated driving, energy efficient and environmentally friendly while performing.

In addition, many individual driving assistant systems based on suitable sensors have been developed for cars. Moreover, car-to-car communication is considered to combine individual cars into more advanced assistance functionality. One particular class of such functionality considered by several researchers from the formal methods and verification community is that of driver assistance on multi-lane roads, with lane-changing and overtaking maneuvers. Overall this constitutes a hybrid systems verification problem, where the car dynamics, discrete

---

This paper is partially sponsored by the EU FET projects SENSATION and CASST-ING as well as the Sino-Danish Basic Research Center IDEA4CPS.

[1] http://recode.net/2015/05/15/googles-homemade-self-driving-cars-are-hitting-the-roads-this-summer/

R. Meyer et al. (Eds.): Olderog-Festschrift, LNCS 9360, pp. 260–277, 2015.
DOI: 10.1007/978-3-319-23506-6_17

or timed controllers as well as road-specific assumptions should imply safety, i.e. absence of collisions.

The California PATH (Partners for Advanced Transit and Highways) project has spurred a series of research towards provable safe lane-maneuvers by developing advancing technologies that connect vehicles to surrounding infrastructure and other vehicles or automates vehicle processes[2]. For example the work by Lygos et al. [9,14], sketch a safety proof for manoeuvres of car platoons including lane change by taking car dynamics into account, but admitting safe collisions, i.e., collisions at a low speed.

More recently, within the German Transregional Collaborative Research Center AVACS – sponsored by the German Research Council (DFG) – Olderog together with colleagues made significant progress on proving safety of a number of automated maneuvers for overtaking on roads of varying complexity, ranging from highways (with uni-directional traffic) [11] to country roads (with two-directional traffic) [10]. The main contribution has been to show that (also here) one can separate the purely spatial reasoning from the underlying car dynamics in the safety proof. In particular, the approach taken by Olderog et. al. introduces a Multi-Lane Spatial Logic inspired by Moszkowski's interval temporal logic [16], Zhou, Hoare and Ravn's Duration Calculus [5] and Schäfer's Shape Calculus [17], and may be summarized by the following characteristics:

- The dynamics of cars are separated from the control laws – thus related to work by Raisch et al [15] and Van Schuppen et al [8].
- Development of the special-purpose multi-lane spatial logic MLSL to allow for easy formulation and verification.
- Design of controllers for lane-change maneuvers.
- Manual verification for proving safety of proposed controllers under general scenarios.

In addition to the Multi-Lane Spatial Logic (MLSL) [13], the formalism of timed automata [1] were used for specifying the protocol for safe lane-changing. However, though in principle possible, tool-support using a timed automata model-checker was never exploited in the above work. In this paper – celebrating the several contributions by Olderog to the area of the modeling and verification of real-time systems in general – we consider a small part of lane-change manoeuvres, namely the existence of a safe-distance controller (assumed in the above work of Olderog et al.). In particular, we aim at demonstrating how the most recent developments of the real-time model-checking tool UPPAAL [12] may be applied. Contrasting with the methodology of Olderog et.al. our method may be characterized as follows:

- Abstract away from dynamics for safety, but reconsidered dynamics for optimization (here to minimize the expected distance between cars).
- Use of a general purpose formalism in terms of various extensions of timed automata and games.

---

[2] http://www.path.berkeley.edu/

– Automatic synthesis of a range of controllers, ranging from safety controllers to optimal controllers, and to optimal yet safe controllers under specific scenarios.
– Extensive use of automated tool support, in terms of model checking, synthesis and optimization as provided by the most recent branch of the UPPAAL tool suite, UPPAAL STRATEGO [6, 7].

The outline of the paper is as follows: in Section 2 we describe the Adaptive Cruise Control problem considered. In Section 3, we present the formalism of (weighted and stochastic) timed automata and games as used in UPPAAL STRATEGO by means of a small Route Choosing Problem. Section 4 details our game model of the Adaptive Cruise Control problem, and Section 5 offers our results in terms of synthesis and analysis of safe, optimal and optimal-yet-safe controllers for the Adaptive Cruise Control problem.

Acknowledgement The authors would like to thank Anders P. Ravn for suggesting the Adaptive Cruise Control (or safety distance control) problem to us.

2   The Problem of Adaptive Cruise Control

We now define the case we will be analyzing using UPPAAL STRATEGO in the remainder of the paper.

Two cars *Ego* and *Front* are driving on a road shown in Figure 1. We are capable of controlling *Ego*, but not *Front*. Both cars can drive a maximum of 20 m/s forward and a maximum of 10 m/s backwards. The cars have three different possible accelerations: $-2\,\text{m/s}^2$, $0\,\text{m/s}^2$ and $2\,\text{m/s}^2$, between which they can switch instantly.

For the car to be safe there should be a distance of at least 5 m between them. Any distance less than 5 m between the cars is considered unsafe.

Ego's sensors can detect the position of *Front* only within 200 meters. If the distance between the cars is more than 200 meters then *Front* is considered to be *far away*. *Front* can reenter the scope of *Ego*'s sensor with arbitrary velocity it desires, as long as the velocity is smaller or equal to that of *Ego*.

We would then like to know the answers to the following questions:

– Is it possible that the cars crash?
  • If so, what is the probability of the cars crashing?
– Can we find a strategy for *Ego* such that the cars can never crash, no matter what *Front* does?

VelocityEgo
AccelerationEgo

VelocityFront
AccelerationFront

Distance

Fig. 1. Distance, velocity and acceleration between two cars.

- Are there more than one of such strategies?
- Which of the safe strategies lets *Ego* travel the furthest?
- How does *Ego* respond to *Front*'s choices under all these different strategies?

3   Stochastic Priced Timed Games

For the synthesis of safe and optimal strategies, we will use (weighted and stochastic) timed automata and games, exploiting the tool UPPAAL STRATEGO [7] being a novel branch of the UPPAAL tool suite that allows to generate, optimize, compare and explore consequences and performance of strategies synthesized for stochastic priced timed games (SPTG) in a user-friendly manner. In particular, UPPAAL STRATEGO comes with an extended query language (see Table 1), where strategies are first class objects that may be constructed, compared, optimized and used when performing (statistical) model checking of a game under the constraints of a given synthesized strategy.

To illustrate the features of UPPAAL STRATEGO, let us look at the example in Fig. 2, providing an "extended" timed automata based model of a car, that needs to make it from its initial position Start to the final position End. In fact the model constitutes a timed *game*, where the driver of the car twice needs to make a decision as to whether (s)he wants to use a high road (H1 and H2) or a low road (L1 and L2). The four roads differ in their required travel-time (up to 100

Table 1. Various types of UPPAAL STRATEGO queries: "**strategy S =**" means strategy assignment and "**under S**" is strategy usage via strategy identifier S. Here the variables NS, DS and SS correspond to non-deterministic, deterministic and stochastic strategies respectively; **bound** is a bound expression on time or cost like x<=100 and n is the number of simulations.

Strategy generators using [6]:	
Minimize objective:	strategy DS = minE (expr) [bound]: <> prop
Maximize objective:	strategy DS = maxE (expr) [bound]: <> prop under NS

Strategy generators using UPPAAL TIGA:	
Guarantee objective:	strategy NS = control: A<> prop
Guarantee objective:	strategy NS = control: A[] prop

Statistical Model Checking Queries:	
Hypothesis testing:	Pr[bound](<> prop)>=0.1 under SS
Evaluation:	Pr[bound](<> prop) under SS
Comparison:	Pr[bound](<> prop1) under SS1 >= Pr[<=20](<> prop2) under SS2
Expected value:	value E[bound;n](min: prop) under SS
Simulations	simulate n [bound] { expr1, expr2 } under SS

Symbolic model checking queries:	
Safety:	A[] prop under NS
Liveness:	A<> prop under NS
Infimum of value:	inf { condition } : expression
Supremum of value:	sup { condition } : expression

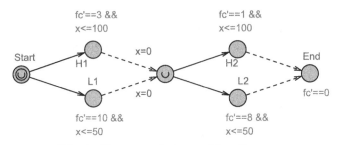

Fig. 2. The route choice problem for a car.

minutes respectively 50 minutes as reflected by the invariants on the clock x). Also the roads differ in fuel-consumption reflected by the difference in the rate of the continuous variable fc (representing the total amount of fuel consumed). Whereas the choice of road is up to the driver of the car to control (indicated by the solid transitions), the actual travel-time of the road is uncontrollable (indicated by the dashed transitions) reflecting the uncertainty of the amount of traffic on the particular day. In one scenario, the objective of the car it to choose the combination of roads that will ensure the shortest overall travel-time even in the most hostile traffic situation on the four roads. Under this interpretation, Fig. 2 represents a timed game. However, it may also be seen as a stochastic priced timed game (SPTG), assuming that the travel-times of the four roads are chosen by uniform distributions, and the objective of the control strategy is to minimize the expected overall travel-time, or the expected overall fuel-consumption (e.g. the rate or fuel-consumption fc'==3 on the first high road H1 indicates that the cost variable fc grows with rate 3 in this location).

We are interested in synthesizing strategies for various objectives. Being primarily concerned with fuel-consumption, the query

```
strategy Opt = minE (fc) [<=200] : <> Car.End
```

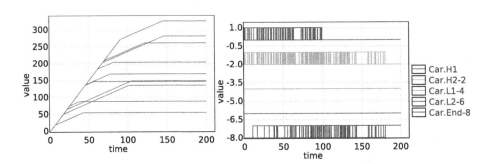

(a) fc trajectory samples. Fuel consumption on the vertical axis

(b) Road choice samples.

Fig. 3. Evaluation of strategy Opt via simulation.

will provide (by reinforcement learning[3]) the strategy Opt, that minimizes the expected total fuel-consumption, learning from runs which are maximally 200 time units long. The relativized query E[<=200 ; 1000] (max: fc) under Opt, generates 1000 runs of length 200 time units and then averages the maximum value of fc from each run. this is used to estimate the expected cost to be 200.39. Figure 3a summarizes 10 random runs according to Opt illustrating fuel-consumption. None of the runs had a fuel consumption of 400 indicating that we always choose the energy-efficient roads. In Figure 3b we see that this is actually the case as the simulations always choose to go to locations H1 and H2, which models the energy-efficient roads.

Now, assume that the task *must* be completed before 150 time-units. From Fig. 3 it can be seen that the strategy Opt unfortunately does not guarantee this, as there are a few runs which exceeds 150 before reaching End. However, the query

strategy Safe = control: A<> Car.End and time<=150

will generate the most permissive (non-deterministic) strategy Safe that guarantees this bound but unfortunately with a high expected total fuel-consumption of 342.19. However, the relativized learning query

strategy OptSafe = minE (fc) [<=200] : <> Car.End under Safe

will provide a sub-strategy OptSafe that minimizes the expected total fuel-consumption – here found to be 279.87 – subject to the constraints of Safe. Figure 4 summarizes 10 random runs according to SafeOpt, incidating that only road L1 is never choosen. Also, the failed model checking of E<> Car.H2 and time>=51 and Car.x==0 under Safe reveals that the high road H2 may only be choosen in case the first phase is completed before 50 time-units, confirming the observations from the simulations.

In general, as shown in the overview Fig. 5, UPPAAL STRATEGO will start from a SPTG $\mathcal{P}$. It can then abstract $\mathcal{P}$ into a timed game (TGA) $\mathcal{G}$ by simply ignoring prices and stochasticity in the model. Using $\mathcal{G}$, UPPAAL TIGA [2] may now be used to (symbolically) synthesize a (most permissive) strategy σ meeting a required safety or (time-bounded) liveness constraint ϕ. The TGA $\mathcal{G}$ under σ (denoted $\mathcal{G}|\sigma$) may now be subject to additional (statistical) model checking using classical UPPAAL [3] and UPPAAL SMC [4]. Similarly, the original STGA $\mathcal{P}$ under σ may be subject to statistical model checking. Now using reinforcement learning[6], we may synthesize near-optimal strategies that minimizes (maximizes) the expectation of a given cost-expression *cost*. In case the learning is performed from $\mathcal{P}|\sigma$, we obtain a sub-strategy σ^o of σ that optimizes the expected value of *cost* subject to the hard constraints guaranteed by σ. Finally, given σ^o, one may perform additional statistical model checking of $\mathcal{P}|\sigma^o$.

---

[3] The reinforcement learning uses machine learning techniques to learn strategies from sets of randomly generated runs. See [6] for more details.

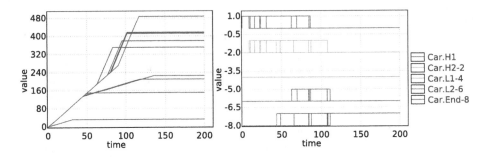

(a) `fc` trajectory samples.   (b) Road choice samples.

Fig. 4. Evaluation of strategy `OptSafe` via simulation.

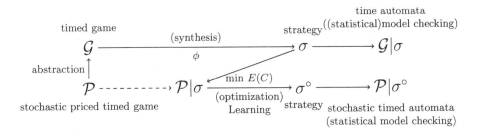

Fig. 5. Overview of UPPAAL STRATEGO

4   Modeling the Adaptive Cruise Control

In this section we introduce a model of the scenario described in Section 2.

We model the velocities and accelerations of the cars, where both cars can choose the acceleration they wish to have in each second. There are two sets of variables modeling the velocity and distance between the cars. The first set is a discrete set of variables. These variables are updated each second using the method `updateDiscrete()` seen in Listing 1. Note that when `distance > maxSensorDistance` we keep the distance constant, as the sensor cannot read how far away *Front* is anymore.

The second set measures the velocity and distance in a continuous way. The equations in Listing 2 define the rate of growth for continuous variables, we see that these are updated as expected. The rate of `rDistance` is defined using the function `distanceRate`, this function is presented in Listing 3.

What we see in Listing 3 is that when *Front* is *far away* from *Ego* we keep the distances between the cars constantly at the `maxSensorDistance` by setting the rate of `rDistance` to zero like with the discrete distance. If *Front* is not *far away* the rate of the distance is the expected `velFront - velEgo`, thus the relative velocity of the cars.

```
void updateDiscrete(){
    int oldVel, newVel;
    oldVel = velocityFront - velocityEgo;
    velocityEgo = velocityEgo + accelerationEgo;
    velocityFront = velocityFront + accelerationFront;
    newVel = velocityFront - velocityEgo;
    if (distance > maxSensorDistance) {
        distance = maxSensorDistance + 1;
    } else {
        distance += oldVel + (newVel - oldVel)/2;
    }
}
```

Listing 1. The code updating the discrete variables: `updateDiscrete()`.

```
rVelocityEgo' == accelerationEgo && rVelocityFront' ==
accelerationFront && rDistance' ==
distanceRate(rVelocityFront,rVelocityEgo, rDistance) &&
    D' ==
rDistance
```

Listing 2. The differential equations controlling the continuous variables.

Figure 6 validates by simulation that the discrete `distance` is correctly tracking the continuous trajectory of `rDistance`: the updated discrete values lie exactly on the continuous curve; the same applies to the difference of velocities.

Four parallel automata model and control the scenario: the system component, the models of *Ego* and *Front*, and the monitor component. The system component controls the system and enforces the discretisation of time, the monitor component monitors the state of the system and updates the hybrid variables correspondingly. We do not show the monitor component here as it is simply one location with the equations in Listing 2 as its invariant.

```
double distanceRate(double velFront, double velEgo,
    double dist) {
    if (dist > maxSensorDistance) return 0.0;
    else return velFront - velEgo;
}
```

Listing 3. The function `distanceRate()` defining how `rDistance` grows.

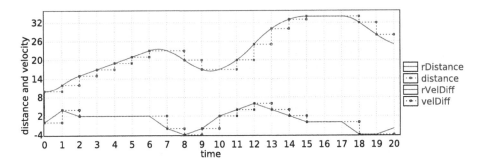

Fig. 6. Continuous and discrete distance (above) and velocity (below).

4.1   System Component

The *System* component, seen in Fig. 7 controls the discretisation of time, and also makes sure that *Ego* chooses its acceleration before *Front*. This means that when *Ego* chooses its acceleration for the next second it does not know what the acceleration of *Front* will be in the next step. It is also responsible for updating the discrete variables via the function `updateVelocities()`.

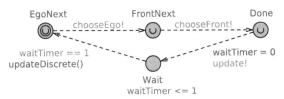

Fig. 7. The system component of the model: the component first lets *Ego* choose an acceleration, and then *Front*, i.e. *Ego* chooses an acceleration without knowing which acceleration *Front* will choose. The accelerations are chosen at discrete time instances every second.

Note that the only location in the *System* component which is not urgent (no time can pass in an urgent location) is the `Wait` location. Thus the cars have to choose their acceleration urgently.

When the system component has finished a loop and is in the `UpdateFront` location it will send an `update` broadcast before going to the `Wait` location, this signal is used by *Front* to determine weather it is *far away*.

4.2   The Model of *Ego* and *Front*

The two cars are modeled with a component each. The basic models of the cars are the same, however *Front* has extra behavior for the case when it is *far away*. We describe the model of *Ego* and in the end we describe the differences in the component for *Front*.

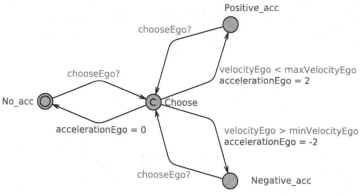

Fig. 8. Model of *Ego*.

The component for *Ego* can be seen in Fig. 8. There are four locations, the three locations which are not committed[4]; No_acc, Positive_acc and Negative_acc denotes the choice of acceleration in the current second, the last, choice is the location from which the choice is made. We also keep track of the acceleration via the variable **accelerationEgo**.

When the car gets a signal chooseEgo? it will go to the committed location Choose. From this location *Ego* will choose which acceleration to have in the next round. The location is committed to make sure *Ego* chooses it's acceleration before *Front*.

In this component we also enforce that the car cannot go faster than it's top speed (both backwards and forwards). This is done via the guards velocityEgo < maxVelocityEgo and velocityEgo > minVelocityEgo which ensure that it is only possible to have a positive acceleration if the current velocity is less than the top speed, and similarly it is not possible to have a negative acceleration if the current velocity is greater than the maximum negative speed.

The model of *Front* can be seen in Fig. 9. We see that it has the same set of places as *Ego* and the same set of variables. In addition to these it has the location Faraway and two committed locations. *Front* will go to the location Faraway if the distance between the cars is greater than **maxSensorDistance**, and it gets the update signal from the **System** component. The next time it gets the chooseFront signal there is fifty percent chance it will stay *far away* and fifty percent chance it will reenter the sensor distance. If it reenters the sensor distance it will choose which velocity it wants via the select statement i:int[minVelocityFront, maxVelocityFront].

---

[4] When a component is in a committed location, the next transition *must* be from this location, without any delay.

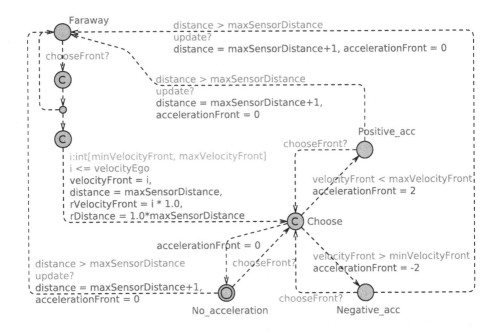

Fig. 9. Model of *Front.*

5 Synthesis and Analysis

With the model defined we can start analyzing it. UPPAAL STRATEGO offers a set of different analysis tools which view the model in different ways.

UPPAAL STRATEGO can view the model as a game with a number of different players.

Half player game In a half player game all choices are stochastic. The part of UPPAAL STRATEGO concerned with this kind of games is UPPAAL SMC. In half player games the kind of questions asked is typically, *what is the probability to reach this state?*. In UPPAAL STRATEGO there has to be a timebound, so the question is then, *what is the probability of reaching this state in t time units?*.

One player game In one player games all choices are nondeterministic. The part of UPPAAL STRATEGO which is concerned with this kind of games is traditional UPPAAL. In one player games we ask questions such as *is it possible to reach this state?*.

One and a half player game In one and a half player games some choices are nondeterministic and others are stochastic. UPPAAL STRATEGO is the first version of UPPAAL to include support for one and a half player games. A question asked in a one and a half player game could be *what is the best strategy to maximize the expected value of x?*. As with half player games

there has to be a timebound in UPPAAL STRATEGO, so the question will be *what is the best strategy to maximize the expected value of x in t time units?*
Two player game In two player games all choices are nondeterministic, one set of choices are made by the controller, the other set is made by the environment. The part of UPPAAL which handles two player games is UPPAAL TIGA. Questions in two player games could be *does there exist a strategy that makes sure I always reach this state, no matter which choices the environment makes?*

5.1   No Strategy

We can now analyze the model described in Section 4 using UPPAAL STRATEGO. The first question we could ask is if the cars will always be at least five meter from each other no matter what happens. This is a one player question. We can use the following UPPAAL query to answer the question:

$$A[] \ distance > 5$$

This is not true and UPPAAL gives us a trace which tells us that if *Ego* starts by accelerating forward and *Front* starts by accelerating backwards then the cars will collide.

A next question could then be what the probability that the distance between the two cars is less than five after 100 seconds. This is a half player question and is answered using the UPPAAL SMC query:

$$Pr[<=100] \ (<> \ distance <= 5)$$

This tells us that the probability of a crash if both cars drive randomly is in the interval $[0.856, 0.866]$ with a confidence of 95%. In Fig. 10 we can see that the probability of a crash is greatest in the beginning of the run and then decreases. We can also see that the average crash happens around 20 seconds after the cars starts moving.

We can also do a set of random simulations of the model to get a more nuanced view on how the model behaves. A set of simulations can be seen in Fig. 11, on the horizontal axis we see the time and on the vertical we see the distance between the cars.

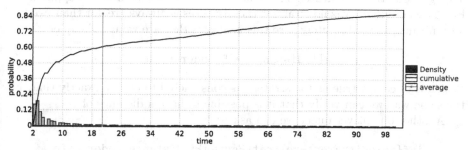

Fig. 10. The cumulative probability and the probability density of a crash.

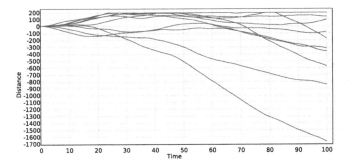

Fig. 11. A set of ten random simulations over the half player version of the model.

5.2   Safe Strategy

Clearly a probability of 85% for a crash is not acceptable. To remove the risk of a crash we consider the model as a two player game where we control *Ego* and the antagonistic environment controls *Front*. We can then ask for a strategy which makes sure that no matter what *Front* does we will avoid a crash. We can request such a strategy using the query:

```
control: A[] distance > 5
```

With UPPAAL STRATEGO it is possible to save a strategy in a named variable, we will do this as we would like to use the strategy later. The query then looks like this:

```
strategy safe = control: A[] distance > 5
```

This will give us the most permissive strategy which makes sure that the distance between the cars is kept greater than 5. The most permissive strategy is a strategy which in every state suggests the biggest set of actions possible to the controller. We can now do a set of simulations under this strategy. This is done using the query:

```
simulate 10 [<=100] rDistance under safe
```

This gives us the plot in Fig. 12. What we can see is that here, as opposed to Fig. 11 where there was no strategy, the distances is always greater than five for all ten runs, to be absolutely sure this is always the case we use the query:

```
A[] distance > 5 under safe
```

Which is of course true as the strategy is constructed to ensure exactly that, in the same way we can verify that it is possible to have a distance of 6.

Another one player query we can ask is:

```
inf{velocityFront-velocityEgo==v}: distance under safe
```

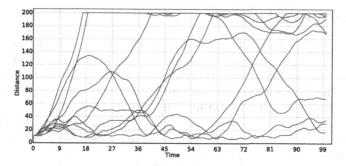

Fig. 12. A set of ten random simulations over the half player version of the model subject to a strategy which guarantees that distance is greater than five.

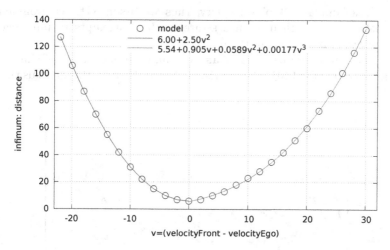

Fig. 13. Smallest distance possible under the `safe` strategy as a function of speed difference computed using `inf{velocityFront-velocityEgo==v}`: `distance under safe` for each v value. Connecting lines are from linear regression analysis.

this will return the smallest (inf) distance seen in any state where

$$\texttt{velocityFront-velocityEgo==v,}$$

thus the smallest distance at the relative velocity v possible under the safe strategy. In Fig. 13 we see the result of running this query with v at different values. When the velocity difference v is negative, the safe distance dependency is exactly quadratic: this is the distance the *Front* car would cover towards *Ego* if it kept accelerating and *Ego* were not able to keep up. Thus the faster the cars are driving towards each other the greater the distance should be to ensure the cars are safe. When v is positive the safe distance slope is not as steep, but it follows a similar worst case scenario that the *Front* may immediately start accelerating towards *Ego*, except it will take longer to make the velocity difference negative.

5.3   Safe and Fast Strategy

In Section 5.2 we generated the most permissive strategy which made sure the cars was safe. However the most permissive strategy is the union of all strategies, thus we have the opportunity of choosing the best strategy according to some measure. In Listing 2 we see that the rate of D is defined as D' == rDistance. This means that the value of D is the accumulated distance during the run. We can choose a strategy which minimizes D, this will mean that *Ego* will try to stay close to *Front*, but it will still stay far enough away that it will be safe. We can learn such a strategy using the query

```
strategy safeFast = minE (D) [<=100]: <> time >= 100 under safe
```

This is a one and a half player query, thus we assume the environment to be stochastic. We can then learn a strategy which attempts to minimize the expected value of D, as done in the query.

Using the learned strategy `safeFast` we can then make a set of simulations with `simulate 10 [<=100] rDistance under safeFast`, this gives us the plot in Fig. 14.

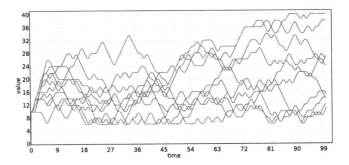

Fig. 14. A set of ten random simulations over the half player version of the model subjected to the strategy `safeFast` which optimizes the value of D, while still guaranteeing that the distance is greater than five.

We see that under this strategy the distance between the cars are much smaller, thus we now have a strategy that is not only safe but which also attempts to make the distance between the cars small.

We can also directly compare the two strategies `safe` and `safeFast` using the query:

```
Pr[rDistance <= maxSensorDistance + 1] <> time >= 100 under safe.
```

This query asks what the probability of reaching `time >= 100` is in runs which are bounded by `rDistance`. As `rDistance` never is greater than maxSensorDistance, we know that all runs will reach `time >= 100`. Thus the query reports

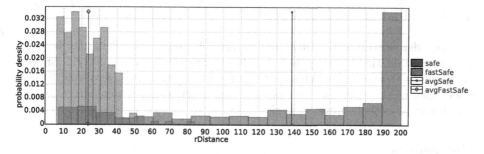

Fig. 15. The probability density distribution over `rDistance` at `time >= 100` thus after 100 time units under the strategies `safe` and `safeFast`. The (dark) red bars for `safe` and the (light) green bars for `safeFast`.

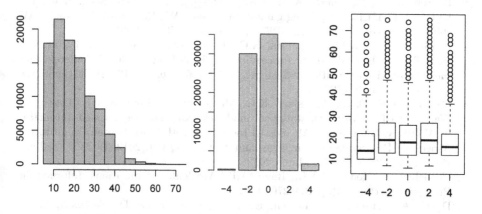

(a) Distance histogram.   (b) Velocity histogram. (c) Distance over velocity.

Fig. 16. Statistics from 1000 fixed-time-step simulations under `safeFast` strategy: overall 100000 data points about distance and velocity difference.

correctly that the probability is 1. However we can then look at what `rDistance` was when we reached `time >= 100`.

In Fig. 15 we see the probability density distribution for the value of `rDistance` when `time >= 100` generated using the query above. The red part shows the distribution under `safe` and the green part shows the distribution under `safeFast`.

What we can see is that under `safe` *Ego* just stays far away from *Front*, as this is safe. This is of cause boring as *Ego* will then never really move forward. Under `safeFast` on the other hand we see that *Ego* stays relatively close to *Front* thus minimizing D just like we intended.

Figure 16 summarizes the characteristics of the `safeFast` strategy of individual simulation points. The histograms show that the distance is limited by 70 metres and most of the time the cars are close to each other. Interestingly the velocity is also limited to a narrow range meaning that *Ego* manages to mimic the speed of *Front* and only rarely the speed difference greater than 2m/s.

6 Conclusion

We have demonstrated that safe and optimal distance controllers may be synthesized automatically using the recently emerged UPPAAL STRATEGO branch of the UPPAAL tool suite. What remains for future work that we would like to undertake, is to synthesize the complete safe overtaking protocol suggested by Olderog et al in their previous work.

References

1. Alur, R., Dill, D.L.: A theory of timed automata. Theor. Comput. Sci. **126**(2), 183–235 (1994)
2. Behrmann, G., Cougnard, A., David, A., Fleury, E., Larsen, K.G., Lime, D.: UPPAAL-tiga: time for playing games!. In: Damm, W., Hermanns, H. (eds.) CAV 2007. LNCS, vol. 4590, pp. 121–125. Springer, Heidelberg (2007)
3. Behrmann, G., David, A., Larsen, K.G., Håkansson, J., Pettersson, P., Yi, W., Hendriks, M.: Uppaal 4.0. In: Proceedings of the 3rd International Conference on the Quantitative Evaluation of Systems, QEST 2006, pp. 125–126. IEEE Computer Society, Washington (2006)
4. Bulychev, P.E., David, A., Larsen, K.G., Mikučionis, M., Poulsen, D.B., Legay, A., Wang, Z.: UPPAAL-SMC: statistical model checking for priced timed automata. In: Wiklicky, H., Massink, M. (eds.) Proceedings 10th Workshop on Quantitative Aspects of Programming Languages and Systems, QAPL 2012. EPTCS, Tallinn, Estonia, vol. 85, pp. 1–16, March 2012
5. Chaochen, Z., Hoare, C.A.R., Ravn, A.P.: A calculus of durations. Information Processing Letters **40**(5), 269–276 (1991)
6. David, A., Jensen, P.G., Larsen, K.G., Legay, A., Lime, D., Sørensen, M.G., Taankvist, J.H.: On time with minimal expected cost!. In: Cassez, F., Raskin, J.-F. (eds.) ATVA 2014. LNCS, vol. 8837, pp. 129–145. Springer, Heidelberg (2014)
7. David, A., Jensen, P.G., Larsen, K.G., Mikučionis, M., Taankvist, J.H.: UPPAAL STRATEGO. In: Baier, C., Tinelli, C. (eds.) TACAS 2015. LNCS, vol. 9035, pp. 206–211. Springer, Heidelberg (2015)
8. Habets, L.C.G.J.M., Collins, P.J., van Schuppen, J.H.: Reachability and control synthesis for piecewise-affine hybrid systems on simplices. IEEE Transactions on Automatic Control **51**(6), 938–948 (2006)
9. Haddon, J.A., Godbole, D.N., Deshpande, A., Lygeros, J.: Verification of hybrid systems: monotonicity in the AHS control system. In: Alur, R., Sontag, E.D., Henzinger, T.A. (eds.) HS 1995. LNCS, vol. 1066. Springer, Heidelberg (1996)
10. Hilscher, M., Linker, S., Olderog, E.-R.: Proving safety of traffic manoeuvres on country roads. In: Liu, Z., Woodcock, J., Zhu, H. (eds.) Theories of Programming and Formal Methods. LNCS, vol. 8051, pp. 196–212. Springer, Heidelberg (2013)
11. Hilscher, M., Linker, S., Olderog, E.-R., Ravn, A.P.: An abstract model for proving safety of multi-lane traffic manoeuvres. In: Qin, S., Qiu, Z. (eds.) ICFEM 2011. LNCS, vol. 6991, pp. 404–419. Springer, Heidelberg (2011)
12. Larsen, K.G., Pettersson, P., Yi, W.: Uppaal in a nutshell. International Journal on Software Tools for Technology Transfer **1**(1–2), 134–152 (1997)
13. Linker, S., Hilscher, M.: Proof theory of a multi-lane spatial logic. In: Liu, Z., Woodcock, J., Zhu, H. (eds.) ICTAC 2013. LNCS, vol. 8049, pp. 231–248. Springer, Heidelberg (2013)

14. Lygeros, J., Pappas, G.J., Sastry, S.: An approach to the verification of the center-TRACON automation system. In: Henzinger, T.A., Sastry, S.S. (eds.) HSCC 1998. LNCS, vol. 1386, pp. 289–304. Springer, Heidelberg (1998)
15. Moor, T., Raisch, J., O'Young, S.: Discrete supervisory control of hybrid systems based on l-complete approximations. Discrete Event Dynamic Systems **12**(1), 83–107 (2002)
16. Moszkowski, B.: A temporal logic for multilevel reasoning about hardware. Computer **18**(2), 10–19 (1985)
17. Schäfer, A.: A calculus for shapes in time and space. In: Liu, Z., Araki, K. (eds.) ICTAC 2004. LNCS, vol. 3407, pp. 463–477. Springer, Heidelberg (2005)

Author Index

Printed in the United States
By Bookmasters